Composites Science and Technology

Series Editor

Mohammad Jawaid, Department of Chemical and Petroleum Engineering,
United Arab Emirates University, Al Ain, United Arab Emirates

This book series publishes cutting edge research monographs comprehensively covering topics in the field of composite science and technology. The books in this series are edited or authored by top researchers and professional across the globe. The series aims at publishing state-of-the-art research and development in areas including, but not limited to:

- Conventional Composites from natural and synthetic fibers
- Advanced Composites from natural and synthetic fibers
- Chemistry and biology of Composites and Biocomposites
- Fatigue damage modelling of Composites and Biocomposites
- Failure Analysis of Composites and Biocomposites
- Structural Health Monitoring of Composites and Biocomposites
- Durability of Composites and Biocomposites
- Biodegradability of Composites and Biocomposites
- Thermal properties of Composites and Biocomposites
- Flammability of Composites and Biocomposites
- Tribology of Composites and Biocomposites
- Applications of Composites and Biocomposites

Review Process

The proposal for each volume is reviewed by the main editor and/or the advisory board. The chapters in each volume are individually reviewed single blind by expert reviewers (at least two reviews per chapter) and the main editor.

Ethics Statement for this series can be found in the Springer standard guidelines here - https://www.springer.com/us/authors-editors/journal-author/journal-author-helpdesk/before-you-start/before-you-start/1330#c14214

Chandrasekar Muthukumar · Mohammad Jawaid ·
Senthil Kumar Krishnasamy ·
Senthil Muthu Kumar Thiagamani · Carlo Santulli
Editors

Flammability Performance of Biocomposites and Bionanocomposites

Springer

Editors
Chandrasekar Muthukumar
Department of Aerospace Engineering
Hindustan Institute of Technology
and Science
Chennai, Tamil Nadu, India

Mohammad Jawaid
Laboratory of Biocomposite Technology
Universiti Putra Malaysia
INTROP
Serdang, Malaysia

Senthil Kumar Krishnasamy
Department of Mechanical Engineering
PSG Institute of Technology and Applied
Research
Coimbatore, Tamil Nadu, India

Senthil Muthu Kumar Thiagamani
Department of Mechanical Engineering
Kalasalingam Academy of Research
and Education
Krishnankoil, Tamil Nadu, India

Carlo Santulli
Geology Division (SST)
Università degli Studi di Camerino
Camerino, Macerata, Italy

ISSN 2662-1819 ISSN 2662-1827 (electronic)
Composites Science and Technology
ISBN 978-981-95-2388-7 ISBN 978-981-95-2389-4 (eBook)
https://doi.org/10.1007/978-981-95-2389-4

This Springer imprint is published by the registered company Springer Nature Singapore Pte Ltd.
The registered company address is: 152 Beach Road, #21-01/04 Gateway East, Singapore 189721, Singapore

Contents

About the Editors

Dr. Chandrasekar Muthukumar is presently working as an Associate Professor in the Department of Aerospace Engineering, Hindustan Institute of Technology and Science, Chennai, India. He graduated with a Bachelor's degree in Aeronautical Engineering from Kumaraguru College of Technology, Coimbatore, India. His Master's degree in Aerospace Engineering was from Nanyang Technological University-TUM ASIA, Singapore. He received his Ph.D. in Aerospace Engineering from Universiti Putra Malaysia (UPM), Malaysia. His Ph.D. was funded through a research grant from the Ministry of Education, Malaysia. During his association with the UPM, he obtained an internal research fund from the University worth 16,000 MYR and 20,000 MYR, respectively. His research interests include Fiber Metal Laminate (FML), natural fibers, biocomposites, aging, and their characterization. He has published 150+ research articles in reputed SCI journals, authored, and co-authored book chapters and conference proceeding articles. He was featured in the Elsevier's Top 2% Scientist list consecutively for the year 2025 and 2024.

Dr. Mohammad Jawaid is currently working as Distinguished Professor at Chemical and Petroleum Engineering Department, United Arab Emirates University, UAE. He has more than 25 years of experience in teaching, research, and industries. His area of research interests includes hybrid composites, lignocellulosic reinforced/filled polymer composites, advance materials: graphene/nanoclay/fire retardant, modification and treatment of lignocellulosic fibers and solid wood, biopolymers and biopolymers for packaging applications, nanocomposites and nanocellulose fibers, and polymer blends. So far, he has published 100 books, more than 100 book chapters, more than 500 peer-reviewed international journal papers, and several published review papers under top 25 hot articles in science direct during 2013–2020. He also obtained 6 Patents and 6 Copyrights/Industrial design. H-index and citation in Scopus are 104 and 44864 and in Google scholar, and H-index and citation are 122 and 60087. 40 Ph.D. and 17 Master's students graduated under his supervision in 2014–2024 in the fields of hybrid composites, green composites, nanocomposites, natural fiber-reinforced composites, nanocellulose, etc. He secured various research grants at university, national, and international levels on polymer composites of around

USD 3 Million. He also delivered plenary and invited talks in international conferences related to composites in India, Turkey, Malaysia, Thailand, United Kingdom, France, Saudi Arabia, Egypt, China, Germany, Indonesia, Spain, UAE, Portagal, etc. Besides that, he is also a member of technical committees of several national and international conferences on composites and material science. He is founding Series Editor of Springer Three Book Series. He worked as guest editor of special issues of Several Peer reviewed Journals. Besides that, he is also reviewer of several high-impact international peer-reviewed journals of Elsevier, Springer, Wiley, Saga, ACS, RSC, Frontiers, etc. Dr. Mohammad Jawaid also received Several Award at National and International Level. He is also Winner of Newton-Ungku Omar Coordination Fund: UK-Malaysia Research and Innovation Bridges Competition 2015. He is also a Professional member of ACS, SPE, APA, MySET, etc. He is Fellow and Chartered Scientist of The Institute of Materials, Minerals & Mining (IOM3), UK. He is also Fellow of International Engineering and Technology Institute(IETI), Hongkong. He is also Clarivate 1% Highly cited Researcher-2023.

Senthil Kumar Krishnasamy, Ph.D. is an Associate Professor in the Department of Mechanical Engineering at PSG Institute of Technology and Applied Research, Coimbatore, Tamil Nadu, India. He earned a bachelor's degree in mechanical engineering at Anna University, Chennai, India, in 2005. He earned a master's degree in CAD/CAM at Anna University, Tirunelveli, in 2009. He earned a Ph.D. in mechanical engineering at Kalasalingam University in 2016. From 2010 to 2018, Dr. Krishnasamy worked in the Department of Mechanical Engineering at Kalasalingam Academy of Research and Education (KARE), India. He completed a postdoctoral fellowship at Universiti Putra Malaysia, Serdang, Selangor, Malaysia, and King Mongkut's University of Technology North Bangkok (KMUTNB) under the research topics of experimental investigations on mechanical, morphological, thermal, and structural properties of kenaf fiber/mat epoxy composites and sisal composites and fabrication of eco-friendly hybrid green composites on tribological properties in a medium-scale application.

Senthil Muthu Kumar Thiagamani, Ph.D. is an Associate Professor in the Department of Mechanical Engineering at Kalasalingam Academy of Research and Education (KARE), Tamil Nadu, India. In 2004, he earned a Diploma in Mechanical Engineering at the State Board of Technical Education and Training, Tamil Nadu. In 2007, he graduated from Anna University, Chennai, with a Bachelor's degree in Mechanical Engineering. In 2009, he obtained his Master's degree in Automotive Engineering at Vellore Institute of Technology, Vellore. In 2018, he earned a Ph.D. in Mechanical Engineering, specializing in Biocomposites, at KARE. He completed postdoctoral research in the Materials and Production Engineering Department at the Sirindhorn International Thai-German Graduate School of Engineering (TGGS), KMUTNB, Thailand, in the year 2019. He started his academic career in 2010 as Lecturer in Mechanical Engineering at KARE.

Carlo Santulli, Ph.D. is an Associate Professor at the School of Science and Technology in Università di Camerino, Italy. Here, he lectures on materials science, natural fibers and biocomposites, waste management, and environmental sustainability. He owns a M.Sc. in chemical engineering (Università La Sapienza, 1991), an M.A. in arts (modern history) (Università La Sapienza, 2001), an M.Sc. in Environmental Decision Making (EDM) (Open University, 2004), and a Ph.D. in materials science and engineering (University of Liverpool, 2000). His research interests over 35+ have been composites, natural fibers, biopolymers, agro- and food waste management, and biomimetics. He is a referee for over 80 journals in various fields concerning materials science and sustainability. He has been an invited Professor in various universities in India, Malaysia, and across Europe.

Introduction to Flammability Tests and Fire Retardant Mechanisms of the Biocomposites and Bionanocomposites

M. Ramesh, S. Ganeshkumar, A. Felix Sahayaraj, and J. Maniraj

Abstract This research review article provides a comprehensive overview of flammability tests and fire retardant mechanisms specific to biocomposites and bionanocomposites. As the demand for sustainable materials in various industries grows, the investigation of fire safety becomes imperative. The article begins by elucidating the fundamental principles of flammability tests, addressing key methodologies employed to assess the fire resistance of biocomposites and bionanocomposites. Subsequently, the discussion delves into the intricate mechanisms underlying the fire retardancy of these bio-based materials. Emphasis is placed on exploring the inherent properties and synergistic effects of incorporating fire retardants within these composite structures. Through a systematic review of current research findings, this article aims to consolidate knowledge in the field, aiding researchers, practitioners, and industries in making informed decisions regarding the application of fire safety measures in the context of sustainable and environmentally friendly materials. The synthesis of information presented in this review contributes to advancing the understanding of flammability characteristics and fire mitigation strategies for biocomposites and bionanocomposites.

Keywords Biocomposites · Bionanocomposites · Flammability tests · Fire retardant mechanisms · Sustainable materials · Environmental fire safety

M. Ramesh (✉) · A. F. Sahayaraj · J. Maniraj
Department of Mechanical Engineering, KIT-Kalaignarkarunanidhi Institute of Technology, Coimbatore, Tamil Nadu, India
e-mail: mramesh97@gmail.com

S. Ganeshkumar
Department of Mechanical Engineering, Sri Eshwar College of Engineering, Coimbatore, Tamil Nadu, India

C. Muthukumar et al. (eds.), *Flammability Performance of Biocomposites and Bionanocomposites*, Composites Science and Technology,
https://doi.org/10.1007/978-981-95-2389-4_1

1 Fundamentals of Flammability Tests

In the realm of materials science, understanding the fundamental principles of flammability tests is paramount for ensuring the safety and reliability of products. This section aims to unravel the intricacies of flammability tests, particularly when applied to biocomposites and bionanocomposites [1, 2]. Flammability tests serve as a crucial tool to assess the propensity of a material to ignite and sustain combustion. For biocomposites and bionanocomposites, which are gaining prominence as sustainable alternatives, the need for a comprehensive evaluation of their fire resistance is more pressing than ever. These tests typically involve exposing the materials to controlled fire conditions, measuring parameters such as ignition time, flame spread, and heat release rate [3, 4]. The methodologies employed in flammability tests for these bio-based composites vary, ranging from standard testing protocols to customized approaches tailored to the unique characteristics of these materials. Standard tests, such as the cone calorimeter and the horizontal burning test, provide valuable insights into the fire behavior of biocomposites and bionanocomposites under specific conditions. Researchers and industry practitioners rely on these methodologies to gauge the materials' performance and compliance with safety standards. Moreover, the importance of considering real-world scenarios cannot be overstated [5, 6]. Flammability tests must simulate conditions that these materials might encounter in practical applications, ensuring that the results accurately reflect their behavior in diverse environments [7]. The mechanism of flammability in polymers is presented in Fig. 1 [8].

2 Methodologies for Assessing Fire Resistance

In the realm of assessing fire resistance in biocomposites and bionanocomposites, various methodologies play a crucial role in providing insights into the materials' ability to withstand and mitigate the impact of fire. These methodologies are essential for understanding the complex dynamics involved in the combustion processes of bio-based materials. One prominent method employed in evaluating fire resistance is the cone calorimeter test. This method involves exposing the composite samples to a defined radiant heat source, measuring parameters such as heat release rate, smoke production, and gas emissions [9]. The data obtained from the cone calorimeter test offers valuable information about the material's reaction to fire, aiding researchers in identifying potential weaknesses and areas for improvement. Another significant approach is the limiting oxygen index (LOI) test, which measures the minimum concentration of oxygen required to sustain combustion. This method is particularly relevant in assessing the flammability of biocomposites and bionanocomposites, as it helps determine the materials' susceptibility to ignition under different oxygen levels [10]. Understanding the LOI values provides critical insights into the inherent fire resistance of these materials and guides the development of fire

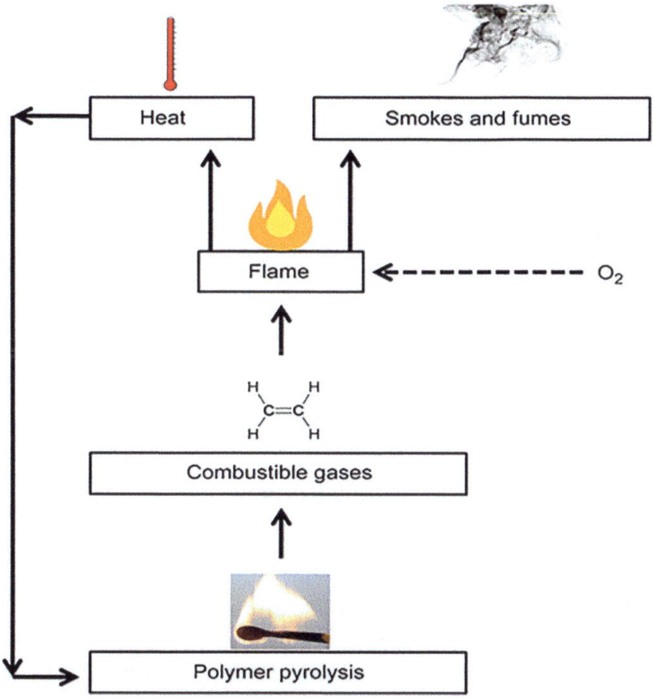

Fig. 1 Mechanism of flammability in polymers [8]

retardant strategies. Furthermore, thermogravimetric analysis (TGA) stands out as a method to assess the thermal stability and decomposition characteristics of bio-based composites. By subjecting the materials to controlled temperature increases, TGA enables the identification of degradation patterns and the determination of key temperature thresholds related to fire susceptibility. These methodologies, along with others such as the microscale combustion calorimetry and fire propagation apparatus tests, collectively contribute to a comprehensive understanding of the fire resistance of biocomposites and bionanocomposites (Fig. 2) [11]. The utilization of multiple testing methods allows researchers to triangulate data, ensuring a more robust evaluation of the materials' performance in diverse fire scenarios. Ultimately, these methodologies empower the scientific community to advance the development of fire-resistant bio-based materials for a sustainable and safer future [12].

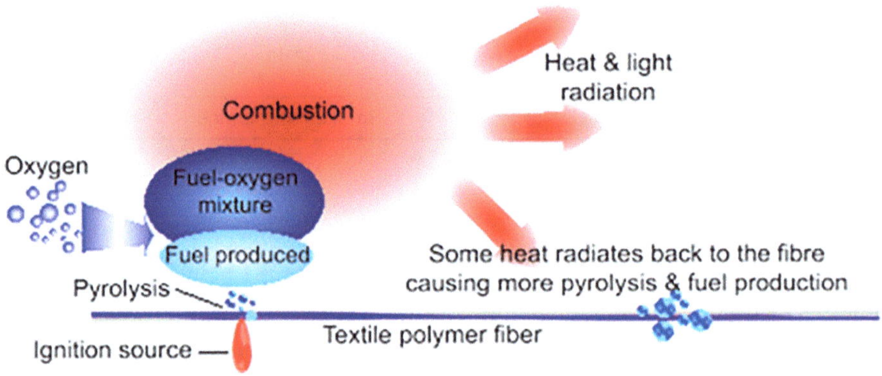

Fig. 2 Fire cycle [11]

3 Inherent Properties of Biocomposites and Bionanocomposites

In the exploration of flammability tests and fire retardant mechanisms within the realm of Biocomposites and Bionanocomposites, a critical examination of their inherent properties unveils fundamental aspects that contribute to their unique fire behavior [13]. Biocomposites, derived from renewable resources such as wood, natural fibers, or agricultural residues, possess distinctive characteristics that impact their flammability profile. The composition, morphology, and thermal stability of these materials play pivotal roles in determining their response to fire. One crucial aspect is the composition of biocomposites, where the ratio of organic components to inorganic fillers significantly influences their flammability [14]. The presence of natural fibers, for instance, may impart a higher degree of combustibility due to the organic nature of these materials. Researchers delve into understanding how different combinations and concentrations of components affect the thermal degradation and ignition susceptibility of biocomposites. Moreover, the morphology of these materials adds another layer of complexity to their inherent properties. The arrangement and distribution of fibers or particles within the composite structure can impact heat transfer, flame propagation, and overall fire resistance. Researchers investigate the microstructure to elucidate how variations in morphology contribute to the thermal performance of biocomposites and bionanocomposites [14]. Thermal stability is a key parameter in assessing the inherent fire behavior of these bio-based materials. Understanding the temperature at which decomposition occurs and the release of flammable gases takes place is essential in predicting their response to fire exposure. Researchers employ sophisticated techniques, such as TGA and differential scanning calorimetry (DSC), to scrutinize the thermal stability of biocomposites and identify potential vulnerabilities. By comprehensively examining the inherent properties of biocomposites and bionanocomposites, researchers aim to unravel the intricacies that govern their flammability characteristics. This knowledge serves as

Table 1 Flammability, ignition and thermal properties of bionanocomposites

Bionanocomposite	Flammability test	Ignition time (s)	Flame spread (cm)	Heat release rate (kW/m^2)
Nanocellulose-PLA	Cone calorimeter	15.2	18.5	450
Chitosan-silver nanoparticles	Horizontal burning	12.8	14.2	380
Starch-zinc oxide	LOI test	21.5	22.0	320
Cellulose nanofiber-graphene	Cone calorimeter	17.3	19.8	410
PLA-nanoclay	Cone calorimeter	14.6	16.5	390
PCL-silver nanowires	Horizontal burning	13.2	15.6	360
Gelatin-TiO_2 nanocomposite	LOI test	19.8	20.5	340
Chitin-carbon nanotubes	Cone calorimeter	16.5	18.0	430
Soy protein-Fe_3O_4	Cone calorimeter	14.9	17.2	400
Polyhydroxyalkanoate-CuO	Horizontal burning	12.5	14.8	370
Lignin-silver nanoparticles	LOI test	20.2	21.8	310
Cellulose nanocrystal-SiO_2	Cone calorimeter	18.0	20.2	420
Polyethylene-GO	Horizontal burning	11.8	13.5	350
Alginate-Fe_2O_3 nanocomposite	LOI test	22.1	23.5	300
PVA-montmorillonite	Cone calorimeter	15.8	17.8	410
Lignin-copper nanoparticles	Horizontal burning	13.6	16.0	380
Chitosan-carbon nanofibers	LOI test	20.5	21.2	330
Gelatin-silver nanowires	Cone calorimeter	16.2	18.6	440
PLA-MWCNT	Horizontal burning	12.2	14.5	360
Cellulose nanofiber-ZnO	LOI test	19.2	20.0	350

a foundation for developing targeted fire safety strategies and designing materials with enhanced resistance to combustion [15]. The flammability, Thermal properties of bionanocomposites are exhibited in Table 1.

4 Synergistic Effects of Fire Retardants

In the realm of fire safety for biocomposites and bionanocomposites, a critical aspect lies in understanding the synergistic effects achieved through the incorporation of fire retardants. This subheading explores the intricate interplay between the inherent properties of bio-based materials and the supplementary influence of fire retardant additives. Biocomposites and bionanocomposites, derived from renewable resources, possess unique characteristics that can be both advantageous and challenging in the context of fire resistance [16]. The integration of fire retardants aims to enhance these materials' ability to withstand combustion, offering a multi-faceted approach to fire safety. The synergistic effects manifest as a result of the collaboration between the inherent properties of the base material and the specific mechanisms triggered by the introduced fire retardants. One key consideration is the role of fire retardants in altering the thermal degradation pathways of biocomposites. These additives can modify the decomposition kinetics, delaying the release of flammable gases and hindering the propagation of flames. Additionally, the interaction between fire retardants and the structural components of biocomposites can create a barrier effect, limiting the access of oxygen and heat to critical areas. Synergy is also observed in the way fire retardants influence the char formation during combustion [16]. The char acts as a protective layer, shielding the underlying material from further thermal degradation. The combination of the inherent charring tendencies of certain biocomposite constituents with the catalytic effect of fire retardants contributes to the formation of a more resilient and insulating char layer. Understanding the synergistic effects is pivotal for optimizing the formulation of fire-retardant biocomposites [17]. Researchers and practitioners can leverage this knowledge to tailor materials with enhanced fire resistance while considering the environmental impact and sustainability aspects. As industries increasingly embrace these bio-based materials, unlocking the synergies between inherent properties and fire retardants becomes a cornerstone in advancing fire safety measures within the realm of sustainable materials [18]. The fire retardants used in composites is presented in Fig. 3 [19].

5 Current Research Findings in Flammability Studies

In the realm of flammability studies for biocomposites and bionanocomposites, current research findings provide crucial insights into the evolving landscape of fire safety within the domain of sustainable materials. Recent investigations have focused on refining testing methodologies to more accurately evaluate the fire resistance of these bio-based composites [20]. Researchers have been exploring innovative approaches that go beyond traditional testing procedures, incorporating advanced analytical techniques to delve deeper into the intricate behaviors exhibited during combustion [21]. One notable area of interest lies in understanding the dynamic interactions between various components within biocomposites and bionanocomposites under fire conditions [22]. Recent studies have shed light on the role of individual constituents, such as biopolymers, nanoparticles, and other additives, in influencing the overall flammability characteristics of these materials. This nuanced exploration aims to uncover synergistic effects and potential trade-offs in fire performance arising from the combination of different elements within the composite structure. Furthermore, the emphasis has been placed on investigating the effectiveness of novel fire retardants specifically tailored for biocomposites and bionanocomposites [23–25]. Researchers are exploring sustainable and environmentally friendly additives that not only enhance fire resistance but also align with the overarching goal of creating eco-conscious materials. This avenue of research seeks to strike a balance between achieving optimal fire safety and ensuring the ecological sustainability of

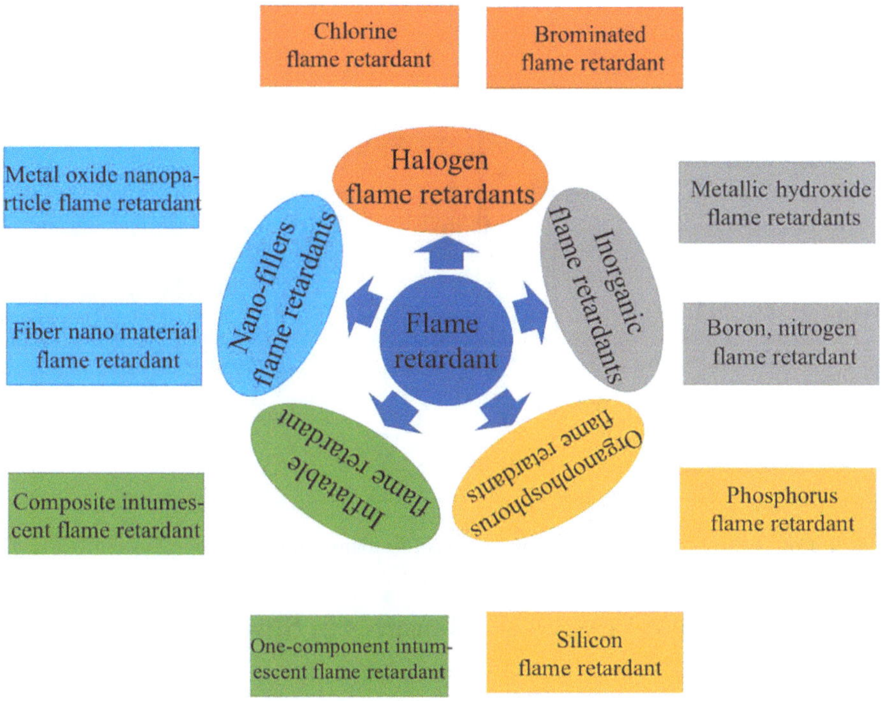

Fig. 3 Fire retardants used in composites [19]

these materials throughout their lifecycle. Recent advancements in material science and nanotechnology have opened up new possibilities for tailoring the flammability characteristics of biobased composites. Nanocomposite formulations, in particular, have gained attention for their potential to impart enhanced flame retardancy without compromising other material properties. These findings underscore the dynamic nature of research in this field, presenting a mosaic of possibilities for advancing fire safety in biocomposites and bionanocomposites. As researchers continue to unravel the complexities of flammability in these materials, the collective knowledge generated contributes significantly to the ongoing evolution of fire mitigation strategies within the realm of sustainable materials. Comparison of ATR-FTIR spectra results of nanobiocomposites is presented in Fig. 4 [26].

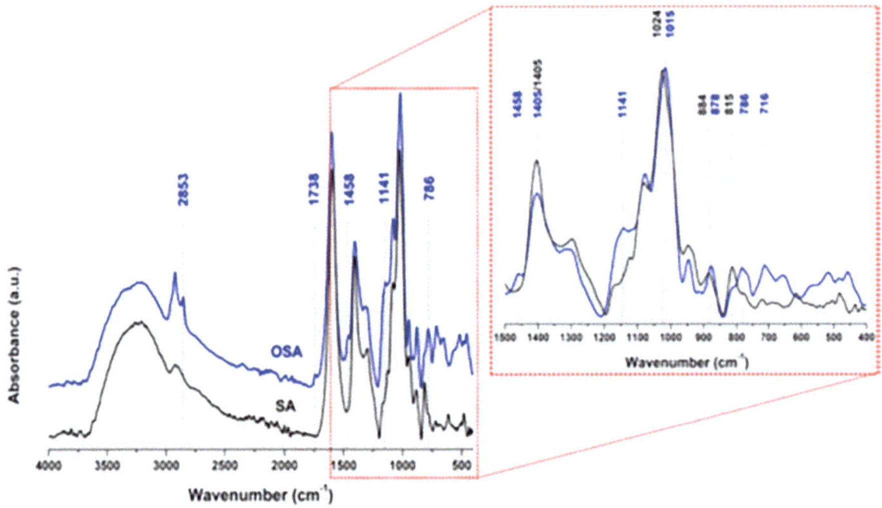

Fig. 4 Comparison of ATR-FTIR spectra results of nanobiocomposites [26]

6 Application of Fire Safety Measures in Sustainable Materials

In the context of biocomposites and bionanocomposites, the application of fire safety measures in sustainable materials represents a crucial stride toward ensuring the overall safety and viability of these bio-based structures [27]. As industries increasingly shift toward eco-friendly alternatives, understanding and implementing effective fire safety measures become paramount. One notable approach in enhancing the fire safety of biocomposites involves the judicious selection and incorporation of fire retardants. These additives play a pivotal role in mitigating the flammability of materials, acting as a barrier against the initiation and propagation of flames [28]. The choice of a suitable fire retardant is often dictated by the specific properties and composition of the biocomposite or bionanocomposite under consideration. Researchers and practitioners explore a spectrum of fire retardants, ranging from traditional halogenated compounds to more environmentally friendly alternatives, such as phosphorus-based or intumescent agents. Moreover, the application of coatings or surface treatments designed to enhance fire resistance represents another avenue of interest. These coatings can provide an additional layer of protection, preventing or slowing down the combustion process [29]. The selection of coating materials is guided by their compatibility with the base biocomposite and their ability to withstand environmental conditions, ensuring long-term efficacy. In addition to these active measures, designing biocomposites with inherent fire-resistant properties is a promising direction. Researchers investigate the incorporation of inherently flame-retardant components during the manufacturing process, capitalizing on the intrinsic characteristics of certain bio-based materials [30]. This approach aligns

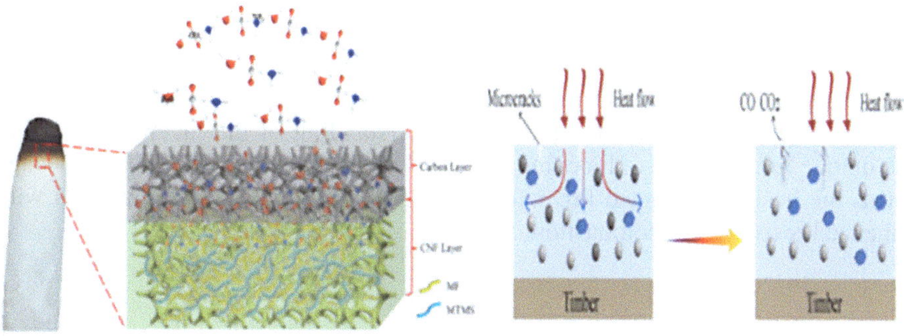

Fig. 5 Flame retardant mechanisms of bionanocomposites [30, 31]

with the broader goal of developing sustainable materials that not only meet structural and performance requirements but also adhere to stringent fire safety standards. As industries continue to adopt biocomposites and bionanocomposites in various applications, the effective implementation of these fire safety measures becomes integral. Researchers and industry professionals alike must collaborate to strike a balance between sustainability and safety, fostering the evolution of materials that not only contribute to a greener future but also ensure robust fire resistance in diverse settings [31]. Flame retardant mechanisms of bionanocomposites are presented in Fig. 5 [30, 31].

7 Summary

7.1 Synthesis of Knowledge

Through the systematic review of current research findings, this article has synthesized a wealth of knowledge in the field, providing a robust foundation for researchers, practitioners, and industries. The amalgamation of information serves as a valuable resource for making informed decisions regarding the application of fire safety measures in the context of sustainable and environmentally friendly materials.

7.2 Fundamental Principles and Methodologies

The elucidation of fundamental principles of flammability tests and the exploration of key methodologies for assessing fire resistance underscore the importance of rigorous testing procedures. Standard tests, such as the Cone Calorimeter and the Horizontal Burning Test, play a crucial role in evaluating the fire behavior of biocomposites and bionanocomposites, ensuring their compliance with safety standards.

7.3 Inherent Properties

The examination of inherent properties, including composition, morphology, and thermal stability, has provided a deeper understanding of the factors influencing the unique fire behavior of biocomposites and bionanocomposites. This knowledge forms the basis for targeted fire safety strategies and the design of materials with enhanced resistance to combustion.

7.4 Synergistic Effects of Fire Retardants

The exploration of synergistic effects between inherent properties and fire retardants has highlighted the collaborative interplay that enhances the fire resistance of bio-based materials. The incorporation of fire retardants modifies degradation pathways, influences char formation, and creates a barrier effect, contributing to a more resilient and insulating char layer.

7.5 Current Research Findings

Recent research findings have showcased the dynamic nature of flammability studies for biocomposites and bionanocomposites. Innovations in testing methodologies, exploration of individual constituent roles, and the development of novel fire retardants emphasize the ongoing evolution of fire safety strategies tailored for sustainable materials.

7.6 Application of Fire Safety Measures

In the practical application of fire safety measures, the judicious selection of fire retardants, the use of coatings, and the design of inherently fire-resistant biocomposites emerge as promising avenues. Striking a balance between sustainability and safety is essential as these materials find increased adoption in diverse industrial applications.

8 Conclusion

The comprehensive exploration of flammability tests and fire retardant mechanisms specific to biocomposites and bionanocomposites has unveiled critical insights into the evolving landscape of fire safety within the domain of sustainable materials. As industries increasingly embrace eco-friendly alternatives, the imperative to understand and implement effective fire safety measures becomes paramount. In conclusion, the collective insights presented in this research review contribute significantly to advancing the understanding of flammability characteristics and fire mitigation strategies for biocomposites and bionanocomposites. As the demand for sustainable materials continues to grow, the integration of effective fire safety measures becomes integral to ensuring a greener and safer future.

References

1. Xu K, Yan C, Du C, Xu Y, Li B, Liu L (2023) Preparation and mechanism of toughened and flame-retardant bio-based polylactic acid composites. Polymers 15(2):300
2. Rajeshkumar L, Ramesh M, Bhuvaneswari V, Balaji D, Deepa C (2023) Synthesis and thermo-mechanical properties of bioplastics and biocomposites: a systematic review. J Mater Chem B 11:3307–3337
3. Ao X, Vázquez-López A, Mocerino D, González C, Wang DY (2023) Flame retardancy and fire mechanical properties for natural fiber/polymer composite: a review. Compos Part B Eng, 111069
4. Ramesh M, Deepa C (2024) Processing and properties of jute (Corchorus olitorius L.) fibres and their sustainable composite materials: a review. J Mater Chem A 12:1923–1997
5. Kanaginahal GM, Tambrallimath V, Murthy M, Mahale RS, Patil A, Pawar SY, Kakkamari PP (2023) Flammability studies of natural fiber-reinforced polymer composites fabricated by additive manufacturing technology: a review. J Inst Eng (India) Ser D, 1–13
6. Rajeshkumar L, Sathish Kumar P, Ramesh M, Sanjay MR, Siengchin S (2023) Assessment of biodegradation of lignocellulosic fiber-based composites—a systematic review. Int J Biol Macromol 253:127237
7. Xia L, Tan C, Ren W, Liu X, Zhang X, Wu J, Yang R (2024) Robust, biodegradable and flame-retardant nanocomposite films based on TEMPO-oxidized cellulose nanofibers and hydroxyapatite nanowires. Carbohydr Polym 324:121495
8. Das O, Kim NK, Hedenqvist MS, Bhattacharyya D (2019) The flammability of biocomposites. In: Durability and life prediction in biocomposites, fibre-reinforced composites and hybrid composites. Woodhead Publishing, pp 335–365
9. Zhang Y, Wang X, Duan Y, Li C, Cheng Y, Tao Y, Wang H (2023) Biological-mechanical pretreated rice straw for constructing foam with mechanically robust and fire-retardant behaviors. Ind Crops Prod 204:117409
10. Bajwa DS, Holt G, Stark N, Bajwa SG, Chanda S, Quadir M (2023) Nano boron oxide and zinc oxide doped lignin containing cellulose nanocrystals improve the thermal, mechanical and flammability properties of high-density poly (ethylene). Polymers 16(1):36
11. Al-Mosawi AI (2016) Flammability of composites. In: Lightweight composite structures in transport, pp 361–371
12. García-Muñoz MA, Valera-Zaragoza M, Aguirre-Cruz A, Carmona-García R, Ramírez-Hernández A, Ramírez-Vargas E, Sánchez-Valdes S (2023) Influence of compatibility in the EVA/starch/organoclay biodegradable nanocomposite on thermal properties and flame self-extinguishing behavior. Polym-Plast Technol Mater 62(17):2318–2333

13. Odera RS, Idumah CI, Ezeani EO, Okpechi VU, Madu IO, Oyeoka HC, Ogbu JE (2023) Novel advancements in flame retardant mechanisms of halloysite nanotubes polymeric nanoarchitectures. Polym-Plast Technol Mater, 1–34
14. Liu SH, Kuan CF, Ke C, Shen M, Chiang CL (2023) A novel effective halogen-free biomass-based chitosan derived flame retardant through microencapsulation technology for improved flame retardancy, thermal stability of polyurethane. Polym Degrad Stabil, 110604
15. Muhammed Raji A, Hambali HU, Khan ZI, Binti Mohamad Z, Azman H, Ogabi R (2023) Emerging trends in flame retardancy of rigid polyurethane foam and its composites: a review. J Cell Plast 59(1):65–122
16. Li Y, Xia Y, Li M, Liu J, Yu M, Li Y (2023) Aramid/alginate blended nonwoven fabrics: thermal degradation properties, flame-retardant mechanism and smoke suppression. Int J Cloth Sci Technol 36:52–67
17. Khan A, Sapuan SM, Yusuf J, Siddiqui VU, Zainudin ES, Zuhri MYM, Rahman AAA (2023) An examination of cutting-edge developments in Bamboo-PLA composite research: a comprehensive review. Renew Sustain Energy Rev 188:113832
18. Abdelkhalik A, Makhlouf G, Ameen H, El-Gamal AA (2023) Preparation of multifunctional flame-retardant coating of cotton fabrics for electrical insulating applications. J Thermoplast Compos Mater, 08927057231203549
19. Shen J, Liang J, Lin X, Lin H, Yu J, Wang S (2021) The flame-retardant mechanisms and preparation of polymer composites and their potential application in construction engineering. Polymers 14(1):82
20. Ye J, Gao Y, Xu Q, Jin Z, Wu G, Wang S, Li Q (2023) P/N/S synergistic flame retardant holocellulose nanofibrils efficiently pretreated from ternary deep eutectic solvents. Chem Eng J 477:147142
21. Wang Y, Wang H, Wang L, Zhu Z (2023) Turning waste into treasure: Effects of waste polyphenylene sulfide (PPS) microfibers as potential synergists on flame retardancy and mechanical property of polylactic acid. React Funct Polym 192:105728
22. Yuan Y, Lin W, Xiao Y, Yu B, Wang W (2023) Advancements in flame-retardant systems for rigid polyurethane foam. Molecules 28(22):7549
23. Xavier JR, Vinodhini SP, Ramesh B, Raja BJ (2023) Flame retardant and anticorrosion behavior of multifunctional epoxy nanocomposite coatings containing graphitic carbon nitride/silanized HfO_2 nanofillers for the protection of steel surface in automobile industry. ACS Chem Health Saf 30(6):428–450
24. Ganapathy T, Ramasamy K, Suyambulingam I, Siengchin S (2023) Synergetic effect of graphene particles on novel biomass–based Ficus benghalensis aerial root/flax fiber–reinforced hybrid epoxy composites for structural application. Biomass Convers Biorefin, 1–38
25. Attia NF, Zakria AM, Nour MA, Abd El-Ghany NA, Elashery SEA (2023) Rational strategy for construction of multifunctional coatings for achieving high fire safety, antibacterial, UV protection and electrical conductivity functions of textile fabrics. Mater Today Sustain 23:100450
26. Quaratesi I, Micu MC, Rebba E, Carsote C, Proietti N, Di Tullio V, Badea E (2023) Cleaner leather tanning and post-tanning processes using oxidized alginate as biodegradable tanning agent and nano-hydroxyapatite as potential flame retardant. Polymers 15(24):4676
27. Dzulkifli MH, Majid RA, Yahya MY (2023) Influence of sodium-montmorillonite nanoclay on flammability, thermal stability, morphology, and mechanical properties of rigid water-blown palm oil-based polyurethane foam. Cell Polym 42(3–4):140–155
28. Zhong J, Wang E, Sun Y, Yin N, Tian S, Ying W, Zhang W (2023) Fabrication of phytic acid/urea co-modified bamboo biochar and its application as green flame retardant for polylactic acid resins. Polymers 15(2):360
29. Fazeli M, Mukherjee S, Baniasadi H, Abidnejad R, Mujtaba M, Lipponen J, Rojas OJ (2024) Lignin beyond the status quo: recent and emerging composite applications. Green Chem 26:593–630
30. Yue X, Zhang S, He J, Wang Z (2023) Fabrication of flame retarded cellulose aerogel with hydrophobicity via MF/MTMS double cross-linking. J Nat Fibers 20(1):2133053

31. Yin P, Cai H, Liao W, Deng X, Wu J, Qu H, Ding Y (2023) Design and preparation of flame-resistant geopolymer coatings for timber. J Mater Sci 58(34):13865–13874

Fire Retardant Properties of the Epoxy Resin-Based Biocomposites and Bionanocomposite

Emel Kuram

Abstract Epoxy resins have great modulus, high glass transition temperature, superb adhesion to large variety of fibers and chemical resistance. But the fact that epoxy resin cannot be recycled or dispersed due to permanent cross-linked structure causes them not environmentally friendly. Therefore, reinforcing biopolymers or synthetic polymers via natural fibers in place of synthetic ones within polymeric composites is good option for environmental improvement. There is increasing attention for natural fiber-reinforced composites at industrial implementations as natural fibers are appealing alternative to synthetic fibers. But, since natural fibers possess high flammability, fire behavior of the material needs to be enhanced. Fire retardant properties of polymer composites is a significant matter when environment, public health and safety are concerned. This chapter presents an overview on fundamentals for fire retardancy properties of epoxy resin based biocomposites and bionanocomposites. Tests that are utilized to evaluate flammability of epoxy resin based biocomposites and bionanocomposites and their results are reviewed.

Keywords Fire retardancy · Epoxy · Biocomposites · Bionanocomposites

1 Introduction

Epoxy resins have great modulus, high glass transition temperature, superb adhesion to large variety of fibers and chemical resistance, making them versatile as a structural thermosetting. But, fact that epoxy cannot be recycled, or dispersed due to permanent cross-linked structure causes them not environmentally friendly in comparison to their thermoplastic counterparts [2, 21, 49, 87, 98]. Solving problem on environment caused with thermosets is a major field of interest, and development of biodegradable polymers and composites may give path forward. Therefore, reinforcing biopolymers

E. Kuram (✉)
Department of Mechanical Engineering, Gebze Technical University, Gebze, Kocaeli, Turkey
e-mail: kuram@gtu.edu.tr

C. Muthukumar et al. (eds.), *Flammability Performance of Biocomposites and Bionanocomposites*, Composites Science and Technology,
https://doi.org/10.1007/978-981-95-2389-4_2

or synthetic polymers via natural fibers in place of synthetic ones within polymeric composites is good option for environmental improvement [2, 35, 38, 69, 70].

Bio-based composites are reinforced polymers where one of reinforcement and matrix or both elements are derived by bio-based origin [5]. Protection of environment is one of the most important subjects facing by nowadays generation. New procedures are needed to protect the environment [56]. Increasing in environmental attention and petroleum-based sources shortage have driven researchers to produce sustainable bio-based composite using renewable feedstocks [57, 74]. Usage of natural fiber as reinforcement in composites will not only meet future industrial requirements but will also be environmentally friendly and economical. For these reasons, researchers have concentrated on biocomposite with bio resin and natural fiber recently [57, 74]. In recent years, works concerning biocomposites and bionanocomposites have increased. Environmental awareness and depletion of petro-based resources are main reasons for the development of biocomposites and bionanocomposites. Most of constituents in biocomposites and bionanocomposites depend on agriculture products as resource of raw material. Therefore, their usage offers a solution to waste disposal, decrement in agriculture wastes and thus environment contamination caused by burning. It also presents economic chance to agriculture and rural zones for developing countries [55]. The biggest disadvantage for biocomposites is relatively weak resistance to fire [5, 60]. Natural materials are flammable and undergo thermal decomposition under small temperatures between 200 and 300 °C [14, 22, 60]. Due to its flammable properties, it will restrict implementation of material in industry. It is significant to produce materials having greater fire resistance without influencing moderate mechanical features of materials [56, 60]. Main aim of study in the literature is to produce biocomposites with improved flame retardancy that require higher oxygen amount in order to burn. Biocomposites have flammability characteristic. To increase fire resistance of biocomposites, flame retardants that are not easily ignited are loaded to biocomposites. Flame retardants are chemicals with smaller total emission of heat as subjected to source of ignition. They constrain and delay fire spread after ignition. Some studies concerning usage of flame retardants in biocomposites have been conducted [34].

Because of rising environmental awareness, depletion of petroleum sources, post-use disposal problems, and introduction of novel regulations and rules by legislative authorities to meet new demands of society, environmental impacts of new products need to be minimized. This issue is particularly relevant in area of composite manufacturing, where petroleum polymers and synthetic fibers are still extensively employed. This concept is directly linked to the effective usage of materials and energy resources, waste management and usage of products and techniques that have less impact on environment. Environmental and economic concerns combined with government rules have spurred research on greener alternatives to glass and petroleum-based composites. Much of the work has concentrated on production of biocomposites and natural renewable materials. This chapter presents an overview on fundamentals for fire retardancy properties of epoxy resin based biocomposites and bionanocomposites. Tests that are utilized to evaluate flammability of epoxy resin based biocomposites and bionanocomposites and their results are reviewed.

2 Biocomposites and Bionanocomposites

Composites are composed of one or higher discontinuous reinforcement phases embedded within continuous matrix phase [5]. These materials are unique since they combine material features in a way that does not exist in nature [77].

Biocomposites is described as: biodegradable polymers reinforced via natural fibers; biodegradable polymers reinforced by man-made fibers; non-biodegradable or petroleum-based polymers reinforced via natural fibers [17, 95]. Biocomposites have many pros over traditional materials, such as lightness, lower production price, recyclability and renewability, and supporting people well-being and ecosystem health [17]. Owing to these properties, recently biocomposites have attracted large interest in various applications, from household appliances to automotive sector [60]. Biocomposites can be classified into completely and partially biocomposites with respect to fiber and matrix (Fig. 1). In completely biocomposites, natural fibers are utilized with natural polymeric matrix. However, in partially biocomposites, natural polymers are reinforced via synthetic fibers or synthetic polymers are reinforced with natural fibers [46].

Bionanocomposites are the combination of polymers with a reinforcement having at least a dimension in nanometer size [52].

Natural fibers (biofibers) are renewable fibers derived from agricultural waste, animal, mineral and plant sources. They could be utilized as reinforcements in the biocomposites production. These materials are promising as new generation reinforcements since they present good features such as lightness, lower price, toughness [55], biodegradability, easy availability (abundant supply), environmental friendliness, low carbon emissions, lower energy consumption and renewability [45]. Thanks to these properties, there is a rising trend toward employing biofibers as filler or reinforcement within polymer composites. Flexibility during process, greatly specific stiffness and lower price makes them good options to producers. Usage of such materials into composites has risen because they are relatively inexpensive, can be

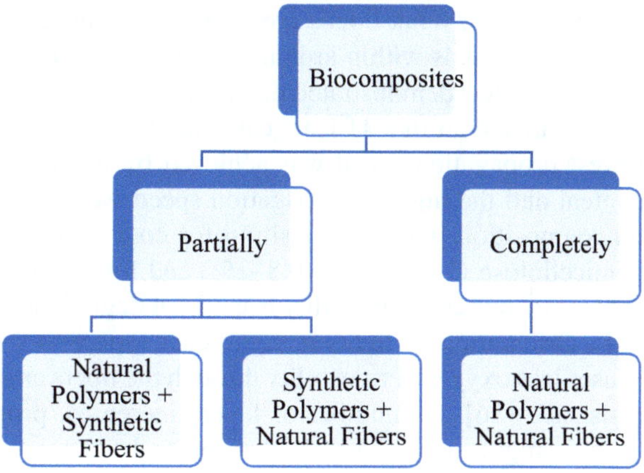

Fig. 1 Biocomposites definition

recycled, and compete well regarding strength per weight of material [16], which causes decrement of negative effects on environment, provides economical materials and creates a path for cleaner production. Natural fiber-reinforced composites have gained attention owing to several pros such as dermal non-toxicity, biodegradability, and strength. In desire to mitigate climate alteration caused by greenhouse gas emissions, biodegradable polymers are being investigated as the best choice for polymers in composites other than their non-renewable synthetic ones [71]. With use of natural fiber, weight of petrochemical raw material employed has been reduced. Cellulose, lignin, hemicellulose and pectin are chief constituents of natural fibers [5]. Each fiber consists of a primary cell wall, three secondary cell walls, and a lumen throughout, this leads to rounded, elongated, hollow cross-sectional nature; therefore, unlike synthetic fibers, it possesses very complicated layered structure [51]. Chemical composition of natural fibers, which bases onto plant species, age, soil properties and climate, is very important for their performances and implementations. Every component could cause various kinds of sensitivities to natural fibers. For instance, lignin chiefly controls fire degradation [5]. Biofibers having higher cellulose are more flammable than those having greater hemicellulose, whereas char occurrence is normally greater at higher lignin amount. Biofiber with higher silica or ash demonstrated superior fire resistance. Additionally, the crystallinity and polymerization of biofibers influence fire resistance [55].

One of the disadvantages for natural fibers as comparison to carbon or glass fibers is flammability [25]. Natural fibers are very vulnerable to flame; it has become increasingly significant to increase flame retardant requirement of composites to comply with safety necessities for natural fiber based composites. Fire behavior for natural fibers bases on fiber components, percentage weight and morphology (i.e., fiber orientation) [26]. Main components of natural fibers controlling flammability are lignin and cellulose [27]. Cellulose decomposition between 260 and 350 °C leads to production of combustible volatiles and gases, non-flammable gases, char and tars [36]. Hemicellulose decomposes between 200 and 260 °C however higher non-flammable gas and lower tar than cellulose are produced [15]. Lignin begins to decompose from approximately 160 °C to approximately 400 °C, during lignin decomposition, weak bonds are broken at smaller temperatures, whereas cleavage of stronger bonds within aromatic rings occurs at greater temperatures [29]. Cone calorimetry test demonstrated that heat release rate (HRR) of bast (hemp, flax) fibers was smaller than that of leaf (cabuya, abaca) fibers [44]. It was reported that the lowest propagation speed was achieved by linen fiber because of its lower lignin content and the highest propagation speed was achieved by banana fiber owing to its composition [64]. Composition for cotton fiber is chiefly cellulose (85–90%), hemicellulose (1–3%), ash (0.8–2%) and lignin (0.7–1.6%) [43]. Higher cellulose content increases flammability because of degradation starting at higher temperature [44]. Orientation of fibers is significant aspect to take into consideration since it causes less oxygen permeability through the fibers and consequently increased flame resistance [26]. As number of layers increases, propagation speed reduces since there is higher material to be burned and fiber amount is greater [64]. Gieparda et al. [33] modified natural fiber chemically via silanization to decrease flammability of epoxy laminates. Flammability of reinforcement fibers under the influence of natural fiber modifications was reduced by 400%. Flame retardant effectiveness of silanes is owing to silicon presence in their chemical structure [33]. Natural based composites change thermal inertia and cause increased fuel quantity in comparison to inert reinforcement [82]. These changes generally result in decreased ignition time and increased total heat release. However, some favorable influences could also be seen, such as char yield production, which could protect underlying polymer to some extent and decrease decomposition rate [24]. Biocomposites must be flame retardant to meet the required specifications for implementations in industries such as transportation and construction. Fireproofing of biocomposites could be obtained either via adding flame retardant to matrix or via grafting or depositing suitable molecules onto natural fibers surface. In last decades, significant efforts have been carried out to modify natural fiber [11].

Heat shielding by forming condensation phase protective layer is one of mechanical manifestations for flame retardant [55]. Fly ash particles that will migrate onto material surface to accumulate and behave as thermal barrier as epoxy is at its molten case. Additionally, part of carbon ring in epoxy can be catalyzed to oxidize during combustion to form hybrid structure consisting of silicon oxidized carbon, behaves as insulating residual carbon layer, rising residue rate carbon of pyrolysis products, hence preventing heat transfer to underlying materials and limits fire spread [6, 28, 31]. Research suggests that competition between absorbent influence and physical barrier determines fire resistance of fly ash/epoxy composite [63].

Natural fiber epoxy composites are low weight, strong and can be employed in several implementations where fire response is dangerous matter, including aerospace, automobile and building [39]. Consequently, awareness and use of traditional fire retardants regarding their response or performance during fire is of great importance. Epoxy based biocomposites have been extensively studied and are taken into consideration as alternative to synthetic fiber-reinforced traditional composites. Among published studies on epoxy based biocomposites, only few researchers have investigated the fire retardant properties of these biocomposites. This chapter reviews literature on fire retardant properties concerning epoxy resin based biocomposites and bionanocomposites.

3 Polymer Combustion, Natural Fiber Combustion and Fire Retardancy

3.1 Polymer and Natural Fiber Combustion

Most of polymers utilized are depended on petroleum origin and therefore greatly combustible/flammable. Polymers are greatly combustible owing to composing chiefly carbon and hydrogen in chemical structure. Combustion reaction contains two parameters: one or higher combustible substances (reducing agent) and combustive substance (oxidizing agents, mainly oxygen in air). Process usually starts when polymer temperature increases enough to result in polymer bonds scissions (polymer degradation, pyrolyse) [4, 47]. Volatile part of resulting polymer fragments diffuses to air and causes flammable gas mixture (fuel) [47]. The release of flammable gases mixed with atmospheric air promotes violent combustion of the substrate and, as a result, decomposition of materials [4]. The gas mixture ignites and releases heat as autoignition temperature (temperature at which activation energy for combustion reaction is reached) is obtained [47]. Decomposition and formation of flammable gases when mixed with ambient oxygen air cause generation of ignitable mixture via polymers through exposure to sufficient heat sources and through autoignition [41, 72]. After combustion, there is the likelihood of self-sustaining combustion cycle to keep volatile amount in flammability limit. Therefore, heat, fuel and O_2 sources are three main components needed to sustain fire [72]. In short, the decomposition of the polymer depends chiefly on flammability, fire spread and heat release properties. Adequate sources of fuel, heat and oxygen are required at every step for combustion, and these resources can be ambient or self-triggered (through material combustion, release of volatile gases and particles serves as source for further combustion and creates cyclic process) [4]. Additionally, chemical composition for polymers and conditions under which combustion takes place directly affect fire nature that develops [41, 72]. As alternative, fuel could ignite at smaller temperature (flash point) when it reacts with intense energy external source [47].

Lifespan of combustion cycle for polymer fire bases on amount of heat released during fuel combustion. When heat release reaches certain level, decomposition of the solid phase proceeds, resulting in the release of further combustibles. This aids maintain and continue combustion cycle, ultimately leading to formation of a fire triangle (Fig. 2) [47].

Heating can result from the contribution of thermal energy by external heat source, chemical process induced in material, or exothermicity for initiated combustion reaction. Energy amount needed to initiate combustion within polymers varies depending on polymer physical properties. Semi-crystalline thermoplastic softens, melts and

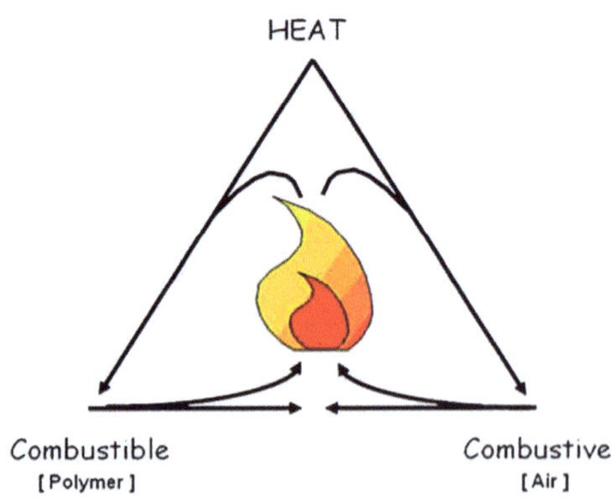

Fig. 2 Fire triangle [47]. Copyright 2008 Elsevier B.V., reprinted by permission of Elsevier B.V.

drips during heating. Energy stored through polymeric material during process bases on heat storage capacity, fusion enthalpy and crystallinity degree. However, heating stage directly causes polymer decomposition owing to lack of melting point for amorphous thermoplastics and most thermosets [47].

Polymer thermal decomposition is the endothermic phenomenon requiring energy input. Energy introduced into system have to be greater than binding energy between covalently connected atoms (C–C polymers, value is 200–400 kJ/mol). In general, thermal decomposition is due to heat and oxygen effects combination [47].

Like other thermoset resins, epoxy decomposes when subjected to great temperature (300–400 °C), releasing smoke, soot, heat and toxic volatiles by matrix of epoxy [68].

Exposure of natural fibers to heat or fire results in thermal decomposition and rupture, every time started with hemicellulose decomposition at small temperature followed with sudden decrease in mass due to pyrolysis of lignin and cellulose. Lignin decomposition is final constituent to decompose at the greatest temperatures. All ruptures cause charcoal, heat and gases. Both fibers and matrix begin to melt under thermal decomposition [48].

It was declared that flax fiber specimens burned more strongly than glass fiber counterparts (Fig. 3) because of larger content of available fuel as well as the constantly deteriorating fiber structure that would otherwise restrict release of volatile gases for combustion. Flame spread rate for flax fiber specimens was higher. Additional fuel within flax fiber specimens accelerates propagation of flame front by providing higher energy to feed back in pyrolysis and combustion [14].

3.2 *Fire Retardancy*

Fire (flame) retardancy is significant to determine the flammability properties. It is fact that materials are less likely to ignite or burn less efficiently if they are flammable [55]. Several tests could be conducted to evaluate fire behavior of biocomposites: cone calorimetry, limiting oxygen index (LOI) testing, vertical burning testing and micro-scale combustion calorimetry [5]. Fire retardancy for plastics can be determined with analyzing parameters measured via cone calorimeter involving heat release rate (HRR), peak HRR (PHRR), time to ignition (TTI), mass loss rate (MLR), specific extinction area (SEA) and total smoke parameter (TSP). Burning rate takes heat release rate and mass loss rate for specimens [3]. Average burning rate can be estimated by following formula:

$$\text{Average Burning Rate(mm/ min)} = \frac{\text{Damaged(Burned)Length(mm)}}{\text{Time(min)}}$$

Flammability is critical issue in some industries, especially in transportation in which confined spaces present fires an important danger [14]. Babrauskas and Peacock [3] describe flammability by reaction process of a material to fire and divide

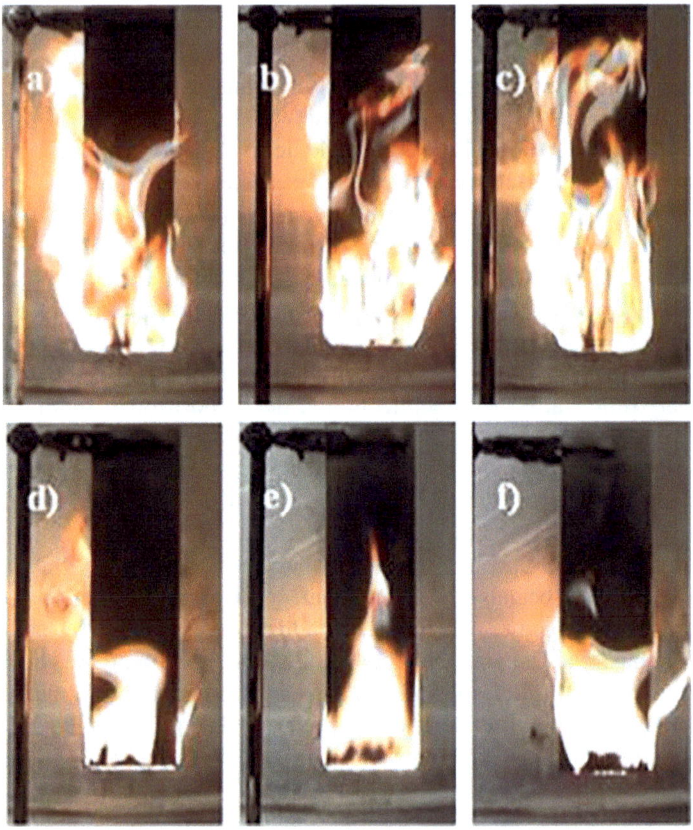

Fig. 3 Vertical burning test specimens 180 s in testing. **a** Flax twill weave 2, **b** flax twill weave 1, **c** flax unidirectional fabric (UNI), **d** glass twill weave, **e** glass plain weave and **f** glass UNI [14]. Copyright 2012 Elsevier Ltd., reprinted by permission of Elsevier Ltd.

it into following parts: ignitability, flame spread, heat release and combustion products. Combustion is catalytic exothermic reaction sustained via internally created heat and free radicals [67]. Bourbigot and Duquesne [7] describe flammability by several parameters: ignitability, burning rate (including heat release and mass loss factors), spread rate and combustion products. Second description presents greater comprehensive set of parameters to describe flammability [14]. Ignitability of a material includes ignition time of material, ignition temperature, and critical heat flux of ignition to take place. They are significant factors to take into consideration when taking fire preventive measures, because preventing ignition inherently prevents fire itself. Spread rate described with term flammability are flame propagation, pyrolysis, burn-out and smoulder. It is vital to understand how quickly fire will spread through a structure, and one of purposes of fire retardant is to slow or entirely stop the process [3].

Main combustion products studied are smoke and toxic species emitted from materials through thermal decomposition and combustion. Most fire deaths result from inhalation of flammable gases and particles (mostly carbon monoxide) [67], and therefore monitoring and prevention of production needs to be taken into consideration. However, formation of seen smoke in early steps of a fire could also give early warning or smoke alarms for human nearby [14]. Charring is incomplete combustion of solid as exposed to great heat. Therefore, charring eliminates oxygen and hydrogen so that only carbon remains inside char [84].

In the literature, fire retardancy for various natural fiber-reinforced epoxy composites was investigated. It was declared that flax based composites ignited earlier, released more heat, and structure was importantly deformed during combustion than synthetic (glass) based counterparts. Flax fibers were sensitive to ignition and combustion. Flame extinguishment time was also higher in flax fiber based composites, although visually they appeared to burn more strongly. But it is to be expected as flax fibers present more fuel for combustion at smaller heat release rate [14]. Linseed and hemp reinforced biocomposites depended on bio-based epoxy resin presented no burning drops when burning. Regarding flame retardant properties, composites had smaller capability to spread fire compared to natural fibers. Linseed fibers could not withstand any of flame retardancy tests conducted. Hemp fiber withstood small burner for 15 s. When composite was characterized entirely, linseed fiber specimens could withstand small burner for 15 s, changing from class F to class E. For hemp, further characterization and methods were required to distinguish class B, C or D. Experimental model followed in the classification of a material is shown in Fig. 4. Fire resistance classification for bio-based nonwoven epoxy composite is given in Table 1. On one hand, existence of epoxy and composite as an entire had greater fire resistance, causing flame propagation slower than in situations where bio-resin was not included within specimen. On other hand, the most striking point was that, although all fibers were generally viewed in same way, hemp showed to have smaller fire transfer than linseed, with a similar fire behavior expected, and its use for this purpose was preferred [89].

In the literature, influence of lychee peel onto fire resistance for epoxy composite was investigated. Flame retardant properties tended to rise with addition of lychee peel. 20 wt% lychee peel presented LOI of 21.5% having burning rate of 23.45 mm/min. Lychee peel is carbon-rich source of biomass that is byproduct of food sector. When it is burned, cinder block is occurred, and it is possible that this is biochar (Fig. 5). Its surface possesses porous honeycomb structure composing of great carbon concentration [60]. The honeycomb porous structure of biochar filler allows molten polymers to penetrate during processing and creates physical bond, which could result in increased mechanical fit and fire resistance [4, 64]. When lychee peel was burned, lychee peel created layers of carbon-rich charcoal and behaved as heat-resistant films. The film restricted transport of fuel and O_2, causing enhanced combustion properties such as burning rate and limiting the oxygen index (Fig. 5). Shortly, carbon-based fillers decreased flammability for composite with creating protective charcoal layer and absorbing free radicals. Fire resistance for reinforcements obtained from agricultural wastes bases on compatibility via main material. At interface between

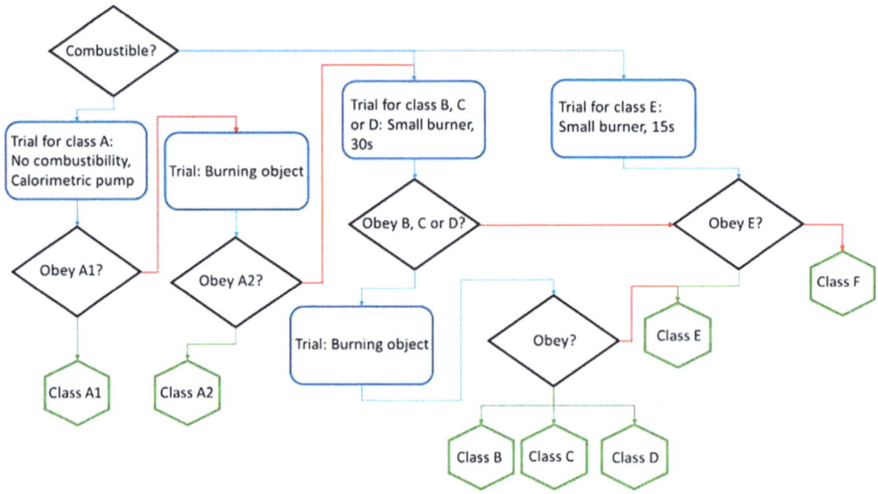

Fig. 4 Scheme for classifying materials based on their fire resistance. Reproduced from Vidal et al. [89]. CC BY 4.0. Copyright 2023

Table 1 Classification of fire resistance for bio-based non-woven epoxy composite

Sample	Classification
Hemp	E
1 layer of hemp 1300 g/m^2	B, C or D
3 layers of hemp 600 g/m^2	B, C or D
Linseed	F
1 layer of linseed 1300 g/m^2	E
3 layers of linseed 600 g/m^2	E

Reproduced from Vidal et al. [89]. CC BY 4.0. Copyright 2023

natural reinforcement and base resin having compatible structure, moderate adhesion and wetting will result in enhanced mechanical features and fire resistance [61, 62].

In the literature, used coffee grounds were blended via epoxy matrix at various contents (30, 40, 50 and 60 wt%). With respect to flame retardant properties for epoxy composites having 30 wt% used coffee grounds, burning rate in accordance with UL94HB was 27.02 $\pm$ 0.29 mm/min. When used coffee grounds content increased, compatibility decreased, causing an unstable structure, which in turn led to the fire retardancy property of material to decrease. Especially in the ratio for 30 wt% used coffee grounds, which had very good compatibility ratio, mechanical features were good and flame retardancy was higher than other ratios [62].

Waste coffee grounds, biomaterial having great amount of carbon, have been proposed as non-toxic organic flame retardant because of great char formation and radical scavenging features, which forming dense char layer through combustion to present fire protection. Coffee grounds contain approximately 40% lignin [86].

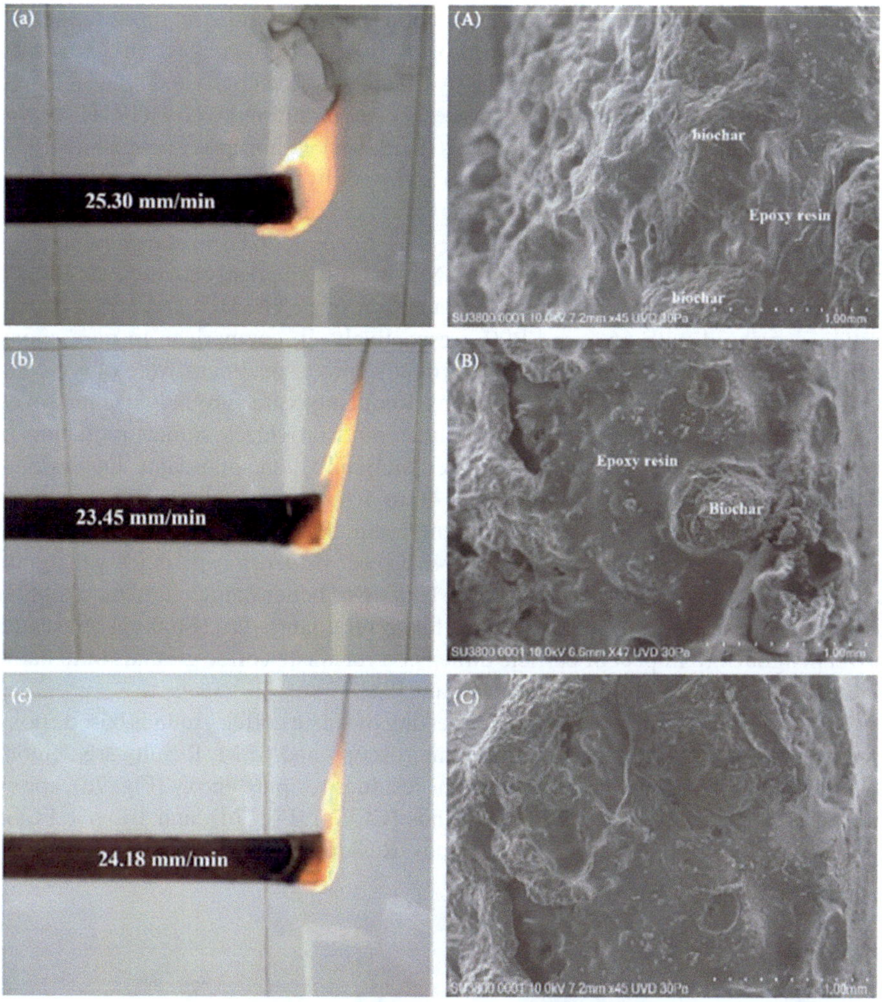

Fig. 5 UL94HB test images for **a** 10 wt%, **b** 20 wt% and **c** 30 wt% lychee peel. SEM images after burning for **A** 10 wt%, **B** 20 wt% and **C** 30 wt% lychee peel. Reproduced from Nguyen [60]. CC BY 4.0. Copyright 2021

Lignin has aromatic structures that can exhibit natural radical scavenging ability [32]. Since methane (CH_4) and carbon monoxide (CO) emissions [97] give more fuel for fire, their decrement through cone-FTIR testings is also evidence of radical scavenging feature for coffee ground. Radical scavenging influence at gas phase could capture combustible volatiles such as oxygen-free radicals from polymer, thus interrupting combustion cycle and creating flame retardant influence. At the solid phase, radical scavenger could react via carbon radicals and create greater solid products, inhibiting combustion reaction [94]. Waste coffee grounds enhanced flame retardancy and decreased toxic gas emissions for epoxy composite, almost completely suppressing highly toxic nitrogen dioxide (NO_2) gas. Toxic gases such as CH_4, hydrogen cyanide (HCN), CO, hydrogen chloride (HCl) and NO_2, released during the combustion and pyrolysis process, were monitored with FTIR. HCN was produced by burning of amine compounds (hardener) within epoxy composite. HCN concentration was decreased from 289 to 155 ppm by 10 wt% coffee ground. CH_4 levels detected at cone calorimeter testings were decreased from 140.3 to 91.0 ppm with loading 10 wt% fine coffee ground (FCG). Loading FCGs to epoxy composites reduced CO emissions by 43%. HCl and NO_2 were also greatly toxic gases formed via combustion of epoxy [65], but after addition of FCGs, HCl and NO_2 concentration decreased with 66% and 90%, respectively. In comparison to pure epoxy residue, greater nitrogen (N) for coffee grounds/epoxy specimens were observed at residue, thus decreasing NO_2 gas emissions. Reducing coffee grounds size improved flame further. Reduced size coffee grounds could provide a similar reduction in peak heat release rate as commonly utilized inorganic flame retardants. Downsizing mean dimension of coffee grounds from 46 to 16 μm employing milling or filtration (Fig. 6) was stated to be efficient in increasing flame retardancy in comparison with as-received waste coffee ground [94]. Lower dimension particles gave good dispersion and big specific surface region leading to better flame retardancy, similar to dimension influences found for other flame retardants [96]. Findings presented novel insights into impact of particle dimension for waste coffee grounds onto flame retardant performance for epoxy composites [94].

Residual chars from cone calorimeter testing of waste coffee grounds based epoxy composites were evaluated with optical microscopy and SEM. Results are demonstrated in Fig. 7a–i. Differences in the char residues for pure epoxy (Fig. 7a), epoxy having 10 wt% coarse waste coffee grounds (CCGs) (Fig. 7d), and 10 wt% FCGs (Fig. 7g) were visible in optical photographs. Residue for pure epoxy specimen was

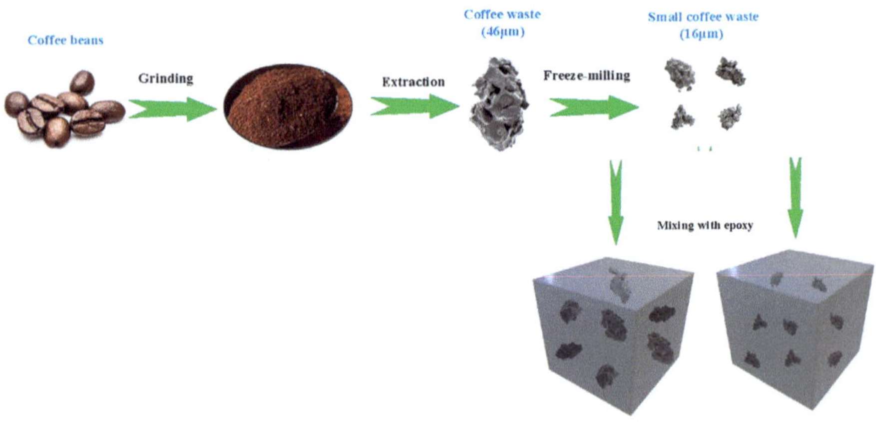

Fig. 6 Downsizing mean dimension of waste coffee grounds and mixing with epoxy procedure. Reproduced from Yang et al. [94]. CC BY 4.0. Copyright 2022

the most fragmented and porous. When 10 wt% CCG was added, residual char was denser, specimen center was completely burned, caused by uniform distribution of CCGs lack (Fig. 7a–c). But residues for epoxy having 10 wt% FCG demonstrated continuous layer of char that did not burn completely. This char layer could form physical barrier during combustion, thus preventing oxygen and heat diffusion and protecting sample. For SEM examinations, the densest sections visually indicated by red circles were removed; images are given in Fig. 7b and c. Chars achieved via epoxy composite having 10 wt% FCG were denser than CCGs (Fig. 7e, f, h, i) [94].

Flammability for sugar palm fiber (SPF, 0, 20, 35 and 50%) hybrid having polyester (PET) yarn reinforced epoxy by loading magnesium hydroxide [$Mg(OH)_2$] as flame retardant was studied by Suriani et al. [84]. Addition of sugar palm increased burning rate because of composition for sugar palm as fuel. Sample with 20% sugar palm fiber (Epoxy/PET/SPF-20) gave the highest burning rate. But loading more sugar palm fiber decreased burning rate of composites. Sample with 35% sugar palm

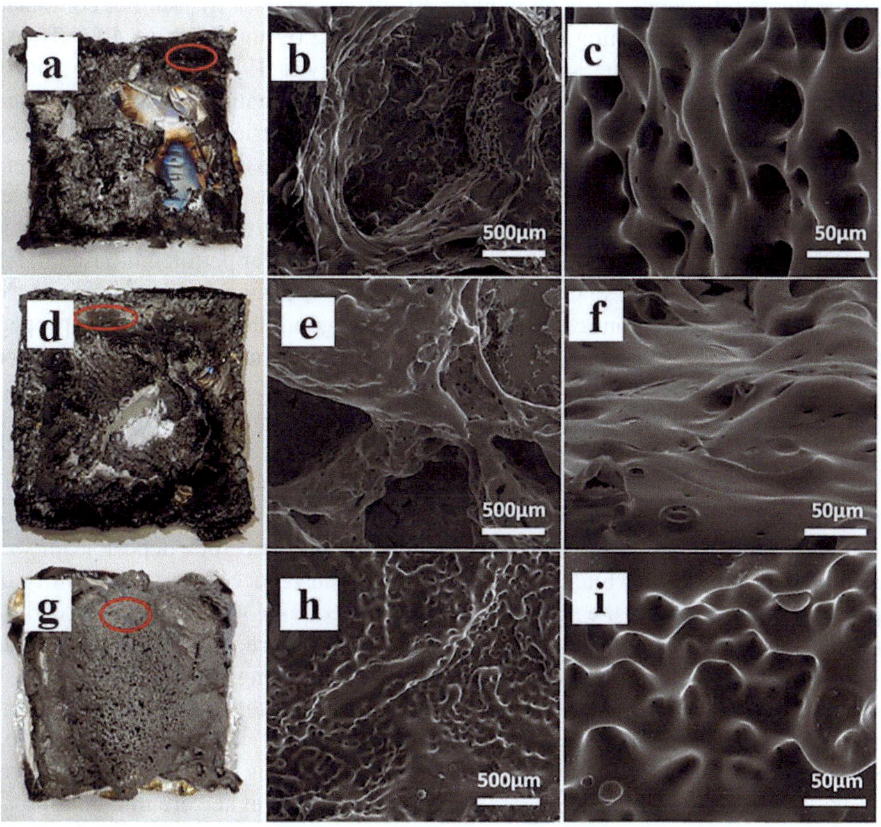

Fig. 7 Residues after cone calorimeter testing. **a–c** Pure epoxy, **d–f** epoxy composite having 10 wt% CCGs. **g–i** epoxy composite having 10 wt% FCGs. Reproduced from Yang et al. [94]. CC BY 4.0. Copyright 2022

fiber (Epoxy/PET/SPF-35) presented the smallest burning rate of 13.25 mm/min. It presented long time for the flame to spread by Epoxy/PET/SPF-35 sample and to form char. Char production is reaction for $Mg(OH)_2$ in burning stage. It is since sugar palm fiber tends to flow in char during burning process, providing greater flame retardant protection [75]. In sample with 50% sugar palm fiber (Epoxy/PET/SPF-50), burning rate increased because of ratio of sugar palm fiber to epoxy exceeding optimal fiber amount. Matrix and fiber were not bonded well, which may contribute to poor bonding. As a conclusion, char producing capacity of composite decreased during burning process. Compatibility of matrix and fiber was influenced by interfacial bonding of matrix and fiber. Weak adhesion for fiber caused weak interfacial bond, leading to poor flammability properties for composites [84].

Nguyen et al. [63] developed epoxy material having good flame retardant properties by addition of fly ash (waste of thermal power plants). The use of fly ash can decrease consumption of existing flame retardants (organic halogen compounds) and therefore meet environmental safety of flame retardant epoxy. Specimens gave higher flame retardancy when employing 30, 40 and 50 wt% fly ash. For combined rate of 30 wt%, the burning rate of material was 22.09 mm/min. At 40% fly ash mixture ratio, burning rate according to UL 94HB was 19.56 mm/min. When fly ash amount was high enough, such as 40 wt%, polymerization of polymer created a more stable and efficient coal layer that acted as a barrier between combustion surface and undissolved material. Flame retardancy increased when fly ash content was increased from 30 to 50 wt%. When fly ash content was low, carbonaceous layer (carbon residue layer) formed too thin to serve as an effective barrier. Results of this work indicated that fly ash would be a suitable candidate for petroleum-based flame retardant additives. Fly ash created a dense ash layer, carbon residue layer created a physical barrier. Carbon residue (coal) layer has good thermal insulation properties and can interfere with material transfer between gas and solid phase, enhance thermal insulation properties, and slow down and prevent combustion of materials [63]. Layer of carbon deposits formed by burning specimen with 30 wt% fly ash was discrete and discontinuous (Fig. 8a). In contrast, when specimen with 50 wt% fly ash was burned, a thick and uniform layer of carbon residue was formed (Fig. 8c). This is because large amount of fly ash with an uneven distribution. As a conclusion, there were holes, cracks and fractures in carbon deposit layer, so carbon deposit layer was no longer continuous (white arrows, Fig. 8b), which reduced protection ability of physical barrier. Composite specimen was developed with 40 wt% fly ash, a carbon residue layer was formed with a continuous dense microstructure (Fig. 8b), no voids were seen. Carbon residue layer demonstrated good bonding ability between fly ash particles. Therefore, compressed carbon layer with good thermal insulation performance was formed. This layer could inhibit mass transfer between gas phase and solid phase, enhance insulation, delay and prevent burning of material. Therefore, epoxy composites with a fly ash content of 40 wt% had higher fire retardant properties than others, with a LOI of 24.4%. When mixing ratio was 50 wt% fly ash, a layer of carbon residue with good thickness was formed, but distribution was uneven, adhesion was not tight. When fly ash content in this gas was low (30 wt%), it was not sufficient to create a carbon layer of required thickness to protect material [63].

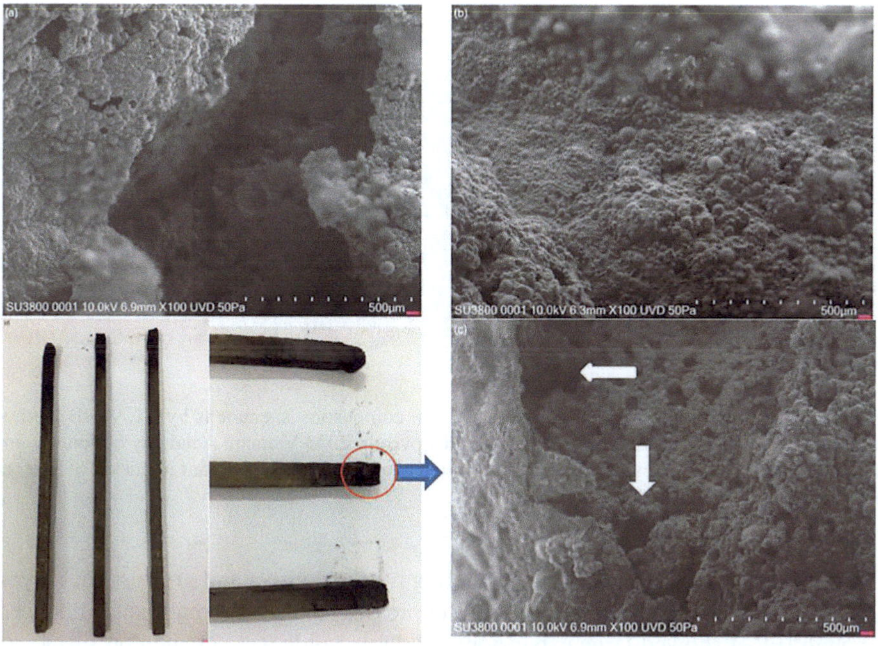

Fig. 8 SEM micrographs for carbon residues from combustion specimens by LOI method **a** 30 wt%, **b** 40 wt% and **c** 50 wt% fly ash, **d** digital photos for fly ash/epoxy composite after LOI tests [63]. Copyright 2023 Vietnam Academy of Science and Technology, Hanoi & Wiley-VCH GmbH, reprinted by permission of Vietnam Academy of Science and Technology, Hanoi & Wiley-VCH GmbH

Figure 9 shows SEM micrographs of char residue layers of fly ash/epoxy composites after combustion test in accordance with UL 94HB. It was clear that fly ash was dispersed on char residue layer surface. Char residue layer for specimen A, C (fly ash/epoxy composite containing 30 wt% and 50 wt% fly ash) was rough and porous. A loose barrier layer could not isolate oxygen and heat well, causing reduced fire performance. Char residue layer of specimen B (40 wt% of fly ash) was uniform, with regular spherical fly ash particles with tight structure on surface. Continuous corrugated layer of char residue formed after combustion could behave as shield that completely covered epoxy. It prevented heat transfer more efficiently than discontinuous layers of char residue, thus greatly increasing fire resistance of fly ash/composites [63].

Chicken eggshell was added as biofiller to epoxy systems at the ratio given in Table 2. After cone calorimeter test, char layers were observed with digital photographs (Fig. 10). IFR-0 created intumescent char layer having some collapse structures and large holes onto surface, releasing great smoke and heat amounts. By addition of chicken eggshell, coating tended to create greater compact and intumescent char layer, which prevented underlying materials from burning, thus efficiently decreasing smoke production and heat release. IFR-3, having 37 wt% intumescent

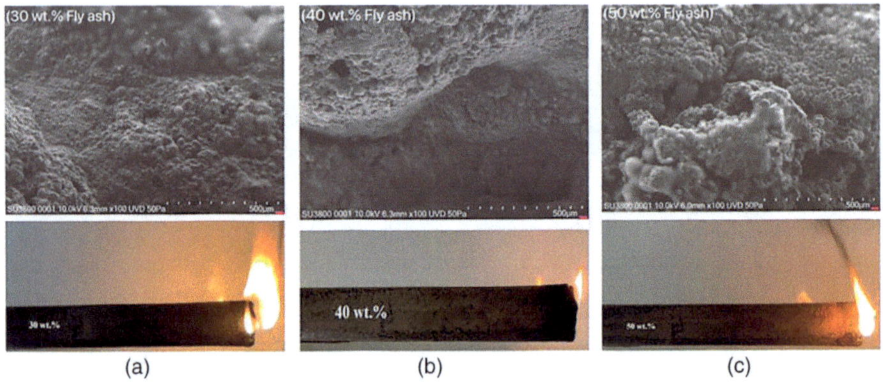

Fig. 9 SEM micrographs for carbon residues from combustion specimens by UL 94HB method **a** 30 wt%, **b** 40 wt% and **c** 50 wt% fly ash [63]. Copyright 2023 Vietnam Academy of Science and Technology, Hanoi & Wiley-VCH GmbH, reprinted by permission of Vietnam Academy of Science and Technology, Hanoi & Wiley-VCH GmbH

fire retardant and 3 wt% chicken eggshell, gave the densest and greatest char layer; this best matches its synergistic smoke suppression and flame retardant influences. Though IFR-4 showed large expansion rate, large voids were created on surface of the char layer because of excess amount of CO_2, and the thermally stable calcium phosphate and calcium metaphosphate had unfavorable impact onto compactness and expansion rate for char layer, therefore synergistic influences onto fire performance were reduced. Thus, moderate chicken eggshell biofiller amount was significant to increase smoke suppression and flame retardant features for intumescent systems [92].

Digital photos for intumescent fire retardant coating after LOI testing are demonstrated in Fig. 11. It was seen that very low char occurred for cured epoxy resin, while specimens having intumescent flame retardant formed continuous, dense and intumescent char layer. As intumescent flame retardant and chicken eggshell biofiller was combined, char layer became more compact and intumescent in comparison to specimen containing only intumescent flame retardant, thus efficiently preventing

Table 2 Compositions for samples [92]

Specimen	Epoxy (wt%)	Polyamide (wt%)	Intumescent flame retardant (wt%)	Chicken eggshell (wt%)
Cured EP	40	20	0	0
IFR-0	40	20	40	0
IFR-1	40	20	39	1
IFR-2	40	20	38	2
IFR-3	40	20	37	3
IFR-4	40	20	35	5

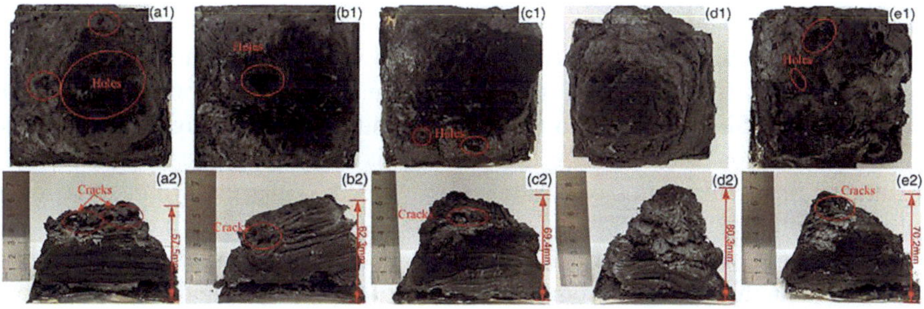

Fig. 10 Digital photos for char residues after cone calorimeter tests. **a** IFR-0, **b** IFR-1, **c** IFR-2, **d** IFR-3 and **e** IFR-4; superscript 1 means top view and 2 means side view [92]. Copyright 2018 Society of Plastics Engineers, reprinted by permission of Society of Plastics Engineers

oxygen and heat transfer. The result was consistent with greater flame retardant efficiency seen from mixture of intumescent flame retardant and chicken eggshell relative to intumescent flame retardant individually. Results demonstrated that addition of chicken eggshell could increase flame retardancy for intumescent coating and synergistic influence between chicken eggshell and intumescent flame retardant depended on chicken eggshell content. Chicken eggshell significantly increased thermal stability and char-formation capability for intumescent coating, with IFR-3 (see Table 2) showing the greatest residual weight of 22.4% at 800 °C. Char residue results revealed that addition chicken eggshell provided greater thermally stable, compact and intumescent char because of higher cross-linking structure remaining at condensed phase, therefore efficiently decreasing mass loss, heat release and smoke formation for coating [92].

Medina et al. [53] developed montmorillonite clay nanoplatelets/cellulose nanofibrils/epoxy (CME) nanocomposites. Neat epoxy ignites rapidly when flame

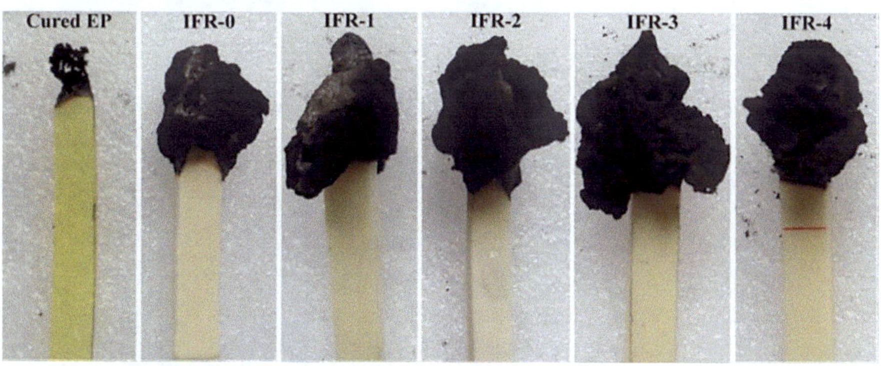

Fig. 11 Digital photos for intumescent fire retardant coatings after LOI testing [92]. Copyright 2018 Society of Plastics Engineers, reprinted by permission of Society of Plastics Engineers

is implemented and undergoes entire combustion having great smoke, melt dripping and no residue. On other hand, nanocomposites do not demonstrate dripping behavior and give some residue after burning even though flame spreads rapidly and reaches clamp. Such residues do not re-ignite and also expand significantly in the thickness direction (>500%) (Fig. 12a and b). This kind of "intumescent" nature is a significant flame retardant mechanism and can even be enhanced with good control of nanostructure [10].

Performance of epoxy matrix having nanocomposites with respect to flammability is smaller than that expected from pure montmorillonite clay nanoplatelets/cellulose nanofibrils nanopaper; where 30 wt% montmorillonite clay nanoplatelets was enough to see self-extinguishing under similar test parameters [12]. Specimen containing 33 wt% montmorillonite clay nanoplatelets burns completely, leaving a relatively small residue. It is probably related to structure of produced composite, inhibiting the montmorillonite clay nanoplatelets/cellulose nanofibrils charring synergy. Dispersion for clay platelets is significant to decrease the clay/polymer composite flammability, mainly via creating clay barriers against volatile substances [76]. Dispersion of montmorillonite clay nanoplatelets is better in montmorillonite clay nanoplatelets/

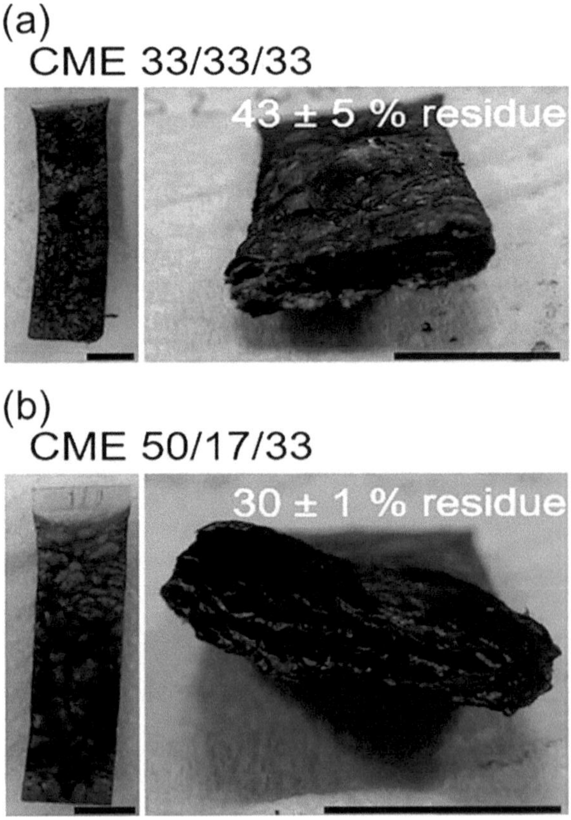

Fig. 12 Residue of **a** CME 33/33/33 and **b** CME 50/17/33 after flaming test. Scale bars are 1 cm [53]. Copyright 2019 American Chemical Society, reprinted by permission of American Chemical Society

cellulose nanofibrils nanopaper than in existing montmorillonite clay nanoplatelets/ cellulose nanofibrils/epoxy nanocomposites [53].

Cellulose nanofiber (CNF) was mixed via montmorrilonite clay platelets (MTM) in colloidal suspension and filtered to produce "clay nanopaper" having strong in-plane orientation for MTM (Fig. 13) [12, 50].

Upon application of a methane flame, pure CNF ignited immediately and burned with strong flames. Flames quickly spreaded along the entire length of specimen. After approximately 9 s, flame went out and afterglow phenomenon was seen. This is flameless combustion (burns red) and entirely consumed remaining CNF residue. Even if no flame was existence, flameless combustion could still pose safety threat. For example, great temperatures during afterglow could cause fire to spread to another ignitable and combustible materials. CNF/clay nanocomposite presented differences at behavior for various clay loadings. 10 and 20% MTM specimens burned entirely, though consistent charred residue remained. Charred remains had same sizes as starting specimen but with increased thickness (Fig. 14). Neither of these two CNF/ MTM composites gave any afterglow. 20% MTM specimens demonstrated increased combustion time (77% in comparison to neat CNF) due to existence of clay reducing flame propagation speed. A threshold in flammability was found when clay amount was increased to 30% or more. Both 30 and 50% MTM specimens demonstrated self-extinguishing behavior. Once ignited by methane flame, progressive flame gradually decreased in dimension and propagation speed as it settles into smaller and smaller portion of specimen, where flame eventually disappeared. Additionally, subsequent application of a methane flame failed to re-ignite specimen. This behavior is significant because potential fire threat could be depressed via self-extinguishing behavior. In addition, content of residue remaining after combustion was extremely great. Occurrence of ordered MTM nanoplatelets into CNF nanofiber network matrix caused a strong improvement of flammability behavior. Features were improved by increasing inorganic material. Visually seen mechanisms during flame exposure were

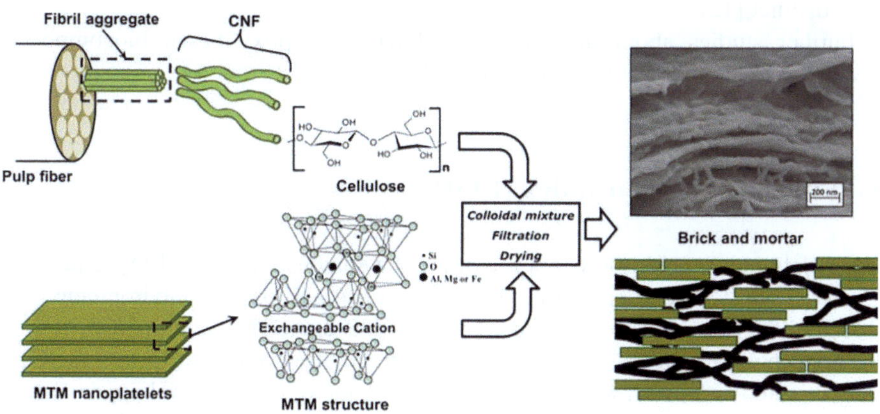

Fig. 13 Schematic view for CNF/MTM elements and final structure [12]. Copyright 2015 American Chemical Society, reprinted by permission of American Chemical Society

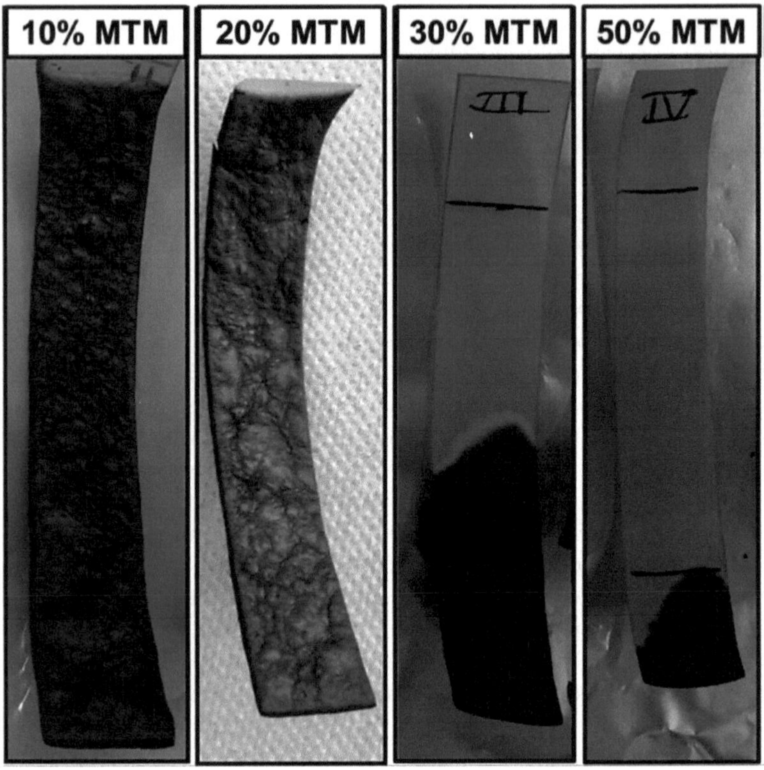

Fig. 14 Residues after vertical flammability testing for nanocomposites [12]. Copyright 2015 American Chemical Society, reprinted by permission of American Chemical Society

essentially similar regardless of clay amount, but last behavior and final conclusion were distinct [12].

Further studies about flammability behavior of epoxy based biocomposites developed in the literature are given in Table 3.

4 Limiting Oxygen Index (LOI)

LOI is laboratory testing process providing the smallest amount of O_2 required to ignite vertically placed part in mixture of nitrogen and oxygen environment. Test procedure can be seen in the standards (ISO 4589-2, ASTM D2863, NES 714). In accordance with standards, in a chimney gas stream flows upwards toward the vertically oriented specimen, while the propane gas flame ignites material upper section. In a result, burning length and time of the sample are evaluated for flammability behavior [4]. In LOI test, the smallest needed oxygen amount to support flaming

Table 3 Flammability behavior for epoxy biocomposites developed in the literature

Fiber	Manufacturing method	Flammability	References
Arenga pinneta fiber/ polyester yarn	Lay-up	Epoxy/arenga pinneta fiber/polyester yarn hybrid composite having 5 wt% $Mg(OH)_2$ (flame retardant) gave smaller burning rate	[83]
Bamboo/kenaf/nanoclay Nanoclay: Halloysite nanotube, montmorillonite or organically modified montmorillonite	Hand lay-up	Flame retardancy enhanced via addition of all kinds of nanoclay (organically modified montmorillonite was the best)	[18]
Banana short fiber and $Al(OH)_3$ filler	Hand lay-up	Addition of $Al(OH)_3$ filler decreased flame propagation rate	[80]
Kenaf with nano oil palm empty fruit bunch filler (3 wt%)	Hand lay-up	Hybrid nanocomposites gave higher flame retardancy compared with epoxy/kenaf composites, relatively little dripping or ignition	[73]
Oil palm empty fruit bunch/expandable graphite	Compression molding	3, 5 and 7 wt% expandable graphite were loaded to 20 wt% natural fiber-reinforced epoxy. Rising expandable graphite loading within composites enhanced fire resistivity	[40]
Pineapple/coir	Hand lay-up	Coir/pineapple/coir layering had greater resistance to burning. Compared to pineapple/ coir/pineapple and bilayer (pineapple/coir) composites, intimately mixed hybrid material exhibited smaller burning rate and total mass loss	[54]
Sisal/coir	Cold pressing	Rising fiber amount increased flammability. Developed hybrid biocomposites was not appropriate for applications where fire response is important	[1]

combustion of material could be determined [5, 94]. Procedure for LOI test is to position the specimen vertically in controlled atmosphere and ignite its upper part via combustor. LOI is minimal amount of O_2 in mixture of nitrogen and oxygen that sustains flame combustion of sample during 3 min or consumes 5 cm length for the specimen as [55]:

$$\text{LOI} = \frac{100 \times [O_2]}{([O_2] + [N_2])}$$

where $[N_2]$ and $[O_2]$ are nitrogen and oxygen concentrations, respectively [55]. Therefore, material that requires more oxygen will exhibit a greater LOI. Additionally, greater index means that material is more resistant to flame (better fire retardant features) [4, 56]. With respect to ISO 4589 standard, LOI is measured onto an 80 mm $\times$ 10 mm $\times$ 4 mm sample positioned vertical direction in middle of glass chimney (Fig. 15). Gas mixture flows up to chimney and is made homogeneous by passing over glass beads layers. Sample top is ignited like a candle after purging column for 30 s [55].

Considering that atmospheric air includes 21% O_2, polymers having LOI values below 21 have a high risk of burning and are called "combustible"; but, materials having LOI greater than 21 are classified as "self-extinguishing" since their combustion could not be sustained in the atmosphere without support for external resource [4, 47]. LOI for epoxy was reported as 23 [91]. LOI of bamboo/kenaf/epoxy composite was measured to be 19.8% and was classified as combustible [18]. With respect to flame retardant features for epoxy composites having 30 wt% coffee grounds, oxygen index was restricted to 20.8% [62]. Nguyen and Nguyen [61] reinforced epoxy matrix by banana fiber at the weight percentage of 10, 15, 20 and 25%. Thermal and flame retardant features were maintained stable, 20 wt% banana fiber provided LOI of

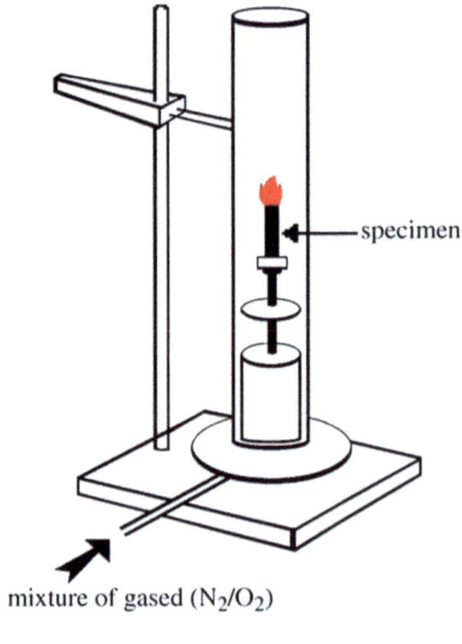

Fig. 15 LOI test [47]. Copyright 2008 Elsevier B.V., reprinted by permission of Elsevier B.V.

Table 4 LOI of epoxy composites

Sample	LOI (%)
Pure epoxy	20
Epoxy having 2 wt% FCGs	20
Epoxy having 5 wt% FCGs	20
Epoxy having 10 wt% FCGs	21
Epoxy having 20 wt% FCGs	22
Epoxy having 30 wt% FCGs	24

20.8% and moderate thermal stability. In existence of banana fiber, sample presented no signs of reduction in flame retardant features in comparison with original epoxy. In accordance with combined proportion of 20% banana fiber, fire resistance rating was the most stable [61]. LOI of cured epoxy resin was 19.2% and this value for intumescent coating having 40 wt% flame retardant rose to 29.1%. Results show that loading of chicken eggshell biofiller enhanced LOI of coatings, and IFR-3 (see Table 2) having 37 wt% intumescent flame retardant and 3 wt% chicken eggshell gave maximal LOI value of 31.5%. But when chicken eggshell amount increased to 5 wt%, LOI for IFR-4 (see Table 2) was lower than that of IFR-3 (see Table 2); showing that excessive chicken eggshell content reduced synergistic effect between chicken eggshell and intumescent flame retardant [92]. At a combined rate of 30 wt% fly ash in epoxy, LOI of material was 23.5%. At 40% fly ash mixture ratio, LOI was 24.4% [63]. LOI values for composites without and with FCG are presented in Table 4. By FCGs, LOI increased from 20 to 24 in comparison to control specimens (epoxy without FCG). The increment in LOI can be owing to creation of char barrier because of FCGs existence in epoxy composites [94].

5 Underwriters Laboratories UL-94 Vertical Flame Test

UL-94 vertical test is ranking measurement for self-extinguishing time of polymeric material [42]. It is a common technique providing rating for sample depended onto its ignition and flame spread for material subjected to small flame. Testing procedure is presented in ASTM D3801 standard. Protocol includes preparation and exposure of 127 mm × 13 mm × 3 mm size sample to controlled flame through 10 s. After flame removal, any combustion action is observed and saved. If sample self-extinguishes, flame is applied for another 10 s then removed. Burning time of flame exposure and afterglow are saved. Qualitative rankings for determining conclusions are no-rating, V0, V1 and V2, as presented in procedure. Materials are classified in categories basing on whether conditions specified in Table 5. If specimen continues to burn after initial flame implementation, it is recorded as no-rating [4]. V0 rating at UL-94 testing represents the best flame retardant for polymers. Schematic representation of UL-94 flammability testing is demonstrated in Fig. 16.

Koyunucu and Ulay [42] developed epoxy based composites containing walnut shell particulates as reinforcing and pumice powder as filler to characterize reducing flammability (Table 6). S7 and S8 composites (40 and 50 wt%) presented V2 rating, was the least flammable sample. Pumice powder was not enough up to 20 wt% to achieve any V rating. Influence of rising wt% pumice percentage onto UL-94 results was found to be promising, and V2 rating was met except for 20 wt% sample. During burning, after-glowing was seen in all 20 wt% specimens and the specimens were flammable. Therefore, all of the samples with 20 wt% pumice powder incorporated

Table 5 Classifications for UL-94 test

Classification (rating)	Specific flaming characteristics
V0	Total flaming combustion time must be out in 10 s or lower. No glow for more than 30 s and no burning material could fall
V1	Total flaming combustion time must be out in 30 s or lower. No glow for more than 60 s and no burning material could fall
V2	Total flaming combustion time must be out in 30 s or lower. No glow for more than 60 s and burning material could fall

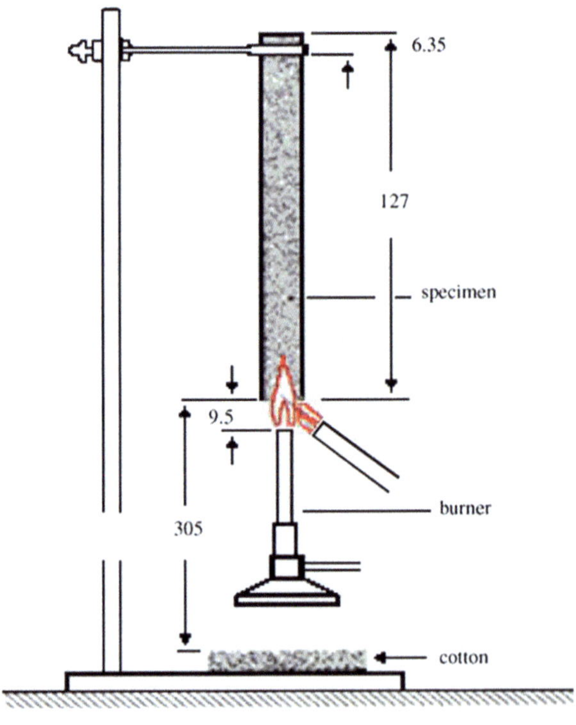

Fig. 16 UL-94 flammability procedure setup [47]. Copyright 2008 Elsevier B.V., reprinted by permission of Elsevier B.V.

failed vertical burning testing. Although no UL class was seen for 20 wt% pumice powder, existence of 40 and 50 wt% pumice significantly reduced burning rate of sample [42].

Results from UL-94 testing showed that cured epoxy resin did not pass UL-94 testing, while all intumescent coatings achieved V0 rating in UL-94 testing [92].

In horizontal flammability tests, it was observed that epoxy burned continuously at approximately 50 mm min^{-1} to the end of the sample, and therefore grade B was given. Flax burned easily in epoxy laminates. For epoxy flax, flame time was 2 min 37 s, burn length was approximately 100 mm and black smoke was observed. In vertical flammability tests, epoxy burned easily with a significant amount of black smoke after the initial application of flame and therefore failed [19].

Incorporation of FCGs decreased burning rate for specimens through UL-94 horizontal testing (Table 7). Free radical scavenging decreased flammable functional groups present at gas phase through combustion, thus delaying onset of combustion and manifested in decrement for combustion rate [94].

Table 6 Compositions of samples

Samples	Sieve mesh dimension for walnut shell (μm)	Walnut shell particles amount (wt%)	Filler amount (wt%)	Filler kind	Sieve mesh dimension for pumice powder (μm)
S1	250	100	–	–	–
S2	630	100	–	–	–
S3	1000	100	–	–	–
S4	250	80	20	Pumice powder	< 250
S5	630	80	20	Pumice powder	< 250
S6	1000	80	20	Pumice powder	< 250
S7	630	60	40	Pumice powder	< 250
S8	630	50	50	Pumice powder	< 250

Reproduced from Koyunucu and Ulay [42]. CC BY 4.0. Copyright 2023

Table 7 Burning length and rate of epoxy composites

Specimen	Burning length (mm)	Burning rate (mm/min)
Pure epoxy	100	45
Epoxy having 2 wt% FCGs	100	42
Epoxy having 5 wt% FCGs	100	39
Epoxy having 10 wt% FCGs	100	35
Epoxy having 20 wt% FCGs	100	28
Epoxy having 30 wt% FCGs	100	20

Reproduced from Yang et al. [94]. CC BY 4.0. Copyright 2022

6 Micro-scale Combustion Calorimetry

It is utilized to determine fire characteristic of material and measures heat release capacity, combination of thermal stability and combustion features for material that could be utilized to categorize flammability. Testing apparatus possesses two separate steps: Pyrolysis and combustion stages. At ASTM D7309-19 standard, protocol involves two selectable pyrolysis modes, i.e., Method-A and Method-B, utilized for controlled thermal and thermal oxidative decomposition, respectively. Specimens with a mass between 0.5–50 mg are pyrolysed into inert gas and volatile effluent is blended by excess O_2 before combustion at Method-A, while at Method-B, pyrolysis takes place within mixture of inert gas and oxygen. Heat release rate from testing is achieved from the O_2 consumption calorimeter [4].

7 Cone Calorimetry

Cone calorimeter is device that works on principle of O_2 consumption of material during the combustion testing [56]. Cone calorimetry test could present valuable information at various steps regarding time to ignition (TTI) (ignition step), heat release rate (HRR) (fire development step) and total heat release (THR) (completely developed fire step) [5]. It is the most extensively implemented bench-scale fire testing. It measures various flammability parameters such as ignition time, mass loss rate, heat release rate, combustion or extinction time, soot yield, smoke production, and CO and CO_2 amounts. Specimen dimension for testing is in range of $0.1 \times 0.1 \times 0.001$–$0.05\ m^3$ and heat flux is in the range of 10 to 100 kW/m^2. Specimens are wrapped within aluminum foil and placed vertically or horizontally under cone heater in accordance with determined standards (ISO 5660, ASTM E1354). Load cell determines weight alteration and pyrolysate is ignited via electric pilot spark igniter. Smoke is collected at equipment hood for evaluation. Cone calorimetry works onto basis of O_2 consumption calorimetry. External parameters such as sample orientation and geometry, specimen thickness, ventilation, ignition resource and temperature influence accuracy of measuring [4].

During experiments, it was noted that the majority of glass fiber specimens retained initial shape, while flax fiber specimens delaminated and bulged outwards, having the greatest expansion occurring at center of specimen. Edges for horizontal burn specimens curve upwards, while cone calorimetry specimens curve in opposite direction, even though both were set horizontally through testing. It shows that direction for incident heat flux is determining parameter on resulting shape alterations. Specimens gave approximately 30% shrinkage in horizontal plane [14].

In the case of pure CNF, when exposed to heat flux, temperature increased and CNF degraded, releasing volatile gases. As concentration of flammable volatiles was enough, ignition occurred and specimen was exposed to almost entire combustion by oxidation. No residue was observed at end of CNF testing. CNF/clay nanocomposites presented much improved fire protection features in comparison to pure CNF. 30 and 50% MTM nanocomposites did not ignite at all through cone calorimetry testing. The conclusion was quite unusual for polymers. Since CNF/30% MTM and CNF/50% MTM brick and mortar compositions were subjected to heat flux, volatile gases released as conclusion of CNF degradation cannot achieve concentration needed for ignition. All nanocomposites gave consistent residue and demonstrated increased specimen thickness as found after vertical testing. Absolute amount of residue rose by clay amount, and last value was greater than initial amount of inorganic phase. At the end of test, it appeared that some of organic degradation products resulting from CNF were existence. Tests in which ignition occurred took approximately 80 s, while specimens that did not ignite were tested for up to 180 s to confirm that samples could not be ignited. MTM concentration, which represented significant changes in fire protection behavior from flammability and cone calorimetry data, was 30 wt% [12].

7.1 Heat Release Rate (HRR)

HRR is described as amount of heat released per unit area of sample as it is exposed to fire and given by kW/m^2. This is not solely a significant factor characterizing fire behavior [18, 22], but also serves as evaluating parameter in defining fire hazard phenomena. In parallel with HRR, formation of smoke, dangerous gases and other kinds of fire dangers also rise [18]. HRR can be employed to express fire intensity. Low HRR peak means higher flame retardant [56]. Heat release rate is a significant indicator for flammability, since it allows quantitative observation for combustion and energy amount to which surrounding material or human are subjected [14]. HRR depends on external heat flux, ventilation degree and degree of destruction of tested material [22, 78]. Total heat release (THR, kJ/m^2) is total amount of calorific value released per unit area after material combustion, and it could be computed by integrating HRR vs. time [4]. THR denotes internal energy of materials, independent of environmental parameters [22, 78].

Heavily woven flax fiber-reinforced composites were the only specimens that exhibited maximal HRR at initial peak through test. It is probably because there is much higher burning rate at surface prior to charring, as looser weave permits pyrolyzed gases from resin to escape to surface more quickly. When glass and flax specimen curves are compared, flax fiber specimens generally exhibit same or greater HRRs as glass counterparts at any given time. It shows that natural fiber specimens burn the same or more strongly during combustion [14].

HRR of epoxy was reported as 1261 kW/m^2 [66]. HRR curves for unaged specimens are depicted in Fig. 17a. Epoxy/flax composite (EpFl) presented typical behavior of intermediate thick non-charring specimen. HRR gradually increased in the first part of testing, with a peak occurring when heat front reached back face of specimen. Since heat could not be removed because of isolating support, pyrolysis rate increased, resulting in HRR peak [13]. Finally, HRR decreased because of fuel depletion. Addition of both flame retardants caused similar influences within epoxy/flax composite. Aluminum diethyl phosphinate (AlPi) had nearly no influence on ignition time of remaining biocomposite, similar to EpFl. 9,10-dihydro-9-oxa-10-phosphaphenanthrene-10-oxide (DOPO) led to slight reduction in TTI, which might be related to influence onto epoxy thermal stability. Wang and Lin [90] found that onset of degradation for epoxy tended to drop by increment in DOPO amount. After ignition, the first peak was seen then HRR dropped significantly because of effect of flame retardant. Both FRs are known to behave chiefly at gas phase, as confirmed with small increment in char residue and decrement in effective heat of combustion (EHC). AlPi decomposes via creation of diethylphosphinic acid released at gas phase and aluminum phosphate remaining at condensed phase [8, 9]. This latter compound can cause 5% increment within residue. DOPO also behaved chiefly at gas phase, as its decomposition releases phosphorus-containing fragments that are likely to trap high-energy radical H° or OH°, causing flame inhibition [30]. As soon as flame retardants were activated, an important decrement in HRR was observed, up to 250–300 kW/m^2. Slowing the burning rate shifted second peak regarding heat feedback on back face to higher period of time. Generally, flame retardants reduced average value of HRR and increased burning time. Aging had slight effect onto composite flammability because neither TTI nor average HRR were changed. Figure 17b shows HRR results for EpFl-AlPi specimens aged at 70 °C within several cases. All curves presented same shape. Main influence for aging was curve shift toward long duration when EWC was great (aging via liquid water). At case of EpFl-AlPi, part of incident heat flux was employed for endothermic vaporization of moisture involved into biocomposite, hence delaying time needed to achieve ignition temperature. In case of EpFl-DOPO specimens, Fig. 17c demonstrates that existence of moisture had no effect onto TTI or the first peak for HRR, remained nearly fixed. Also, THR remained unaltered. Basic difference lied in fact that HRR decreased immediately after first peak of HRR, and second peak for HRR shifted to higher periods of time, about 40 s, in comparison to dry specimen. The shift can be attributed to endothermic release of water, which slows decomposition rate for sample after ignition. Generally, influence of aging onto fire performance for composites was solely important as moisture amount was sufficient (>3%). Impact was chiefly due to the delay in fuel production. But two situations can be distinguished. At condition of EpFl-AlPi, high moisture specimen was achieved by immersing it in liquid water. This type of aging caused water clustering in addition to other adsorption mechanisms. It is assumed that as exposed to a radiant source, water present in clusters can be released easily, thus delaying ignition and subsequently entire HRR curve. Regarding EpFl-DOPO, tested specimen having great moisture amount was achieved after aging at aw = 0.85. At that kind aging, water is assumed to form clusters but is absorbed chiefly in the hydrophilic regions of the matrix, fibers, and DOPO. Therefore, as subjected to a radiant source, the release of water will be more challenging and will mainly take place after ignition [11].

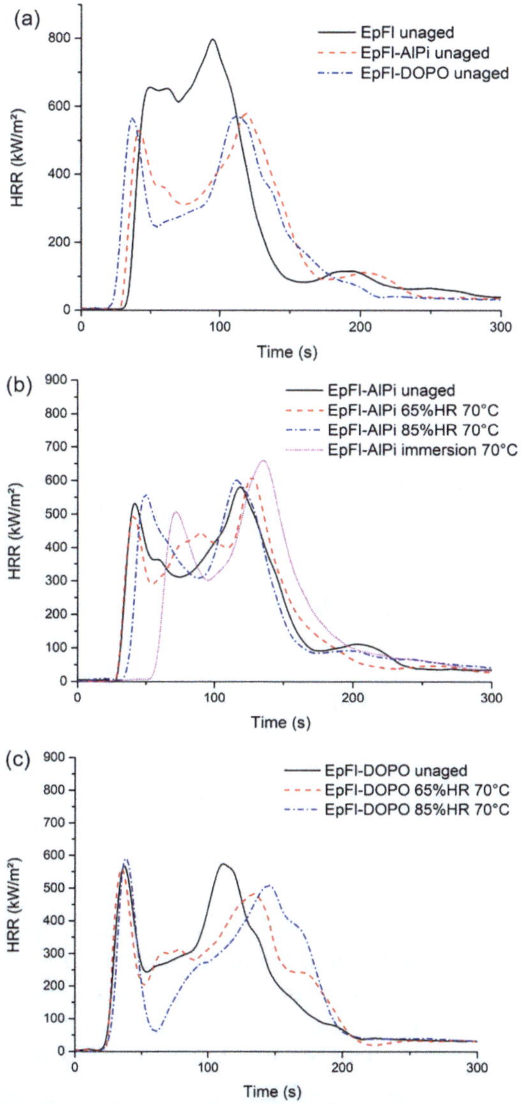

Biocomposites	Matrix (vol%)	Flax (vol%)	Flame retardant (wt% of matrix)
EpFl	70	30	0
EpFl-AlPi	70	30	9.5
EpFl-DOPO	70	30	13.9

Fig. 17 Cone calorimeter curves for **a** unaged biocomposites and **b** EpFl-AlPi aged in various conditions at 70 °C. Reproduced from Campana et al. [11].

7.2 *Peak Heat Release Rate (PHRR)*

Peak for HRR curve as the maximal ignition level is named as heat release rate peak (PHRR) [20, 59]. PHRR is one of the most important factors in evaluating flammability characteristics for material [22]. Essentially, amount of PHRR and the time reached to obtain PHRR are employed to measure fire behavior for material [4]. It represents maximal intensity for the fire, demonstrating fire speed and size [20]. Smaller PHRR represents more fire retardant specimen [59]. Flax fiber based epoxy composites provided higher PHRR than their glass fiber counterparts because of exothermic combustion of fibers [14]. PHRR of bamboo/kenaf/epoxy composite was found to be 680 kW/m^2 [18].

Cured epoxy resin burned rapidly after ignition, demonstrating one peak having PHRR of 1293.3 kW/m^2 at 95 s. With addition of intumescent flame retardant, specimens presented initial increment in HRR until efficient intumescent char was produced then reached PHRR in initial step characterized as thermally thick charring specimens [78]. Compared to cured epoxy resin, PHRR and THR diminished by 75.7 and 41.3% for IFR-0 (see Table 2), 79.4 and 47.2% for IFR-1 (see Table 2), 81.8 and 52.5% for IFR-2 (see Table 2), 86.0 and 62.0% for IFR-3 (see Table 2), 84.5 and 56.3% for IFR-4 (see Table 2), respectively, demonstrating an enhancement in flame retardancy. Compared to IFR-0 (see Table 2), coatings with chicken eggshell gave smaller HRR and THR, while IFR-3 (see Table 2) showed the lowest PHRR and THR. This suggested that chicken eggshell biofiller performed well in increasing flame retardancy for intumescent fire retardant coating during combustion. But synergistic flame retardant influence of chicken eggshell into coating depended on chicken eggshell amount, and the synergistic influence decreased as chicken eggshell increased above 5 wt% in coatings. This result might be attributed to biofiller reacting with volatile phosphorus to create thermally stable calcium metaphosphate and calcium phosphate onto surface of coating, therefore efficiently separating heat and mass transfer between condensed and gas phases, thus improving flame retardancy for coating through combustion. But higher chicken eggshell amount reacted with the acid and decreased release of free radial scavengers including phosphorus, which promoted dehydration and charring of carbon resource and also suppressed chain decomposition of matrix, therefore decreasing synergistic flame retardant effect [92].

Fire growth rate (FGR, kW/(s.m^2)) is estimated via FGR = PHRR/$(\text{PHRR})_{\text{ti}}$, $(\text{PHRR})_{\text{ti}}$ is time reached to achieve PHRR. The faster the FGR, the smaller the time it takes to detect a fire [4].

Research was carried out on influence of the concentration and particle dimension of waste coffee grounds onto toxic gas creation and flame retardancy for epoxy composite. Conclusions showed that decreasing mean dimension of waste coffee grounds from 46 to 16 μm employing milling or filtration could decrease peak and total heat release for epoxy composite with 52.0% and 33.7%, respectively [94], exceeding efficiency of other bio flame retardants such as deoxyribonucleic acid (DNA) [37], seashell [58] and lignin [81]. Fine coffee grounds could reach similar output with aluminum trihydrates, the most extensively employed non-biological flame retardant [94].

Flame retardant index (FRI) was described by ratio of THR × $\left(\frac{\text{PHRR}}{\text{TTI}}\right)$ between pure epoxy and coffee grounds/epoxy composite [94]:

$$\text{FRI} = \frac{\left[\text{THR} \times \left(\frac{\text{PHRR}}{\text{TTI}}\right)\right]_{\text{Epoxy}}}{\left[\text{THR} \times \left(\frac{\text{PHRR}}{\text{TTI}}\right)\right]_{\text{Composite}}}$$

PHRR and THR for pure epoxy were 1876 kW/m^2 and 77.0 MJ/m^2, respectively. At smaller heat flux (35 kW/m^2), PHRR for pure epoxy was found as 1630 kW/m^2 by Shi et al. [79]. Addition of FCGs and CCGs delayed time to achieve the peak heat release rate, showing that specimens could last longer when subjected to high-temperature radiation conditions. In accordance with FRI's grading criteria [88], both FCG/epoxy and CCG/epoxy composites gave good flame retardant influence [94]. In comparison with CCGs, FCGs decreased both PHRR and THR. PHRR and THR diminished with 35.6% and 20.8%, respectively, with addition of 10 wt% FCGs, this was greater than CCGs with similar weight percentage within epoxy composite. As amount of FCGs rose to 20 wt%, decrement in PHRR was 49.6%, and reduction in THR was 29.9%, was greater than weight concentration of FCGs. In contrast, rising amount of CCGs via same content resulted in no important enhancement in PHRR decrement (altered from 21.1 to 32.7% and low increase in THR decrement from 13.0 to 19.5%. This may also have contributed to smaller values for THR and PHRR, as content of flammable epoxy into composites reduced as concentrations of CCGs and FCGs increased. To determine whether loading coffee grounds produced any synergistic effects, it was assumed that heat released via coffee grounds was negligible in comparison with epoxy, so that synergy ratio with respect to THR or PHRR from composites could be expressed with specific THR or PHRR for epoxy [94]:

$$\text{Synergy Ratio}_{(\text{PHRR})} = \frac{\text{PHRR}_{\text{reduction}}}{\text{Epoxy Weight Loading}_{\text{reduction}}} - 1$$

$$\text{Synergy Ratio}_{(\text{THR})} = \frac{\text{THR}_{\text{reduction}}}{\text{Epoxy Weight Loading}_{\text{reduction}}} - 1$$

Greater synergy ratio values indicated a better synergy influence between epoxy and coffee grounds onto flame retardancy, while negative and zero values represented the absence of any synergy. The highest synergy took place at the addition of 2 wt% FCG and gradually reduced as additional loading increased because of worsening of particle dispersion. In general, FCG-epoxy composites showed better flame retardant performance than CCG-epoxy composites at different weight loadings in cone calorimeter tests. Compared to aluminum hydroxide (ATH) [93], the most commonly utilized flame retardant for polymeric materials, fine coffee grounds were efficient

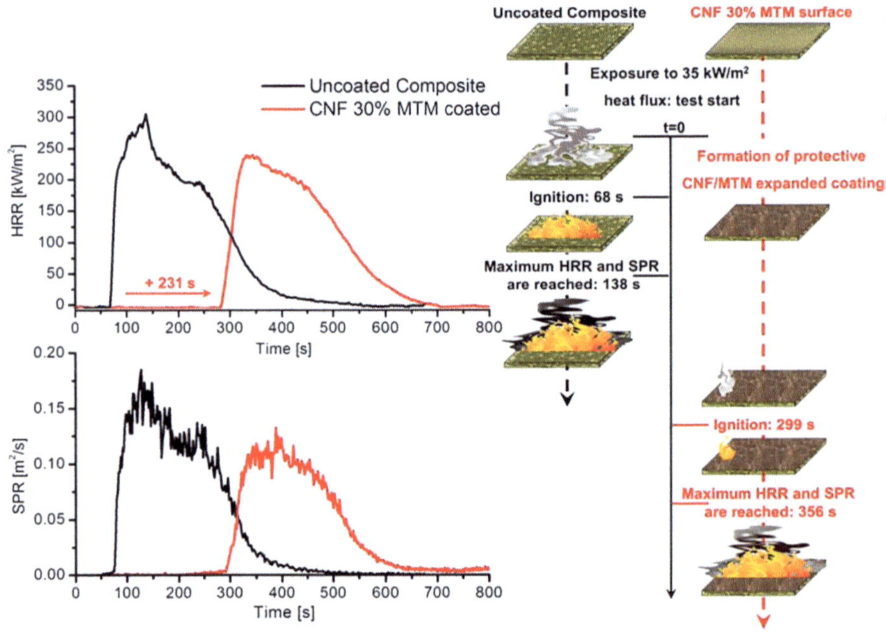

Fig. 18 HRR and SPR graphs for uncoated and clay nanopaper (CNF/30% MTM) coated GF/EP [12]. Copyright 2015 American Chemical Society, reprinted by permission of American Chemical Society

for decreasing total heat release however had lower influence onto diminishing peak heat release rate [94].

Mean heat release peak rate (pkHRR) was found to be 285 kW/m^2. pkHRR and THR were decreased by 18 and 24% compared with unmodified composite (Fig. 18) [12].

7.3 Time to Ignition (TTI)

Time to ignition (TTI) visually determines time required for specimen whole surface to burn with the continuous luminous flame [20]. TTI (in s) describes ignition ease of a material via measuring how quickly flaming combustion takes place as specimen is subjected to incident heat flux (kW/m^2) and at an oxygen-controlled ambient environment. Therefore, material having great TTI value denotes a material that is challenging to ignite. At flame retardant polymer composites sometimes loading of flame retardants causes the advancement of decomposition and hence decreasing TTI. Therefore, lower TTI is not indication that the flame retardancy for material has worsened [4]. TTI and time to PHRR (TPHRR) present time required for sample to ignite and form maximal heat, respectively at radiative heat. In this reason, specimen ignition could be characterized via TTI. TTI depends on external heat flux, ventilation degree and degree of destruction of material [22, 78]. TTI of bamboo/kenaf/epoxy composite was reported as 95 s [18].

Mean TTI of unmodified GF/EP was found to be as 68 s. Composite then began to burn, releasing dark and dense smoke. Usage of protective clay nanopaper onto GF/EP surface caused an important increment in TTI, shifting from 68 to 299 s for uncoated and coated composite, respectively (Fig. 18). From safety perspective, there was greater time (4 min) to escape prior to ignition took place [12].

7.4 Total Smoke Parameter (TSP)

TSP is the parameter for amount of gases and smoke emitted during material ignition. The specimen with greater TSP is a more flammable material [85].

There should be low visual distinction in smoke volume released between glass and corresponding flax specimens through combustion. Considering that flax specimens generally ignite earlier and extinguish later; this causes higher total smoke amount being created. There were solely minor distinctions in total smoke production between natural and synthetic heavy woven specimens [14].

Light absorption values for specimens increased continuously, and addition of intumescent fire retardant demonstrated earlier smoke release at the first decomposition step because APP decomposition at lower temperature released smoke particles. After initial step, a significant decrement in light absorption values of specimens combined with intumescent fire retardant was observed. In comparison to IFR-0 (see Table 2), addition of chicken eggshell reduced light absorption values of coatings during combustion, indicating a reduction in smoke production. IFR-3 (see Table 2), on the other hand, showed the smallest light absorption values among specimens, regarding the smallest smoke formation. Smoke density rating (SDR) value for cured epoxy resin was 78.75% and addition of intumescent fire retardant reduced SDR value of coatings. SDR value of IFR-0 (see Table 2) was 63.3%, which was higher than coatings containing chicken eggshell biofillers. This result showed that addition of chicken eggshell in coatings could efficiently suppress smoke production, which was attributed to production of efficient intumescent char layer onto materials surface that prevented mass and heat transfer. IFR-3 (see Table 2) showed the smallest SDR, and this factor showed same trend with SPR and TSR achieved by cone calorimeter testing. It could be concluded that combination of intumescent fire retardant and chicken eggshell had a synergistic smoke suppression influence within epoxy-based coating, and synergistic influence depended on chicken eggshell amount. Smoke production rate (SPR) and total smoke release (TSR) values of specimens incorporated with intumescent flame retardant were found to be smaller than that for neat cured epoxy. Peak SPR (PSPR) and TSR for IFR-0 (see Table 2) were 51.7 and 37.4% smaller, respectively, than those of cured epoxy resin. In comparison to IFR-0 (see Table 2), PSPR and TSR of coatings diminished by addition of chicken eggshell, indicating increased smoke suppression properties. Moreover, IFR-3 (see

Table 2) presented the smallest PSPR and TSR, and these values were 39.4 and 15.0% smaller than those of IFR-0 (see Table 2), respectively. SPR and TSR results showed that addition of chicken eggshell biofiller gave superb synergistic smoke suppression influence to intumescent fire retardant coating, and synergistic influence depended on chicken eggshell content in coatings. Main reason for this influence was that addition of chicken eggshell created more compact and thermally stable char, which acted as efficient barrier suppressing production rate of smoke precursor, therefore efficiently reducing smoke creation [92].

Results for cone calorimetry and smoke density testings demonstrated that loading of chicken eggshell greatly reduced heat release and smoke formation for specimens concomitant by increment in residual weight attributed to creation of more compact, thermally stable intumescent char against mass and heat transfer through combustion. Synergistic influence onto fire behavior between chicken eggshell and intumescent flame retardant depended on chicken eggshell content, and excessive chicken eggshell content reduced this synergistic effect. Cone calorimetry and smoke density tests showed that chicken eggshell biofiller could greatly reduce smoke production and heat release for coating, and synergistic influence between intumescent fire retardant and chicken eggshell depended on chicken eggshell content. IFR-3 (see Table 2) having 37 and 3 wt% chicken eggshell presented the best synergistic influence, and THR and TSR reduced with 35.3 and 15.0% compared to IFR-0 (see Table 2) without chicken eggshell [92].

Synergistic smoke suppression and flame retardant mechanism for chicken eggshell into epoxy-based intumescent system is given in Fig. 19. Through combustion, ammonium polyphosphate decomposed to produce non-flammable gases (H_2O and NH_3), phosphoric acid and orthophosphates. Orthophosphates and phosphoric acid esterified pentaerythritol (PER) to produce ester mixtures, whereas flammable gases diluted combustible gases and decreased intensity of combustion. As temperature increased, melamine and ester mixtures decomposed to produce precursor for intumescent char through cyclization via release of gaseous products (CO_2, NH_3, and H_2O). Simultaneously, non-flammable gases (CO_2, H_2O and NH_3) caused char precursor to swell and produce intumescent char covered onto matrix surface. As chicken eggshell was incorporated to ammonium polyphosphate-pentaerythritol-melamine (APP-PER-MEL) system, chicken eggshell reacted by phosphoric acid to produce thermally stable calcium metaphosphate and calcium phosphate, which had positive influence on barrier and anti-oxidation feature for char layer, resulting in well synergistic influence between chicken eggshell and intumescent fire retardant at condensed phase. Inert gases (H_2O and CO_2) released from this reaction could dilute flammable gases and promote expansion for carbonized layer. In conclusion, more compact and intumescent char was created after incorporation of chicken eggshell, provided moderate physical barrier to separate conduction of oxygen, heat and flammable gases, therefore efficiently preventing further combustion of underlying materials [92].

Delay in the smoke production rate (Fig. 18, SPR graphs) showed that a dense smoke was not emitted before ignition [12].

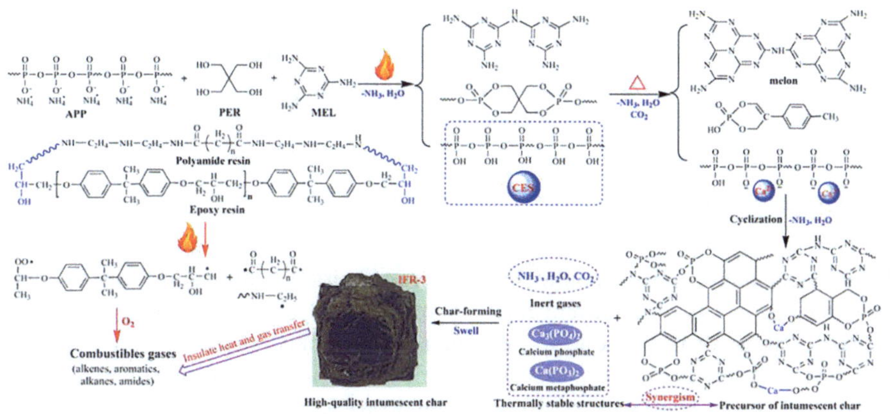

Fig. 19 Synergistic smoke suppression and flame retardant mechanism for chicken eggshell biofiller within epoxy-based intumescent system [92]. Copyright 2018 Society of Plastics Engineers, reprinted by permission of Society of Plastics Engineers

7.5 Mass Loss Rate (MLR)

MLR (g/s) is mass loss amount for sample per unit time through combustion [4]; the lower the MLR, the smaller the weight loss and the more stable the specimen against ignition [23]. Mass loss rate is indicator for influence on structural integrity of material [14].

Mass loss is a very important parameter regarding heat release and smoke production for materials. Cured epoxy gave rapid mass loss and a residue of 2.2% at end of test. With addition of intumescent fire retardant and chicken eggshell, mass loss curves became slower and residual weight of specimens increased highly. From IFR-0 (see Table 2) to IFR-4 (see Table 2), residual weights increased to 17.5%, 20.4%, 23.1%, 29.0% and 24.8%, respectively. Addition of chicken eggshell imparted great residual weight to intumescent coating. Increased residual weight helped to decrease release of smoke precursors and flammable gases through combustion, resulting in lower smoke formation and heat release. Hence, residual weight was in good agreement with smoke production and heat release behavior. IFR-3 (see Table 2) exhibited the highest residual weight having the smallest smoke production and heat release. Reason of this behavior might be that the appropriate amount of the gaseous product released via decomposition of chicken eggshell biofiller in range of 350–600 °C could promote intumescent process of char layer, exhibited positive influence on fire performance. Moreover, the appropriate amount of thermally stable calcium metaphosphate and calcium phosphate, produced via the interaction between ammonium polyphosphate (APP) and calcium carbonate prior to phosphate chains degradation, helped to form high-quality charred layer. But if higher chicken eggshell was incorporated, greater calcium phosphate produced onto the char layer surface not only inhibit diffusion of oxygen, flammable gases and heat, but also inhibit APP decomposition and evaporation, hence inhibiting char layer swelling and synergistic influence on fire performance. All things considered, it could be concluded that addition of chicken eggshell provided superb synergistic smoke suppression and flame retardant influences to intumescent fire retardant coating [92].

8 Conclusions

Epoxy resins have great modulus, great glass transition temperature, superb adhesion to large variety of fibers and chemical resistance. But the fact that epoxy cannot be recycled, or dispersed due to permanent cross-linked structure provides them not environmentally friendly. Because of rising environmental awareness, depletion of petroleum resources, post-use disposal problems, and introduction of novel regulations and rules by legislative authorities to meet new demands of society, environmental impacts of new products need to be minimized. This issue is particularly relevant in area of composite manufacturing, where petroleum polymers and synthetic fibers are still extensively employed. This concept is directly linked to the effective usage of materials and energy resources, waste management and usage of products and techniques that have less impact on environment. Environmental and economic concerns combined with government rules have spurred research on greener alternatives to glass and petroleum-based composites. Much of that work has concentrated on production of biocomposite and other natural renewable material. However, since natural fibers have high flammability, the fire behavior of the material needs to be enhanced. Fire retardant properties of polymer composites is significant issue when environment, public health and safety are concerned. This chapter presents an overview on fundamentals for fire retardancy properties of epoxy resin based biocomposites and bionanocomposites. Tests that are utilized to evaluate flammability of epoxy resin based biocomposites and bionanocomposites and their results are reviewed. Banana, chicken eggshell, coir, flax, fly ash, hemp, kenaf, linseed, lychee peel, oil palm empty fruit bunch, pineapple, pumice, sisal, sugar palm, walnut shell and waste coffee grounds were employed to develop epoxy based biocomposites in the literature. Montmorillonite, cellulose nanofibril and nano oil palm empty fruit bunch were employed to develop epoxy based bionanocomposites. Flammability of these developed biocomposites and bionanocomposites were studied and following results were achieved:

- Epoxy decomposes when subjected to great temperature (300–400 °C), releasing heat, soot, smoke, and toxic volatiles [68].
- Flax based composites ignited earlier, released more heat, and their structure was significantly deformed during combustion than synthetic (glass) based composites [14].
- Hemp and linseed reinforced biocomposites gave no burning drops when burning [89].
- Flame retardant features tended to increase with addition of fly ash and lychee peel but reduced when used coffee grounds content increased [63, 60, 62].

- Fire resistance for reinforcements obtained from organic agricultural wastes bases on their compatibility via main material.
- Downsizing mean dimension of coffee grounds from 46 to 16 μm was concluded to be efficient in increasing flame retardancy in comparison with as-received waste coffee grounds [94].
- Moderate chicken eggshell biofiller content was significant to increase flame retardant and smoke suppression features [92].
- Neat epoxy ignites rapidly when flame is implemented and undergoes entire combustion having great smoke, melt dripping and no residue. On other hand, montmorillonite clay nanoplatelets/cellulose nanofibrils/epoxy nanocomposites do not demonstrate dripping and give residue after burning even though flame spreads rapidly and reaches clamp [53].
- Flame retardancy enhanced with montmorillonite, halloysite nanotube and organically modified montmorillonite nanoclay addition [18].
- Addition of chicken eggshell biofiller enhanced LOI value of coatings, sample having 37 wt% intumescent flame retardant and 3 wt% chicken eggshell gave maximal LOI of 31.5% [92].

Flammability nature of developed epoxy based biocomposites and bionanocomposites was investigated, however further research is required to evaluate flammability characteristics of epoxy based biocomposites and bionanocomposites. Extensively employed natural fibers should be incorporated to epoxy and flammability of these biocomposites should be studied in detail. Epoxy based biocomposites and bionanocomposites have flammability characteristic and are prone to fire damage. In future, innovative approach should be developed to employ bio-resources waste as an easily available and a cheaper fire retardant material.

Acknowledgements Author would like to thank the interventional neurologist Prof. Hasan Hüseyin Karadeli, who conducted operation and therapy after brain disease she suffered in August 2019. Author wants to dedicate current chapter to her family and Prof. Hasan Hüseyin Karadeli.

References

1. Akash KG, Girisha NSVG, Rao KVS (2017) A study on flammability and moisture absorption behavior of sisal/coir fiber reinforced hybrid composites. IOP Conf Ser Mater Sci Eng 191:012003. https://doi.org/10.1088/1757-899X/191/1/012003
2. Alsuwait RB, Souiyah M, Momohjimoh I, Ganiyu SA, Bakare AO (2023) Recent development in the processing, properties, and applications of epoxy-based natural fiber polymer biocomposites. Polymers 15:145. https://doi.org/10.3390/polym15010145
3. Babrauskas V, Peacock RD (1992) Heat release rate: the single most important variable in fire hazard. Fire Saf J 18:255–272. https://doi.org/10.1016/0379-7112(92)90019-9
4. Babu K, Rendén G, Mensah RA, Kim NK, Jiang L, Xu Q, Restás Á, Neisiany RE, Hedenqvist MS, Försth M, Byström A, Das O (2020) A review on the flammability properties of carbon-based polymeric composites: state-of-the-art and future trends. Polymers 12:1518. https://doi.org/10.3390/polym12071518

5. Bahrami M, Abenojar J, Martínez MÁ (2020) Recent progress in hybrid biocomposites: mechanical properties, water absorption, and flame retardancy. Materials 13:1–46. https://doi.org/10.3390/ma13225145
6. Bao Q, Wang B, Liu Y, Wang Q, Yang Z (2020) Epoxy resin flame retarded and toughed via flexible siloxane chain containing phosphaphenanthrene. Polym Degrad Stab 172:109055. https://doi.org/10.1016/j.polymdegradstab.2019.109055
7. Bourbigot S, Duquesne S (2007) Fire retardant polymers: recent developments and opportunities. J Mater Chem 17:2283–2300. https://doi.org/10.1039/b702511d
8. Braun U, Schartel B (2008) Flame retardancy mechanisms of aluminium phosphinate in combination with melamine cyanurate in glass-fibre-reinforced poly(1,4-butylene terephthalate). Macromol Mater Eng 293:206–217. https://doi.org/10.1002/mame.200700330
9. Buczko A, Stelzig T, Bommer L, Rentsch D, Heneczkowski M, Gaan S (2014) Bridged DOPO derivatives as flame retardants for PA6. Polym Degrad Stab 107:158–165. https://doi.org/10.1016/j.polymdegradstab.2014.05.17
10. Camino G, Costa L, Luda di Cortemiglia MP (1991) Overview of fire retardant mechanisms. Polym Degrad Stab 33:131–154. https://doi.org/10.1016/0141-3910(91)90014-I
11. Campana C, Léger R, Sonnier R, Ienny P, Ferry L (2022) Effect of hygrothermal ageing on the mechanical and fire properties of a flame retardant flax fiber/epoxy composite. Polymers 14:3962. https://doi.org/10.3390/polym14193962
12. Carosio F, Kochumalayil J, Cuttica F, Camino G, Berglund L (2015) Oriented clay nanopaper from biobased components—mechanisms for superior fire protection properties. ACS Appl Mater Interfaces 7:5847–5856. https://doi.org/10.1021/am509058h
13. Carvel R, Steinhaus T, Rein G, Torero JL (2011) Determination of the flammability properties of polymeric materials: a novel method. Polym Degrad Stab 96:314–319. https://doi.org/10.1016/j.polymdegradstab.2010.08.010
14. Chai MW, Bickerton S, Bhattacharyya D, Das R (2012) Influence of natural fibre reinforcements on the flammability of bio-derived composite materials. Compos Part B 43:2867–2874. https://doi.org/10.1016/j.compositesb.2012.04.051
15. Chapple S, Anandjiwala R (2010) Flammability of natural fiber-reinforced composites and strategies for fire retardancy: a review. J Thermoplast Compos Mater 23:871–893. https://doi.org/10.1177/0892705709356338
16. Chaudhary AK, Gope PC, Singh VK (2015) Water absorption and thickness swelling behavior of almond (*Prunus Amygdalus* L.) shell particles and coconut (*Cocos Nucifera*) fiber hybrid epoxy-based biocomposite. Sci Eng Compos Mater 22:375–382. https://doi.org/10.1515/secm.2013.0317
17. Chee SS, Jawaid M, Sultan MTH, Alothman OY, Abdullah LC (2019) Accelerated weathering and soil burial effects on colour, biodegradability and thermal properties of bamboo/kenaf/epoxy hybrid composites. Polym Testing 79:106054. https://doi.org/10.1016/j.polymertesting.2019.106054
18. Chee SS, Jawaid M, Alothman OY, Yahaya R (2020) Thermo-oxidative stability and flammability properties of bamboo/kenaf/nanoclay/epoxy hybrid nanocomposites. RSC Adv 10:21686–21697. https://doi.org/10.1039/d0ra02126a
19. Crossley R, Schubel P, Stevenson A (2014) Furan matrix and flax fibre as a sustainable renewable composite: mechanical and fire-resistant properties in comparison to phenol, epoxy and polyester. J Reinf Plast Compos 33:58–68. https://doi.org/10.1177/0731684413502108
20. Cusack PA, Hornsby PR (1999) Zinc stannate-coated fillers: novel flame retardants and smoke suppressants for polymeric materials. J Vinyl Add Tech 5:21–30. https://doi.org/10.1002/vnl.10302
21. Daniyan IA, Mpofu K, Adeodu AO, Adesina O (2021) Development of carbon fibre reinforced polymer matrix composites and optimization of the process parameters for railcar applications. Mater Today Proc 38:628–634. https://doi.org/10.1016/j.matpr2020.03.480
22. Das O, Kim NK, Sarmah AK, Bhattacharyya D (2017) Development of waste based biochar/wool hybrid biocomposites: flammability characteristics and mechanical properties. J Clean Prod 144:79–89. https://doi.org/10.1016/j.jclepro.2016.12.155

23. Delichatsios MA, Zhang J (2009) Chapter 19: micro to mesoscale testing and modeling for nanocomposite polymers in fires. In: Wilkie CA, Morgan AB (eds) Fire retardancy of polymeric materials. CRC Press, pp 509–550
24. Dorez G, Taguet A, Ferry L, Lopez-Cuesta JM (2013) Thermal and fire behavior of natural fibers/PBS biocomposites. Polym Degrad Stab 98:87–95. https://doi.org/10.1016/j.polymdegradstab.2012.10.026
25. Dorez G, Ferry L, Sonnier R, Taguet A, Lopez-Cuesta J-M (2014) Effect of cellulose, hemicellulose and lignin contents on pyrolysis and combustion of natural fibers. J Anal Appl Pyrol 107:323–331. https://doi.org/10.1016/j.jaap.2014.03.017
26. Elsabbagh A, Attia T, Ramzy A, Steuernagel L, Ziegmann G (2018) Towards selection chart of flame retardants for natural fibre reinforced polypropylene composites. Compos B 141:1–8. https://doi.org/10.1002/pc.23530
27. El-Sabbagh A, Steuernagel L, Meiners D, Ziegmann G, Toepfer O (2016) Optimization of flame retardant content with respect to mechanical properties of natural fiber polymer composites: case study of polypropylene/flax/aluminum trihydroxide. Polym Compos 37:3310–3325. https://doi.org/10.1016/j.compositesb.2017.12.020
28. Fang S, Hu Y, Song L, Zhan J, He Q (2008) Mechanical properties, fire performance and thermal stability of magnesium hydroxide sulfate hydrate whiskers flame retardant silicone rubber. J Mater Sci 43:1057–1062. https://doi.org/10.1007/s10853-007-2241-2
29. Ferdous D, Dalai AK, Bej SK, Thring RW (2002) Pyrolysis of lignins: experimental and kinetics studies. Energy Fuels 16:1405–1412. https://doi.org/10.1021/ef0200323
30. Gaan S, Neisius M, Cuchere O, Liang S, Mispreuve H (2014) Flame retardant polyurethanes based on novel phosphonamidate additives. Fire Saf Sci 11:821–831. https://doi.org/10.3801/IAFSS.FSS.11-821
31. Gao C, Huo S, Cao Z (2021) Solid wastes toward flame retardants for polymeric materials: a review. Front Mater 8:712188. https://doi.org/10.3389/fmats.2021.712188
32. García A, Alriols MG, Spigno G, Labidi J (2012) Lignin as natural radical scavenger. Effect of the obtaining and purification processes on the antioxidant behaviour of lignin. Biochem Eng J 67:173–185. https://doi.org/10.1016/j.bej.2012.06.013
33. Gieparda W, Rojewski S, Wüstenhagen S, Kicinska-Jakubowska A, Krombholz A (2021) Chemical modification of natural fibres to epoxy laminate for lightweight constructions. Compos Part A 140:106171. https://doi.org/10.1016/j.compositesa.2020.106171
34. Habibi S, Rashidi A, Bazgir S, Katbab AA, Montazer M (2009) Preparation and flame retardancy of poly(ethylene terephthalate)/montmorillonite nanocomposites. Asian J Chem 21:4881–4888
35. Helaili S, Chafra M, Chevalier Y (2021) Natural fiber alfa/epoxy randomly reinforced composite mechanical properties identification. Structures 34:542–549. https://doi.org/10.1016/j.istruc.2021.07.095
36. Horrocks AR (1983) An introduction to the burning behaviour of cellulosic fibres. J Soc Dyers Colour 99:191–197. https://doi.org/10.1111/j.1478-4408.1983.tb03686.x
37. Isarov SA, Lee PW, Towslee JH, Hoffman KM, Davis RD, Maia JM, Pokorski JK (2016) DNA as a flame retardant additive for low-density polyethylene. Polymer 97:504–514. https://doi.org/10.1016/j.polymer.2016.05.060
38. Jawaid M, Awad S, Fouad H, Asim M, Saba N, Dhakal HN (2021) Improvements in the thermal behaviour of date palm/bamboo fibres reinforced epoxy hybrid composites. Compos Struct 277:114644. https://doi.org/10.1016/j.compstruct.2021.114644
39. Khalifa M, Berndt A, Shamitha C, Pichler S, Lammer H, Wuzella G (2023) Hybridization effect of cellulose paper and postcuring conditions on the mechanical properties of flax fiber reinforced epoxy biocomposite. J Appl Polym Sci 140:e53297. https://doi.org/10.1002/app.53297
40. Khalili P, Tshai KY, Kong I (2017) Natural fiber reinforced expandable graphite filled composites: evaluation of the flame retardancy, thermal and mechanical performances. Compos A 100:194–205. https://doi.org/10.1016/j.compositesa.2017.05.015

41. Kiliaris P, Papaspyrides CD (2010) Polymer/layered silicate (clay) nanocomposites: an overview of flame retardancy. Prog Polym Sci 35:902–958. https://doi.org/10.1016/j.progpolymsci.2010.03.001
42. Koyunucu M, Ulay G (2023) Thermal and flammability behavior of walnut shell reinforced epoxy composites. Polímeros 33:e20230018. https://doi.org/10.1590/0104-1428.20230018
43. Kozlowski R, Wladyka-Przybylak M (2004) Chapter 14: uses of natural fiber reinforced plastics. In: Wallenberger FT, Weston NE (eds) Natural fibers, plastics and composites. Springer, pp 249–274
44. Kozłowski R, Władyka-Przybylak M (2008) Flammability and fire resistance of composites reinforced by natural fibers. Polym Adv Technol 19:446–453. https://doi.org/10.1002/pat.1135
45. Kumar S, Saha A, Bhowmik S (2022) Accelerated weathering effects on mechanical, thermal and viscoelastic properties of kenaf/pineapple biocomposite laminates for load bearing structural applications. J Appl Polym Sci 139:e51465. https://doi.org/10.1002/app.51465
46. Kuram E (2022) Chapter 14: hybrid biocomposites: utilization in aerospace engineering. In: Mazlan M, Sapuan SM, Ilyas RA (eds) Advanced composites in aerospace engineering applications. Springer, pp 281–301
47. Laoutid F, Bonnaud L, Alexandre M, Lopez-Cuesta J-M, Dubois P (2009) New prospects in flame retardant polymer materials: from fundamentals to nanocomposites. Mater Sci Eng R 63:100–125. https://doi.org/10.1016/j.mser.2008.09.002
48. Lee CH, Salit MS, Hassan MR (2014) A review of the flammability factors of kenaf and allied fibre reinforced polymer composites. Adv Mater Sci Eng 2014:514036. https://doi.org/10.1155/2014/514036
49. Lekrine A, Belaadi A, Makhlouf A, Amroune S, Bourchak M, Satha H, Jawaid M (2022) Structural, thermal, mechanical and physical properties of *Washingtonia Filifera* fibres reinforced thermoplastic biocomposites. Mater Today Commun 31:103574. https://doi.org/10.1016/j.mtcomm.2022.103574
50. Liu A, Walther A, Ikkala O, Belova L, Berglund LA (2011) Clay nanopaper with tough cellulose nanofiber matrix for fire retardancy and gas barrier functions. Biomacromol 12:633–641. https://doi.org/10.1021/bm101296z
51. Liu K, Takagi H, Osugi R, Yang Z (2012) Effect of lumen size on the effective transverse thermal conductivity of unidirectional natural fiber composites. Compos Sci Technol 72:633–639. https://doi.org/10.1016/j.compscitech.2012.01.009
52. Medeiros ES, Santos ASF, Dufresne A, Orts WJ, Mattoso LHC (2014) Chapter 11: bionanocomposites. In: Thomas S, Joseph K, Malhotra SK, Goda K, Sreekala MS (eds) Polymer composite, vol 3. Wiley, pp 361–430
53. Medina L, Ansari F, Carosio F, Salajkova M, Berglund LA (2019) Nanocomposites from clay, cellulose nanofibrils, and epoxy with improved moisture stability for coatings and semistructural applications. ACS Appl Nano Mater 2:3117–3126. https://doi.org/10.1021/acsanm.9b00459
54. Mittal M, Chaudhary R (2019) Effect of layering pattern on flammability and water absorption behavior of pineapple leaf/coir fiber reinforced hybrid composites. I-Manager's J Mater Sci 7:44–54. https://doi.org/10.26634/jms.7.1.15204
55. Mngomezulu ME, John MJ, Jacobs V, Luyt AS (2014) Review on flammability of biofibres and biocomposites. Carbohyd Polym 111:149–182. https://doi.org/10.1016/j.carbpol.2014.03.071
56. Mochane M, Mokhena TC, Mokhothu TH, Mtibe A, Sadiku ER, Ray SS, Ibrahim ID, Daramola OO (2019) Recent progress on natural fiber hybrid composites for advanced applications: a review. eXPRESS Polym Lett 13:159–198. https://doi.org/10.3144/expresspolymlett.2019.15
57. Mosiewicki MA, Aranguren MI (2013) A short review on novel biocomposites based on plant oil precursors. Eur Polymer J 49:1243–1256. https://doi.org/10.1016/j.eurpolymj.2013.02.034
58. Moustafa H, Duquesne S, Darwish NA, Youssef AM (2016) Highly flame retardant green composites using seashells. Soc Plast Eng Plast Res Online 2016:1–3. https://doi.org/10.2417/spepro.006334
59. Naguib HM (2018) Environmental-friendly recycled polyester/$Mg(OH)_2$ nanocomposite: fire retardancy and thermal stability. Polym Testing 72:308–314. https://doi.org/10.1016/j.polymertesting.2018.10.026

60. Nguyen TA (2021) Biocomposites developed with litchi peel based on epoxy resin: mechanical properties and flame retardant. J Chem 2021:3287733. https://doi.org/10.1155/2021/3287733
61. Nguyen TA, Nguyen TH (2021) Banana fiber-reinforced epoxy composites: mechanical properties and fire retardancy. Int J Chem Eng 2021:1973644. https://doi.org/10.1155/2021/1973644
62. Nguyen TA, Nguyen QT (2021) Hybrid biocomposites based on used coffee grounds and epoxy resin: mechanical properties and fire resistance. Int J Chem Eng 2021:1919344. https://doi.org/10.1155/2021/1919344
63. Nguyen TA, Nguyen XH, Ngo TV (2023) Research on mechanical properties and fire retardancy of epoxy composites reinforced by fly ash of thermal power plant. Vietnam J Chem 61(S3):97–108. https://doi.org/10.1002/vjch.202300048
64. Ortega R, Monzón MD, Ortega ZC, Cunningham E (2020) Study and fire test of banana fibre reinforced composites with flame retardance properties. Open Chem 18:275–286. https://doi.org/10.1515/chem-2020-0025
65. Özmen FK, Üreyen ME, Koparal AS (2020) Cleaner production of flame-retardant-glass reinforced epoxy resin composite for aviation and reducing smoke toxicity. J Clean Prod 276:124065. https://doi.org/10.1016/j.jclepro.2020.124065
66. Pan M, Guan D, Wang T, Huang R, Mu J, Zhang C (2013) Morphology, thermal properties, and fire behavior of epoxy resin nanocomposites containing octaammonium polyhedral oligomeric silsesquioxane-modified montmorillonite. High Perform Polym 25:992–999. https://doi.org/10.1177/0954008313493101
67. Price D, Anthony G, Carty P (2001) Chapter 1: introduction: polymer combustion, condensed phase pyrolysis and smoke formation. In: Horrocks AR, Price D (eds) Fire retardant materials. Woodhead Publishing, pp 1–30
68. Rakotomalala M, Wagner S, Döring M (2010) Recent developments in halogen free flame retardants for epoxy resins for electrical and electronic applications. Materials 3:4300–4327. https://doi.org/10.3390/ma3084300
69. Ranjan C, Ahmed Z, Kumar SR, Kumar A, Kumar A, Kumar K (2021) Fabrication and strength analysis of rice straw fibers reinforced epoxy biodegradable composite. Mater Today Proc 46:331–335. https://doi.org/10.1016/j.matpr.2020.08.299
70. Reddy BM, Reddy PV, Reddy YVM, Reddy BCM, Reddy RM (2022) Mechanical, thermal and morphological properties of raw cordia dichotoma fiber reinforced epoxy composites. Curr Res Green Sustain Chem 5:100264. https://doi.org/10.1016/j.crgsc.2022.100264
71. Rwawiire S, Tomkova B, Militky J, Jabbar A, Kale BM (2015) Development of a biocomposite based on green epoxy polymer and natural cellulose fabric (bark cloth) for automotive instrument panel applications. Compos B 81:149–157. https://doi.org/10.1016/j.compositesb.2015.06.021
72. Saba N, Jawaid M, Paridah MT, Al-Othman OY (2016) A review on flammability of epoxy polymer, cellulosic and non-cellulosic fiber reinforced epoxy composites. Polym Adv Technol 27:577–590. https://doi.org/10.1002/pat.3739
73. Saba N, Jawaid M, Alrashed MM, Alothman OY (2019) Oil palm waste based hybrid nanocomposites: fire performance and structural analysis. J Build Eng 25:100829. https://doi.org/10.1016/j.jobe.2019.100829
74. Sahoo SK, Mohanty S, Nayak SK (2017) Mechanical, thermal, and interfacial characterization of randomly oriented short sisal fibers reinforced epoxy composite modified with epoxidized soybean oil. J Nat Fibers 14:357–367. https://doi.org/10.1080/15440478.2016.1212757
75. Sain M, Park SH, Suhara F, Law S (2004) Flame retardant and mechanical properties of natural fibre-PP composites containing magnesium hydroxide. Polym Degrad Stab 83:363–367. https://doi.org/10.1016/S0141-3910(03)00280-5
76. Samyn F, Bourbigot S, Jama C, Bellayer S (2008) Fire retardancy of polymer clay nanocomposites: is there an influence of the nanomorphology? Polym Degrad Stab 93:2019–2024. https://doi.org/10.1016/j.polymdegradstab.2008.02.013
77. Sapuan SM, Sanyang L, Sahari J (2015) Chapter 9: development and properties of sugar palm fiber reinforced polymer composites. In: Thakur VK, Kessler MR (eds) Green biorenewable biocomposites: from knowledge to industrial applications. Apple Academic Press, pp 289–322

78. Schartel B, Hull TR (2007) Development of fire-retarded materials—interpretation of cone calorimeter data. Fire Mater 31:327–354. https://doi.org/10.1002/fam.949
79. Shi Y, Yu B, Zheng Y, Yang J, Duan Z, Hu Y (2018) Design of reduced graphene oxide decorated with DOPO-phosphanomidate for enhanced fire safety of epoxy resin. J Colloid Interface Sci 521:160–171. https://doi.org/10.1016/j.jcis.2018.02.054
80. Shivamurthy B, Thimmappa BHS, Monteiro J (2020) Sliding wear, mechanical, flammability, and water intake properties of banana short fiber/$Al(OH)_3$/epoxy composites. J Nat Fibers 17:337–345. https://doi.org/10.1080/15440478.2018.1492489
81. Song P, Cao Z, Fu S, Fang Z, Wu Q, Ye J (2011) Thermal degradation and flame retardancy properties of ABS/Lignin: effects of lignin content and reactive compatibilization. Thermochim Acta 518:59–65. https://doi.org/10.1016/j.tca.2011.02.007
82. Sonnier R, Taguet A, Ferry L, Lopez-Cuesta J-M (2018) Towards bio-based flame retardant polymers. In: Springer briefs in molecular science. Springer International Publishing, Cham, Switzerland. ISBN 9783319670829
83. Suriani MJ, Ruzaidi CM, Nair DS (2020) Flammability and mechanical properties of Arenga Pinnata fibre/PET reinforced epoxy for fire retardant composites. IOP Conf Ser Mater Sci Eng 864:012002. https://doi.org/10.1088/1757-899X/864/1/012002
84. Suriani MJ, Sapuan SM, Ruzaidi CM, Nair DS, Ilyas RA (2021) Flammability, morphological and mechanical properties of sugar palm fiber/polyester yarn-reinforced epoxy hybrid biocomposites with magnesium hydroxide flame retardant filler. Text Res J 91:2600–2611. https://doi.org/10.1177/00405175211008615
85. Suzanne M, Ukleja S, Delichatsios M, Zhang J, Karlsson B (2014) Fundamental flame spread and toxicity evaluation of fire retarded polymers. Fire Saf Sci 11:846–859. https://doi.org/10.3801/IAFSS.FSS.11-846
86. Taleb F, Ammar M, Ben Mosbah M, Ben Salem R, Moussaoui Y (2020) Chemical modification of lignin derived from spent coffee grounds for methylene blue adsorption. Sci Rep 10:11048. https://doi.org/10.1038/s41598-020-68047-6
87. Uddin MF, Sun CT (2010) Improved dispersion and mechanical properties of hybrid nanocomposites. Compos Sci Technol 70:223–230. https://doi.org/10.1016/j.compscitech.2009.09.017
88. Vahabi H, Kandola BK, Saeb MR (2019) Flame retardancy index for thermoplastic composites. Polymers 11:407. https://doi.org/10.3390/polym11030407
89. Vidal J, Ponce D, Mija A, Rymarczyk M, Castell P (2023) Sustainable composites from nature to construction: hemp and linseed reinforced biocomposites based on bio-based epoxy resins. Materials 16:1283. https://doi.org/10.3390/ma16031283
90. Wang CS, Lin CH (1999) Synthesis and properties of phosphorus-containing epoxy resins by novel method. J Polym Sci Part A Polym Chem 37:3903–3909. https://doi.org/10.1002/(SICI)1099-0518(19991101)37:21%3c3903::POLA4%3e3.0.CO;2-X
91. Wang X, Kalali EN, Wang D-Y (2015) Renewable cardanol-based surfactant modified layered double hydroxide as a flame retardant for epoxy resin. ACS Sustain Chem Eng 3:3281–3290. https://doi.org/10.1021/acssuschemeng.5b00871
92. Xu Z, Chu Z, Yan L, Chen H, Jia H, Tang W (2019) Effect of chicken eggshell on the flame-retardant and smoke suppression properties of an epoxy-based traditional APP-PER-MEL system. Polym Compos 40:2712–2723. https://doi.org/10.1002/pc.25077
93. Yang W, Wu S, Yang W, Yuen AC-Y, Zhou Y, Yeoh G, Boyer C, Wang CH (2020) Nanoparticles of polydopamine for improving mechanical and flame-retardant properties of an epoxy resin. Compos B 186:107828. https://doi.org/10.1016/j.compositesb.2020.107828
94. Yang W, Chang W, Zhang J, Yeoh GH, Boyer C, Wang CH (2022) Effects of waste coffee grounds on the mechanical properties, flame retardancy and toxic gas production of epoxy composites. Mater Des 224:111347. https://doi.org/10.1016/j.matdes.2022.111347
95. Yazdi MFA, Zakaria R, Mustaffar M, Majid MZA, Zin RM, Ismail M, Yahya K (2014) Bio-composite materials potential in enhancing sustainable construction. Desalin Water Treat 52:3631–3636. https://doi.org/10.1080/19443994.2013.854105

96. Zhang L, Liang S, Chen Z (2018) Influence of particle size and addition of recycling phenolic foam on mechanical and flame retardant properties of wood-phenolic composites. Constr Build Mater 168:1–10. https://doi.org/10.1016/j.conbuildmat.2018.01.173
97. Zhang W, Li X, Yang R (2011) Pyrolysis and fire behaviour of epoxy resin composites based on a phosphorus-containing polyhedral oligomeric silsesquioxane (DOPO-POSS). Polym Degrad Stab 96:1821–1832. https://doi.org/10.1016/j.polymdegradstab.2011.07.014
98. Zorko D, Tavčar J, Bizjak M, Šturm R, Bergant Z (2021) High cycle fatigue behaviour of autoclave-cured woven carbon fibre-reinforced polymer composite gears. Polym Testing 102:107339. https://doi.org/10.1016/j.polymertesting.2021.107339

Phenolic and PLA Based Natural Fiber Reinforce Composite: A Fire Performance Analysis

Durgesh Kumar Mishra, Nitish Kumar, and Santosh Kumar

Abstract In order to produce products based on composites, researchers have recently made a significant effort to find environmentally safe and biodegradable reinforcements. The increasing usage of biomaterials in recent times has significantly reduced carbon footprints and greenhouse gas emissions. As a result, the use of natural fiber reinforcement in polymer matrices is highly recommended due to its exceptional ability to balance processing convenience and performance with cost. The low fire resistance and delayed flammability behavior of these materials, limit the application of natural fiber composite for different potential applications. Consequently, it follows that in order to create natural composite with improved thermal degradation, biodegradation, and fire resistance, it is critical to comprehend the flammability and degradation of natural composite as well as the testing procedures used to evaluate them. Hence, in this review, we will present the detailed analysis of fire performance of phenolic and PLA based reinforced polymer composite. In particular, UL-94 horizontal and vertical flammability tests and cone calorimetric analysis was presented the utmost dominant methods for evaluating the fire performance behavior of the natural fiber-reinforced composites.

Keywords Flammability · Fire retardant behavior · Phenolic · PLA · Natural Fiber

---

D. K. Mishra (✉)
Department of Aeronautical Engineering, Hindustan Institute of Technology and Science, Kelambakkam, Tamil Nadu, India
e-mail: mishradurgesh34@gmail.com

N. Kumar
Department of Mechanical and Industrial Engineering, Manipal Institute of Technology, Manipal Academy of Higher Education, Manipal, Karnataka, India

S. Kumar
Department of Mechanical Engineering, COEP Technological University, Pune, Maharashtra, India

C. Muthukumar et al. (eds.), *Flammability Performance of Biocomposites and Bionanocomposites*, Composites Science and Technology,
https://doi.org/10.1007/978-981-95-2389-4_3

1 Introduction

The use of plant fibers as reinforced materials for polymer-based composites has gained increased attention in recent years, primarily because of its widespread availability, inexpensive cost, high specific strength and modulus, and biodegradability [3, 4]. Classification of natural fiber composite is illustrated in Fig. 1. Plant fiber-reinforced composites (PFRCs) have intriguing mechanical and superior physical properties, such as heat isolation and sound absorption when compared to carbon fiber or glass fiber-reinforced composites [30], which make them suitable for use as interior structures in the transportation or civil construction industries [9, 10, 12, 14]. The flame retardant (FR) requirements, however, are extremely stringent for the interior components of those applications. Because of their extremely combustible nature, FR pretreatments are therefore crucial for PFRCs. To increase the flame retardancy of PFRCs, numerous studies have been conducted. The most popular technique for achieving flame retardancy in composites is the introduction of FRs [20]. The additives can exist physically or react chemically within the composites, depending on their composition [15]. The best fillers for fiber-reinforced composites are those that don't contain halogens, which burn to produce harmful fumes, and can prevent breaking down during composite processing. When using flame retardants, it is important that they work well with the matrix or fibers and don't alter the material's excellent intrinsic qualities, especially its modulus and fracture toughness. Factors that affecting the fire performance behavior of natural fiber-reinforced composites is given in Fig. 2 [7].

One of the main causes of the lack of focus on the flammability characteristics of natural fiber-reinforced composites is the time and expense involved in the various

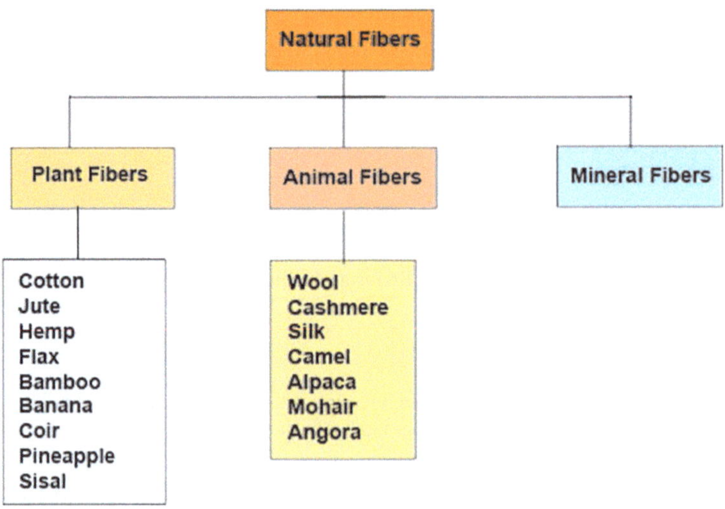

Fig. 1 Classification of natural fiber polymer composite

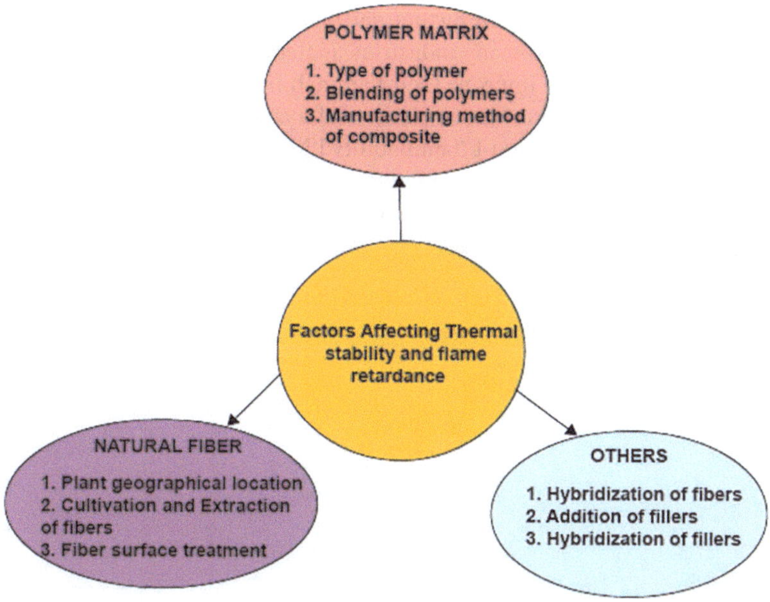

Fig. 2 Factors affecting the fire performance behavior of natural fiber-reinforced composites

tests needed to measure fire performance [23]. In addition, several tests necessitate the implementation of sophisticated equipment and unique arrangements in the home in order to obtain the necessary data. Developing a model to predict the thermal structural correlation and accurately representing the complicated real-life scenario in the experimental space is a challenging task that warrants more investigation Ramesh et al. [34]. The purpose of this chapter is to examine the fire performance of PLA and phenolic based polymeric composite materials Rahman et al. [3].

Following the introduction, the chapter provides readers with an overview of the thermal properties of polymers and the flammability characteristics of natural fibers. An overview of the fire-retardant phenomenon and its many application techniques follows. The fire performance of natural fiber-reinforced polymeric composite materials for building materials and vehicle interiors is covered in the section that follows. The concluding remarks mark in the end of the chapter.

2 Flame Retardancy Behavior of Natural Fiber-Based Composites

The technique that permits the substance to burn slowly is known as flame retardation. Because natural fibers are present in the polymer biocomposites, it is very likely that they will ignite easily. Thus, flame retardants are added to the natural fiber-based biocomposites to extend their burning time and flame resistance. Because natural fiber biocomposites are used in so many industries, including aerospace, automotive, and construction, it is crucial to improve their flammability behavior and fire resistance. Given that the topic at hand is biocomposites safety, it's critical to comprehend how the composites behave in terms of flammability [38, 42]. The several burning phases of natural fiber-reinforced biocomposites are depicted Khandelwal et al. [17] in Fig. 3 and the various fire testing methods are their standards are shown in Fig. 4.

Fig. 3 Burning stages of natural fiber composite

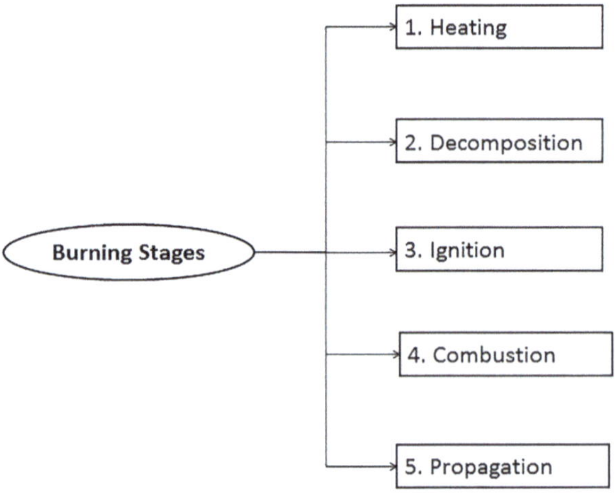

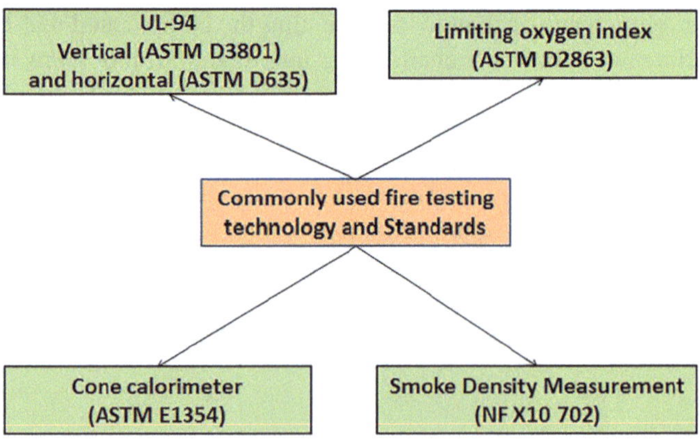

Fig. 4 Different types of commonly used fire technology and standards

3 Outline of PLA Composite

Biodegradable polymer made from renewable resources, polylactic acid (PLA) finds use in a variety of fields, including packaging, biomedical textile fibers, and technological goods. Because of its intrinsic properties, PLA has a significant position in the biopolymer sector, making it a highly attractive contender for future improvements [28]. While specialized end-use qualities are needed for certain applications, PLA unfortunately has several drawbacks. Consequently, adding micro-and/or nanofillers, reinforcing fibers, and specialized additives to the PLA matrix is thought to be a potent way to achieve desired end-use features and significant property improvements. Industrial processes are used to produce poly(lactic acid) or polylactide (PLA), respectively, by polymerizing lactic acid (LA) or lactide (the cyclic dimer of lactic acid, as an intermediate) Lau et al. [21]. Apart from being biocompatible and biodegradable (NB: under regulated composting circumstances), PLA is also a thermoplastic aliphatic polyester made from non-fossil renewable natural resources through the fermentation of polysaccharides or sugar, such as those derived from corn, potatoes, cane molasses, sugar beet, etc. [18].

3.1 Outline of Phenolic Composite

Phenolic composites are a type of materials that incorporate reinforcing fibers, including glass, carbon, or natural fibers, with the strength and durability of phenolic resins. These materials are great at withstanding heat and chemicals, which makes them useful for many different kinds of applications. Formaldehyde and phenol are the sources of these artificial polymers. Their strong chemical stability, low flammability, and excellent heat resistance define them. The mechanical properties of the composite are determined by the selection of reinforcing fibers. Typical choices are Carbon fiber: Offers remarkable strength and stiffness but can be more expensive; glass fiber: Offers a reasonable strength-to-weight ratio and is reasonably priced. Natural fibers: Lightweight and more sustainable composites can be made with jute, flax, or bamboo. To prepare the reinforcing fibers, they might be layered, woven, or cut into the required shape. Through the use of methods like vacuum bagging or resin transfer molding, the fibers are impregnated with a phenolic resin. In order to start the curing process and create a robust and long-lasting material, the impregnated composite is heated under pressure [36]. Phenolic resins are appropriate for use in severe situations since they do not degrade at high temperatures. They can withstand a variety of substances, such as solvents, acids, and alkalis Gholampour et al. [11]. Flame-retardant qualities are inherent in phenolic composites. Excellent strength, stiffness, and impact resistance are achieved by the combination of reinforcing fibers and phenolic resin. Phenolic composites show very little dimensional changes in response to moisture or temperature changes. Phenolic composites are a dependable and adaptable material choice for many different types of applications [16]. Because of their special mix of qualities, they are a desirable option for applications needing high performance, toughness, and resilience to adverse environments.

4 Analysis on PLA-Based Composite

Recently, [8] presented an attempt to increase the flame retardancy of the biocomposites by including chitosan (CS) and ammonium polyphosphate (APP) into polylactic acid (PLA) through melt blending. Cone calorimetry testing (CONE), the UL-94 vertical burning test, and the limiting oxygen index (LOI) were used to assess the flammability of the composites. According to the findings, the PLA sample with 2% CS and 5% APP passed the UL-94 V-0 rating, reduced the peak heat release rate to 425.6 kW/m^2, and reached the maximum LOI value of 33.1. An oragnophosphonate flame retardant (FR) was used to fire retard fabrics made of a blend of thermoplastic (polypropylene (PP) / poly lactic acid (PLA) and natural (flax) polymeric fibers [31]. The corresponding thermoplastic composites were prepared by melt pressing the fire-resistant flax/PP and flax/PLA textiles. To obtain a V-0 rating, the UL-94 test was utilized as a standard, and cone calorimetric Table 1 characteristics were employed for a thorough examination of fire performance. The findings indicate that in order to receive a V-0 rating in the UL-94 test, flax/PP and flax/PLA composites must have a minimum of 0.9 and 0.6% phosphorus (P), respectively.

Moreover, the thermal stability and flammability of coconut fiber/PLA green composites were investigated by [13]. The coconut fibers and PLA matrix interfacial adhesion strength was improved by plasma treatment of the fiber surface, and the green composite was created using the commingled yarn method. There was an improvement in the mechanical characteristics, including Young's modulus and tensile strength. When the plasma treatment was applied, the shrinkage of the coconut fiber/PLA composites generally decreased as the weight percentage of fiber increased. Despite being more environmentally friendly and sustainable than synthetic materials, natural fiber composites have a high flammability [41]. This work explored the synergistic barrier effect of silica, melamine cyanurate, and aluminum diethylphosphinate. Additionally, flame retardant ramie fiber-reinforced poly (lactic acid) composites were developed. The composites achieved UL94 V0 certification

Table 1 Cone calorimetric data of flax/PLA composites [31]

Sample	Time to ignition (s)	Peak heat release rate (kW/m^2)	Total heat release rate (MJ/m^2)	Total smoke released (L)
Flax/PLA	66 ± 2	283 ± 4	49 ± 2	61 ± 20
Flax/PLA-0.4P	71 ± 5	236 ± 9	41 ± 1	194 ± 26
Flax/PLA-0.5P	148 ± 7	223 ± 49	19 ± 5	1210 ± 364
Flax/PLA_0.6P	–	31 ± 2	8 ± 1	1629 ± 43
Flax/PLA_0.8P	–	39 ± 7	10 ± 1	1591 ± 96

with limiting oxygen index of 30.1% when each flame retardant's mass ratio was 10%. Excellent char layers, the chain reaction of free radicals, and the dilution effect of inert gases produced after combustion all contributed to the material's exceptional flame retardancy shown in Table 2. To create natural fiber-reinforced PLA/TPS biocomposites with flame retardancy in an affordable manner, novel multifunctional additives including phosphorus were composed [6]. Glycerol is phosphate, a synthetic polyol containing phosphorus, was utilized as a plasticizer to produce thermoplastic starch with a lower flammability. When added to PLA, this modified starch was found to have increased char-promoting capacity, which improved the flame-retardancy of PLA/TPS blends. The Flame Retardancy Index, or FRI, is a straightforward yet widely applicable dimensionless criterion that was developed from thermoplastic composite cone calorimetry data and used to assess the flame retardancy performance of various polymer composites on a set of trustworthy data [37]. Cone calorimetry's primary characteristics, such as peak heat release rate (pHRR), total heat release (THR), and time-to-ignition (TTI), were utilized to compute a dimensionless measure that quantitatively represents an improvement in the flame retardancy of the thermoplastic composites that were nominated over the neat thermoplastic. Researchers have recently made a significant effort to find environmentally friendly and biodegradable reinforcements for use in the production of composite-based products. Due to the growing usage of biomaterials, there has been a significant decrease in greenhouse gas emissions and carbon footprints [33]. As a result, natural fiber reinforcement in polymer matrices is extensively favored due to its exceptional cost-performance ratio and manufacturing simplicity. The limited fire resistance and delayed flammability behavior of these materials, however, hinder the widespread application of natural fiber polymer composites in industrial areas. It is possible to draw the conclusion that creating biocomposites with improved thermal degradation, fire resistance, and biodegradation properties requires an awareness of the flammability and degradation of biocomposites as well as a comprehension of the testing procedures for these properties.

Knez et al. [19] looked into how certain printing parameters affected the fire characteristics of composites made additively using wood/PLA and neat polylactic acid (PLA) filaments. The ISO 5660-1 cone calorimeter test method was used to examine the 3D-printed specimens' response to fire. The findings demonstrated that the addition of wood flour to the PLA filament had a substantial impact on the specimens'

Table 2 Cone calorimeter data for composites [41]

Sample	Time to ignition (s)	Peak heat release rate (kW/m^2)	Total heat release rate (MJ/m^2)	Total smoke released (m^2/s)
RRP	31	560.3	62.2	20
RRP/3A	39	483.1	58.7	426.6
RRP/3 M	42	401.1	51.8	9.2
RRP/AM	42	451.4	53.8	175.7
RRP/AMS	43	434.3	52.9	219.4

characteristics when they were subjected to fire. It was thought that bamboo fiber would provide an excellent reinforcement option for green composites. Numerous studies have looked into the use of bamboo fibers as reinforcements in structural or fabric forms to create composites. The procedure from the bamboo strips to the fibers combined with polylactide to generate the composite pellets for the subsequent injection molding process was carried out in this study Young et al. [39]. Excellent fire resistance for polymers and their composites was established by phosphorus-based flame retardants, although the mechanical qualities were invariably compromised. DOPO-MA, a diacid derivative containing phosphorus, was created through the reaction of DOPO (9,10-Dihydro-9-oxa-10-phosphaphenanthrene-10-oxide) with maleic acid (MA) [40]. Using a twin-screw extruder, DOPO was functionalized and added to short jute/PLA composites. The active sites that connected PLA or jute to DOPO were made possible by the functionalization of DOPO, ensuring a good interface. Comparing the tensile, flexural, and impact strength after including DOPO-MA to that of DOPO, an improvement was seen. The composites' flammability had decreased and their thermal stability had improved simultaneously, suggesting that DOPO-MA was a very effective fire resistant material.

5 Analysis on Phenolic-Based Composite

Molaba et al. [29] discussed about the flame retardant (FR) flax fiber incorporated phenolic based composite. They have investigated flammability of untreated, chemically treated and FR treated composite. Cone calorimeter technique was used to find out flammability behavior. Results revealed that the FR flex improved effectively the flame retardancy due to decrement in smoke production and peak heat release rate. Tensile strength, elastic modulus and color of the composite were also affected due to high temperature aging.

Li et al's study [24] examined the flame retardants dimethyl methyl phosphonate (DMMP) and nano-montmorillonite (MMT) based on flax fiber-reinforced phosphorus. The limiting oxygen index test and the vertical burning test were used to assess the composites' flammability. The outcomes showed that pre-treating flax fiber with DMMP is an easy and efficient way to increase the flame retardancy of composite materials. By including MMTs into the phenolic resin, more carbonization layers were able to form inside the composites during burning, which worked in concert with the DMMP to retard flame.

The treated and untreated pineapple leaf fiber (PALF) and kenaf fiber (KF) reinforced composite were investigated by Asim et al. [1]. Pure phenolic is very fire resistant, according to UL-94 testing conducted both vertically and horizontally. In phenolic composites, fire resistance dropped with the addition of PALF; fire resistance improved in phenolic composites with the addition of KF. The PALF and KF composites' characteristics did not improve with the fiber treatment.

The treated and untreated coconut leaf sheath fiber included composite was studied by Bharath et al. [5]. For flammability testing, UL-94 vertical and horizontal tests as well as the Limiting Oxygen Index were conducted. The treated fiber had a lower mass loss rate and flame propagation rate, according to the results, while burning untreated fiber requires more oxygen.

The mechanical, physical, and flammability properties of Kenaf Fiber/Unsaturated Polyester/Phenolic Composite were studied by Marliana et al. [27]. Using the LOI and UL-94 fire tests, the flammability characteristics of the composites were ascertained. The UL-94 test shows that the composites have enhanced flame retardancy, although the LOI increased with the PF content.

The effects of fiber length and stress on the physico-mechanical and flammability parameters of phenolic composites reinforced with Cyrtostachys renda fiber were examined by Loganathan et al. [25]. The UL 94 vertical and horizontal did not significantly alter with the addition of CR fiber, although the LOI value did somewhat decrease. For the UL 94, the neat phenolic's Limiting Oxygen Index (LOI) dropped from 29.33 to 26 and 25 for 20 weight percent and 40 weight percent, respectively. Regarding vertical and horizontal classifications, all of the neat phenolic and addition of CR fiber displayed NC and H-B, respectively.

Loganathan et al. [26] looked at how adding multi-walled carbon nanotubes (MWCNT) and Cyrtostachys renda (CR) fiber affected the morphological, physical, mechanical, and flammability aspects of phenolic composites. Based on limiting oxygen index (LOI) and UL-94 horizontal values, all composite materials are classified as self-extinguishing.

The thermal, physical, and flammability characteristics of silane-treated phenolic hybrid composites made of kenaf and pineapple leaf fibers were examined by Asim et al. [2]. The flammability of PALF/KF hybrid composites was not significantly increased by treated hybrid composites.

Sisal fiber-reinforced biobased silicone modified phenolic composites were studied by Li et al. [22]. The optimized modified composites exhibit a 60% reduction in overall heat release. It was discovered that the electrical resistance, impact strength, and flame retardant qualities all met application requirements. Through the use of an innovative, inexpensive biobased silicone modifier, the research offered a practical means of enhancing the performance of these sustainable natural fiber-reinforced composites.

Phenolic composites showing generally good fire resistance because of the inbuilt properties contains in phenolic resins and the capability of natural fibers shown in Table 3.

Table 3 Fire performance details for phenolic/natural fiber composite

Natural fiber type	Peak heat release rate (kW/m^2)	Total smoke production (m^2)	Total heat released (MJ/m^2)	References
Flax fabric	145.5 ± 9.8	2.1 ± 0.3	33.5 ± 8.6	Molaba et al. [29]
Flax fabric	602.36 ± 5.1 to 196.91 ± 31.2	7.43 ± 0.2 to 3.85 ± 0.7	161.33 ± 6.3 to 62.34 ± 6.1	Li et al. [24]
Sisal fiber	559 ± 1 to 342 ± 8	8.9 ± 0.2 to 2.3 ± 0.1	55.9 ± 0.5 to 20.2 ± 0.2	Kim et al. [18]

6 Conclusions

The utilization of plant-based natural fiber composites has several benefits over synthetic fiber-based composites, including reduced cost, lower density, increased sustainability, and renewability. These features allow for a greater range of commercial applications. The fire performance of PLA and phenolic reinforced natural fiber composite has been thoroughly and methodically reviewed in this article. While adding natural fibers makes polymer biocomposites more flammable, adding the right flame retardants or covering the fibers with flame-retardant materials makes the biocomposites more fire-resistant. Cone calorimetric analysis and UL-94 horizontal and vertical flammability tests were the most widely used techniques for analyzing the composites' flame retardancy behavior. Heat release rate, smoke created are the key results derived from these testing. Content of fibers and fillers, type of the flame-retardant and hybridization effects defined the flammability behavior of the biocomposites. Thus, achieving the ideal balance between natural fiber and flame-retardant material can result in a bio-based flame-retardant material.

References

1. Asim M, Jawaid M, Nasir M, Saba N (2018) Effect of fiber loadings and treatment on dynamic mechanical, thermal and flammability properties of pineapple leaf fiber and kenaf phenolic composites. J Renew Mat 6(4):383–393
2. Asim M, Paridah MT, Saba N, Jawaid M, Alothman OY, Nasir M, Almutairi Z (2018) Thermal, physical properties and flammability of silane treated kenaf/pineapple leaf fibres phenolic hybrid composites. Compos Struct 15(202):1330–1338
3. Boban AP, Kumar N, Mishra DK (2024) A review on pineapple and banana natural fiber composite. In: International conference on futuristic advancements in materials, manufacturing and thermal sciences, Springer, pp 25–38
4. Boban AP, Mishra DK, Kumar N, Kumar S, Mahakur VK, Sharma S (2025) Quasi-static indentation, mechanical, and damage evolution of different configurations of laminated banana–pineapple fiber-reinforced composites for structural application. Textile Res J, p 00405175251322952
5. Bharath KN, Basavarajappa S (2014) Flammability characteristics of chemical treated woven natural fabric reinforced phenol formaldehyde composites. Procedia Mat Sci 1(5):1880–1886
6. Bocz K, Szolnoki B, Marosi A, Tábi T, Wladyka-Przybylak M, Marosi G (2014) Flax fibre reinforced PLA/TPS biocomposites flame retarded with multifunctional additive system. Polym Degrad Stab 106:63–73
7. Bourmaud A, Beaugrand J, Shah DU, Placet V, Baley C (2018) Towards the design of high-performance plant fibre composites. Prog Mater Sci 97:347–408
8. Chen C, Gu X, Jin X, Sun J, Zhang S (2017) The effect of chitosan on the flammability and thermal stability of polylactic acid/ammonium polyphosphate biocomposites. Carbohyd Polym 157:1586–1593
9. Faruk O, Bledzki AK, Fink HP, Sain M (2012) Biocomposites reinforced with natural fibers: 2000–2010. Prog Polym Sci 37(11):1552–1596
10. George M, Chae M, Bressler DC (2016) Composite materials with bast fibres: structural, technical, and environmental properties. Prog Mater Sci 83:1–23

11. Gholampour A, Ozbakkaloglu T (2020) A review of natural fiber composites: properties, modification and processing techniques, characterization, applications. J Mater Sci 55(3):829–892
12. Ismail SO, Akpan E, Dhakal HN (2022) Review on natural plant fibres and their hybrid composites for structural applications: recent trends and future perspectives. Compos Part C: Open Access 9
13. Jang JY, Jeong TK, Oh HJ, Youn JR, Song YS (2012) Thermal stability and flammability of coconut fiber reinforced poly (lactic acid) composites. Compos B Eng 43(5):2434–2438
14. Joshi SV, Drzal L, Mohanty A, Arora S (2004) Are natural fiber composites environmentally superior to glass fiber reinforced composites? Compos A 35(3):371–376
15. Kandola BK (2012) In: John MJ, Sabu T, editors. Flame retardant characteristics of natural fibre composites. Nat Polymers, Compos 1:86–117
16. Kerni L, Singh S, Patnaik A, Kumar N (2020) A review on natural fiber reinforced composites. Mater Today Proc 28:1616–1621
17. Khandelwal S, Rhee KY (2020) Recent advances in basalt-fiber-reinforced composites: tailoring the fiber-matrix interface. Compos Part B 192
18. Kim NK, Dutta S, Bhattacharyya D (2018) A review of flammability of natural fibre reinforced polymeric composites. Compos Sci Technol 162:64–78
19. Knez N, Kariž M, Knez F, Ayrilmis N, Kuzman MK (2021) Effects of selected printing parameters on the fire properties of 3d-printed neat polylactic acid (PLA) and wood/PLA composites. J Renew Mat 9(11):1883–1895
20. Ku H, Wang H, Pattarachaiyakoop N, Trada M (2011) A review on the tensile properties of natural fiber reinforced polymer composites. Compos B 42(4):856–873
21. Lau K-T, Hung P-Y, Zhu M-H, Hui D (2018) Properties of natural fibre composites for structural engineering applications. Compos Part B 136:222–33
22. Li C, Wan J, Pan YT, Zhao PC, Fan H, Wang DY (2016) Sustainable, biobased silicone with layered double hydroxide hybrid and their application in natural-fiber reinforced phenolic composites with enhanced performance. ACS Sustain Chem Eng 4(6):3113–3121
23. Li M, Pu YQ, Thomas VM, Yoo CG, Ozcan S, Deng YL, Nelson K, Ragauskas AJ (2020) Recent advancements of plant-based natural fiber-reinforced composites and their applications. Compos Part B 200
24. Li Z, Li Y, Shen Y, Yu T, Wang J (2022) Synergic effects of dimethyl methylphosphonate (DMMP) and nano-sized montmorillonite (MMT) on the flammability and mechanical properties of flax fiber reinforced phenolic composites under hydrothermal aging. Compos Sci Technol 10(230):109487
25. Loganathan TM, Hameed Sultan MT, Ahsan Q, Jawaid M, Naveen J, Md Shah AU, Abu Talib AR, Basri AA, Jaafar CN (2021) Effect of cyrtostachys renda fiber loading on the mechanical, morphology, and flammability properties of multi-walled carbon nanotubes/phenolic bio-composites. Nanomaterials 11(11):3049
26. Loganathan TM, Sultan MT, Ahsan Q, Shah AU, Jawaid M, Talib AR, Basri AA (2021) Physico-mechanical and flammability properties of cyrtostachys renda fibers reinforced phenolic resin bio-composites. J Polym Environ 29:3703–3720
27. Marliana MM, Hassan A, Yuziah MN, Khalil HA, Inuwa IM, Syakir MI, Haafiz MM (2016) Flame retardancy, thermal and mechanical properties of Kenaf fiber reinforced Unsaturated polyester/Phenolic composite. Fibers Polymers 17:902–909
28. Mngomezulu ME, John MJ, Jacobs V, Luyt AS (2014) Review on flammability of biofibres and biocomposites. Carbohydr Polym 111:149–182
29. Molaba TP, Chapple S, John MJ (2018) Flame retardant treated flax fibre reinforced phenolic composites: ageing and thermal characteristics. Fire Mater 42(1):50–58
30. Kumar N, Chellam GK, Ningegowda BM, Mishra DK (2025) Exploration of failure and delamination damage analysis of adhesive-bonded composite joints. Compos: Mech: Comp App Int J 16(2)
31. Pornwannachai W, Ebdon JR, Kandola BK (2020) Fire-resistant flax-reinforced polypropylene/polylactic acid composites with optimized fire and mechanical performances. J Thermoplast Compos Mater 33(7):898–914

32. Rahman MZ (2021) Mechanical and damping performances of flax fibre composites—a review. Compos Part C: Open Access 4
33. Rajeshkumar L, Kumar PS, Boonyasopon P, Rangappa SM, Siengchin S (2024) Flame retardance behaviour and degradation of plant-based natural fiber composites—a comprehensive review. Constr Build Mater 432:136552
34. Ramesh M (2019) Flax (Linum usitatissimum L.) fibre reinforced polymer composite materials: a review on preparation, properties and prospects. Prog Mater Sci 102:109–66
35. Ramesh M (2016) Kenaf (Hibiscus cannabinus L.) fibre based bio-materials: a review on processing and properties. Prog Mater Sci 78–79:1–92
36. Shi X-H, Li X-L, Li Y-M, Li Z, Wang D-Y (2022) Flame-retardant strategy and mechanism of fiber reinforced polymeric composite: a review. Compos B 233:109663
37. Vahabi H, Kandola BK, Saeb MR (2019) Flame retardancy index for thermoplastic composites. Polymers 11(3):407
38. Wu Klingler W, Bifulco A, Polisi C, Huang Z, Gaan S (2023) Recyclable inherently flame-retardant thermosets: chemistry, properties and applications. Compos Part B 110667
39. Young WB, Tsao YC (2015) The mechanical and fire safety properties of bamboo fiber reinforced polylactide biocomposites fabricated by injection molding. J Compos Mater 49(22):2803–2813
40. Yu T, Tuerhongjiang T, Sheng C, Li Y (2017) Phosphorus-containing diacid and its application in jute/poly (lactic acid) composites: mechanical, thermal and flammability properties. Compos A Appl Sci Manuf 97:60–66
41. Zhan J, Wang J, Lin J, Zhao G, Ji S, Li X, Naceur H (2022) Flame-retardant, thermal and mechanical properties of PLA/ramie fiber composites. Polym Compos 43(7):4244–4254
42. Zhao X, Copenhaver K, Wang L, Korey M, Gardner DJ, Li K, Lamm ME, Kishore V, Bhagia S, Tajvidi M (2022) Recycling of natural fiber composites: challenges and opportunities. Resour Conserv Recycl 177:105962

Flammability Properties of the Polyester Resin-Based Biocomposites and Bionanocomposites

Mahmut Ali Ermeydan

Abstract Composites used in construction, rail, and maritime industries frequently use polyester resin cross-linked with styrene. However, the inclusion of biofibers and volatile styrene makes polyester resin-based biocomposites and bionanocomposites susceptible to fire. Strict fire safety laws are also necessary because biocomposite components are heat-and fire-sensitive. Numerous flame-retardant treatments and methods, such as layered silicates, nanofillers, halogenated and non-halogenated flame-retardants, and chemical changes such copolymerization or grafting, have been developed to solve this problem. Green biocomposites have undergone some improvement in their heat resistance by physical and chemical treatment, but they still don't match the strict fire safety standards. In order to improve flame retardancy techniques, find more appropriate and environmentally friendly flame-retardants, and choose suitable natural fibers and biopolymers for the development of fire-safe biocomposite products, a thorough investigation into the thermal breakdown and fire behavior of green biocomposites is required. In this chapter, the research done over the last decades on flammability of polyester biocomposites and some recent research on the reduction of flammability of biocomposites and bionanocomposites are reviewed.

Keywords Flammability · Unsaturated polyester · Fire retardants · Biocomposites

1 Introduction

There has been a shift in focus toward the development of bio-based resources due to the detrimental environmental effects of the widespread use of non-biodegradable and non-renewable petroleum-derived products. Because of its advantages over conventional materials like metals and synthetic composites in terms of the environment, their lightweight nature, superior strength, stiffness, low energy consumption during

M. A. Ermeydan (✉)
Department of Forest Industry Engineering, Bursa Technical University, Yildirim, Bursa, Türkiye
e-mail: mahmut.ermeydan@btu.edu.tr

C. Muthukumar et al. (eds.), *Flammability Performance of Biocomposites and Bionanocomposites*, Composites Science and Technology,
https://doi.org/10.1007/978-981-95-2389-4_4

manufacture, CO_2 neutrality, and soundproofing qualities have raised interest in biocomposites [1, 2]. Unfortunately, despite all the benefits that biocomposites have to offer—such as reduced dependence on petroleum, cost-effectiveness, easier maintenance, improved fuel efficiency through lightweight vehicles, recycling and end-of-life disposal—their use in engineering applications is still relatively uncommon [3]. The difficulties in managing moisture are impeding the commercial use of biocomposites. The difficulties with absorption, low heat stability, poor fire resistance, weak adherence between hydrophilic natural fibers and hydrophobic biopolymers, and low adhesion have all been noted [4]. Nonetheless, these problems have been effectively resolved. Alkali treatment or acetylation or silane modification have been used to reduce moisture absorption, whereas chemical changes or the addition of coupling agents have improved poor adhesion [5]. Attention has been drawn to the problem of low thermal stability and insufficient fire resistance in recent times (Fig. 1) [6].

On the other hand, polyester resins are a very versatile family of thermosetting polymers. They are made of pre-polymer mixes of styrene and unsaturated polyester chains, where styrene acts as a crosslinking agent and diluent during the radical curing process that follows [7]. These resins are widely used in many different industries, such as building construction, automotives, offshore applications, coatings, and equipment [8]. They have many benefits, including superior mechanical, electrical, and chemical resistance, as well as ease of processing. Furthermore, they are economically appealing because of their cheap production and raw material prices. However, regular polyester resin has a low fire resistance and produces a lot of smoke when burned due to its chemical composition and molecular structure. As a result, its use is restricted in several sectors, such the electrical industry. To mitigate the issue of flammability resulting from the increasing commercial use of polyester resins, flame-retardant systems must be developed [9].

The combination of two flammable materials contains several risks and burns almost completely without any precaution. Marliana et al. reported that 100%

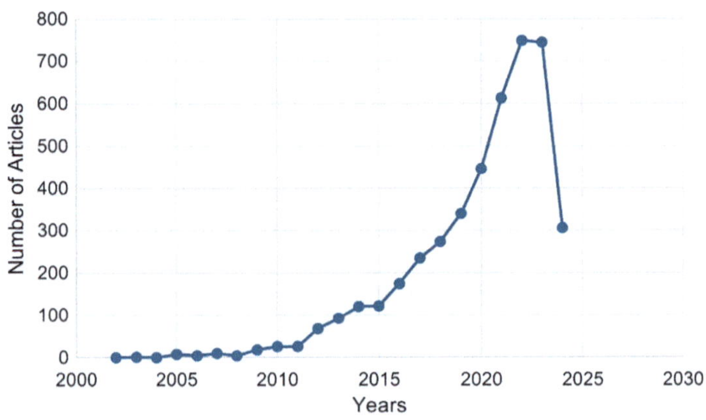

Fig. 1 Number of articles per year on Scopus search using flammability and biocomposites keywords

polyester resin reinforced by 20% kenaf fiber only 6% char yield. By keeping constant the kenaf ratio, Marliana et al. blended phenol formaldehyde resin to the polyester matrix to improve flame resistance [10]. As in the example, several researchers have been carried out to overcome those risks in the past decade. Aluminum dialkylphosphinate [9], iron powder [11], boron compounds [12] are some of the additives to prevent flammability of polyester biocomposites. Here, recent advances of flammability of polyester resin-based biocomposites and bionanocomposites will be summarized under the light of various significant literature.

2 Polyester-Based Biocomposites and Bionanocomposites

Even though it was among the earliest commercially available polymer matrices for composites reinforced with glass fibers, polyester resin is still widely used today. On the other hand, natural fibers have been incorporated as reinforcements to lessen their influence on the environment. These composites continue to be of interest to researchers worldwide, as evidenced by the large number of publications that have been published in the previous decades [13–16]. One of the oldest preliminary studies reported by Burgueño et al. clearly confirmed that biocomposites are on par with E-glass composites in terms of short-term stiffness, but significantly inferior in terms of strength, temperature, and moisture properties [17].

Since then, efforts have been made to improve the properties of polyester biocomposites and hundreds of articles have been published. Many of the articles focused on the improvement of mechanical properties of polyester biocomposites. Misra et al. reported modification of hemp fiber mats by various methods including alkali, silane, or acrylonitrile treatments, as well as matrix modification and compared them with E-glass composite [18]. Following surface treatments, the biocomposites' mechanical and thermal properties increased. However, a hybrid composite of glass and industrial hemp fibers had balanced characteristics [18]. Since then, a number of studies have been carried out with several fibers and matrixes. Specifically, more than 500 hundred articles have been published on unsaturated polyester biocomposites. There is even a bibliographic study published in 2023, which provides gathered, structured, and pertinent information of papers reported between 1982 and 2022 and found in the Scopus database regarding the development and present status of these items by bibliometric analysis [19]. However, there is a limited number of studies on flammability properties of plant fiber/unsaturated polyester resin (UPR) biocomposites or bionanocomposites.

On the other hand, as noted above, natural plant fiber reinforced biocomposites generally provide limited or low strength, are brittle, and are susceptible to moisture. Another solution of improving composite characteristics would be the utilization of cellulosic nanofibers (CNF) [20]. By using nanocellulose several properties of polyester can be improved. Ansari et al. used unsaturated polyester matrix which was reinforced using wood cellulose nanofibers (CNF) without the usage of coupling

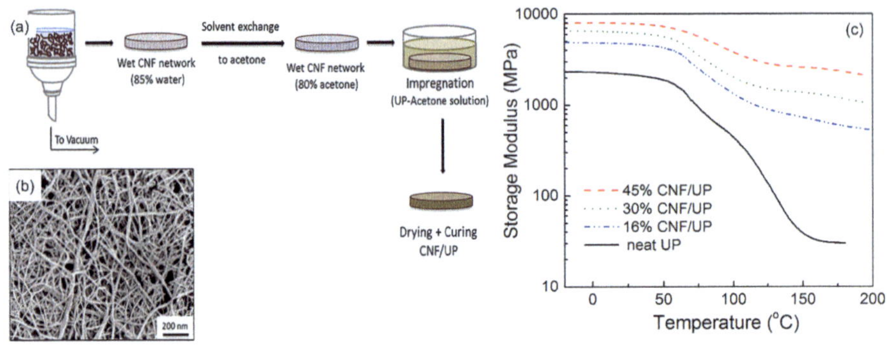

Fig. 2 **a** Preparation procedure illustrated solvent exchange and polyester/acetone solution impregnation on CNF network. **b** The CNF aerogel's surface morphology in relation to the CNF network employed in this investigation. **c** Storage modulus of polyester composites reinforced with CNF. Picture adapted from Ref. [20] with permission from Elsevier (CC-BY License)

agents or surface modification of the nanofibers to consequently form bionanocomposites. Ansari et al. reported that the modulus and strength of composite, as well as its ductility and toughness, are significantly enhanced by the nanostructured CNF network reinforcement. An industrially viable template-based prepreg processing method was used, combining nanoscale CNF dispersion with a high CNF concentration (up to 45 vol%) (Fig. 2). Moisture sorption, dynamic thermal analysis, tensile testing at various humidities, fracture toughness testing, and fractography were all applied to the bionanocomposites. The concentration of CNF caused a significant increase in the glass transition temperature (T_g). At 45 vol% CNF, the modulus and strength of polyester increased by almost three times, while the apparent fracture toughness and ductility doubled [20]. CNF bionanocomposites can also be effective against fire with or without flame retardants, unfortunately until now, there is no research carried out investigating flammability properties of nanocellulose reinforced unsaturated polyester resin.

3 Flammability of Polyester-Based Biocomposites and Bionanocomposites

3.1 Flammability of Unsaturated Polyester Resin

High-performance, environmentally friendly fire-retardant thermoset polymers have been developed in response to the growing demands on structural thermosets, like unsaturated polyester, and the ever-tightening fire safety laws [21]. Additive fire retardants (FR) are typically categorized based on the fire-retardant index (FRI) and the optimum quantity of FRs needed to obtain a UL-94V-0 certification [22, 23]. One of the main issues with unsaturated polyester is its high flammability; in fact, out of all the commonly used commercial resins, polyester has the lowest flammability [21]. The condensation reaction of a diol with a combination of saturated and unsaturated diacids and anhydrides yields UP, and the reaction is continuously driven by the water removal [24]. Through the use of vinyl monomers like styrene as a catalyst, accelerator, reactive diluent, and crosslinker, together with other processing aids, free radical chain growth polymerization creates a three-dimensional network [7].

On the other hand, numerous studies have demonstrated that styrene content, prepolymer chemical structure, and final network structure all affect polyester resin's thermal stability. The bonds closest to the ester group (C–O) undergo homolytic cleavage to start the degradation, forming two free radicals. Subsequently, at around 330 °C, phthalic anhydride and hydroxyester molecules are generated, followed by a reaction with hydrogen at approximately 390 °C. Next, weaker aliphatic carbon–carbon bonds next to the polyester backbone cause the oligostyrene cross-links to depolymerize, resulting in the formation of styrene monomers and oligomers. Additionally, the crosslinked polystyrene can break down. At roughly 380 °C, styrene undergoes depolymerization, creating linear polyester sequences, and subsequent chain scission brought on by heat deterioration. The carbonaceous char's oxidation is the final stage [21].

To create a fire-retardant polyester resin, there are three primary methods. Using fire retardant monomers to synthesis polyester resins that are intrinsically fire-retardant is a promising approach. These monomers typically have the following structural elements: silicon, sulfur, phosphorus, nitrogen, chlorine, and bromine. Copolymerization of polyester resin with reactive fire-retardant comonomers is another effective method. The final type of fire retardants are additives, which are merely physically combined with resin. There are two categories for additive fire retardants: organic and inorganic. P-N complexes, ATH (Aluminum Trihydroxides), nanostructured (newly Mxene), core–shell structured, and charring resins are examples of inorganic fire retardants, whereas halogenated, oligomeric, and polymeric fire retardants are examples of organic fire retardants [21, 25–28].

One of the recent examples of fire-retardant usage has been reported by Liu et al. [29]. As illustrated the effectiveness of fire-retardants in Fig. 3, Liu et al. reported utilization of a practical flame-retardant system which was formed using aluminum diethylphosphinate (ADP) and dimelamine pyrophosphate (DMPY) combination [29]. They found that polyester resin thermosets with fire retardants demonstrated outstanding synergistic effect and water resistance. Polyester/15wt% fire retardant with a LOI of 27.9% met UL-94V-0 rating requirement [29].

3.2 *Flammability of Polyester Resin Biocomposites*

Under the light of the information given so far, it is very clear that two components of polyester biocomposites are flammable materials [21, 30]. To overcome the flammability problem of polyester based biocomposites and achieve technical requirements of several sectors, researchers have been mainly focused on the use of additives so far.

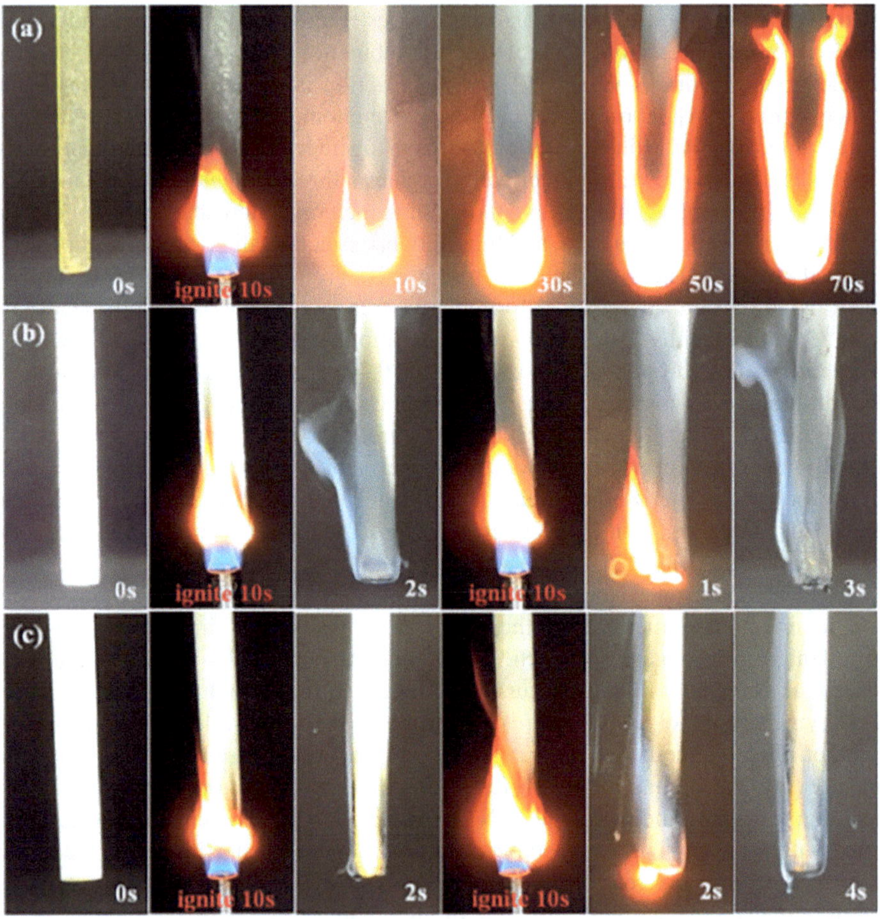

Fig. 3 Screenshots from the UL-94 testing of the polyester (**a**), polyester/15wt% fire retardant (**b**), water treated polyester/15 wt% fire retardant (**c**)

On the other hand, there has been a limited number of studies until recent years, but fortunately interest in the biocomposite topic also attracted attention to flammability properties of the composite to reveal potential or to seek novel options.

One of the initial studies was carried out by Kumar et al. on fire resistance of hybrid oil palm/glass fibers reinforced polyester resin biocomposites [31]. They noted that by combining glass fibers and oil palm biofibers for hybrid reinforcement, fire-resistant reinforced composite materials can be created. When fire retardant fillers like magnesium carbonate, aluminum trihydrate, zinc borate, and brominated phosphate ester were added to the formulations, the LOI values drastically improved over 40, where only glass or natural fiber composite had LOI of about 24 [31].

About ten years later Manfredi et al. reported thermal degradation and fire resistance of composites made of natural or glass fibers, modified acrylic resins (Modar) or unsaturated polyester [32]. Fire resistance of biocomposites was observed by TGA/DTG and cone calorimeter. They found that sisal fiber-reinforced composite with the Modar matrix exhibited the highest fire risk and, overall, the weakest fire resistance. On the other hand, flax fiber composites created a long-lasting but slow-growing fire, while jute fiber composites displayed a quick-growing but short-lived fire. Compared to biocomposites, glass fiber composites demonstrated greater flame resistance. On the other hand, increased CO and CO_2 emissions were noted. Compared to polyester matrix composites, the modified acrylic resin requires a longer ignition time and thermal stability (TGA), because this resin had the ability to create char, so the smoke hazard of a fire was obviously lower for the Modar matrix composites [32]. Closely, the same research group published similar results by Rodríguez et al. with same matrixes [33].

In the following years, which means in the past decade, research on polyester biocomposites has gained pace. In 2016, Marliana et al. reported phenolic resin (PF) mixed in different ratios with unsaturated polyester (UP) resin to create a homogenous blend that has better flame resistance [10]. 20% kenaf was used to strengthen the blend. Thermogravimetric analysis was used to examine the thermal characteristics of the composites (TGA). The degradation temperature at 50% weight loss indicated that the thermal stability of UP/PF kenaf composites varies in tandem with the PF component. The TGA results indicate that the char yield of the composites increases linearly with PF concentration. The limiting oxygen index (LOI) and UL-94 fire tests were used to assess the composites' flammability characteristics. The UL-94 test showed that the composites have enhanced flame retardancy, although the LOI increased with the PF content. As the PF component increased, the char yield of the UP/PF kenaf fiber composite increased from 9.2 to 21.1%. Besides, it was observed that a sharp rise in the LOI from 34 to 41% when the percentage of PF resin is raised from 0 to 40% [10]. As previously mentioned, the 100% unsaturated polyester composite burned for more than 50 s after the flame was removed from the specimen, so it was not given a UL-94V-rating. The outcomes demonstrated that UP is a polymeric substance that readily catches fire and fails the UL-94 test. Nonetheless, the composites with PF resin contents of 10, 20, 30, and 40% were able to attain V-1 grade due to their combustion times being under 30 s [10].

In 2018, Borazan and Gokdai found by LOI test that the pinecone/polyester composites' flame-retardancy improved slightly due to the 8% boron oxide additive [12]. In the same year, Chu et al. [34] reported a fascinating study to enhance the mechanical and fire safety characteristics of unsaturated polyester resin (UPR) composites reinforced with ramie fabric. A silane coupling agent (TMSAP) comprising nitrogen and phosphorus was created and applied to the surface of the ramie fabrics. The fabric/UP resin composites were then formed using an oven vacuum bag method or manual lay-up after ammonium polyphosphate (APP) was added to UP resin. With just 10 weight percent of APP, composites could reach UL-94V-0 in the vertical burning test (Table 1).

Furthermore, the composite showed a notable reduction in fire hazards as measured by a cone calorimeter. For instance, the notable reductions in the total

Table 1 Cone calorimeter test results and UL-94 results for composites made of ramie and UP resin with or without APP additive

Composite	TTI (s)	PHRR (kW/m^2)	THR (MJ/m^2)	Time to PHRR (s)	FGI (kW/m^2 s)	UL-94
RF/UPR	48	754	90.3	150	5.0	NR[a]
RF/UPR@APP$_{10}$	36	482	77.4	138	3.5	NR
RF/UPR@APP$_{15}$	39	451	70	155	2.9	NR
RF/UPR@APP$_{20}$	46	436	60	136	3.2	V0
CF/UPR	48	550	72.2	141	3.9	NR
CF/UPR@APP$_5$	52	489	70.3	135	3.6	NR
CF/UPR@APP$_{10}$	58	449	54.4	140	3.2	V0
CF/UPR@APP$_{15}$	72	406	49.5	105	3.9	V0

a *NR* No rating

heat release (THR) and peak heat release rate (PHRR) of 39.8% and 40.5%, respectively, were accompanied by a more compact char layer and fewer harmful pyrolysis gasses (Fig. 4) [34].

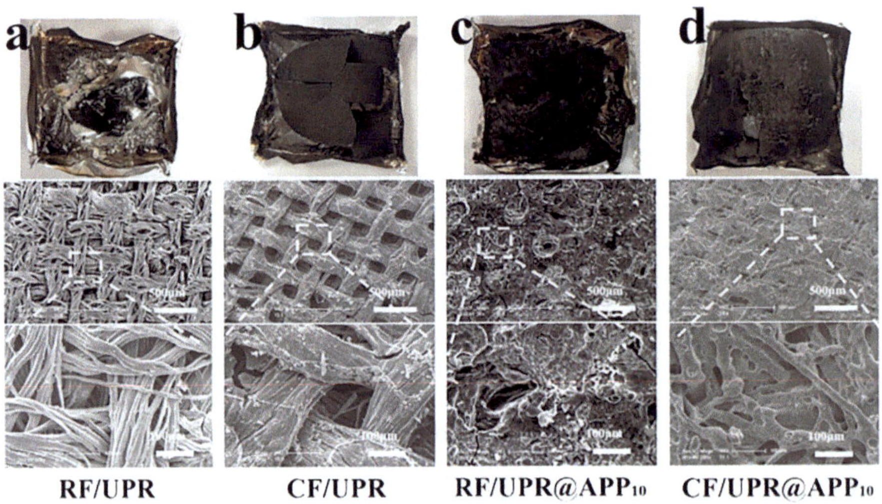

Fig. 4 Char residues for **a** unmodified ramie/UP resin, **b** modified ramie/UP resin, **c** unmodified ramie/UPR@APP$_{10}$ and **d** modified ramie/UPR@APP$_{10}$

In 2017, Yu et al. reported a different approach rather than adding fire retardants in polyester resin, they coated ramie fibers [35]. Layer-by-layer (LbL) assembly was used to modify ramie fabrics with bio-based electrolytes, such as flame-retardant cationic electrolyte melamine and cationic chitosan and anionic phytic acid. Then, using a hand lay-up technique, coated ramie reinforced unsaturated polyester resin composites were formed. The composites' char residues were significantly improved, and during burning, the production of flammable organic volatiles was inhibited. Furthermore, as the LbL assembly number increased, the composites' flame retardancy increased as well, as seen by the significantly lower peak heat release rate and total heat release values found in cone calorimetric measurements. The composites' limiting oxygen index value increased from 26.0 to 34.5% [35]. Zhang et al. also utilized phytic acid to coat hemp fibers and used only 3% of coated hemp for polyester composite [36]. The LOI value of coated hemp/UP biocomposites increased from 18.9 to 22.1% when 3 weight percent of hemp was included, besides the apparent heat release rate (AHRR) and total heat release (THR) values declined to 401.9 kW/m^2 and 150.6 MJ/m^2, respectively. However, thermogravimetric analysis showed that the char yield did not change much in the coated hemp/UP biocomposites, but had better thermal stability as shown by thermogravimetric analysis [36].

In 2018, Salim et al. utilized sodium water glass as a binding and fire-retardant agent in nonwoven kenaf acrylic based crosslinked polyester resin [37]. However, the interference of water glass on the polymer's curing reaction prevented the formation of a stiff polyester, which limited the improvement of fiber to matrix bonding. As the amount of water glass rose, the fracture toughness in some way also increased. Acrodur resin's uncured functional group increases fracture toughness by contributing to its ability to absorb energy and ductility. The performance of the composite as a flame retardant, therefore, was found to be very promising depending on the water glass loading. When the flame was removed from the sample having about 20% water glass, the flame spontaneously went out during the fire test, yielding an outstanding UL94 classification of V-0. The increased LOI of the composites as the amount of water glass in the composites increases quantitatively proved the improvement in flame-retardant efficiency of the composites having water glass. The composites had a good compromise between flame-retardant qualities and composite strength in order to be employed in the automobile industry and they concluded that this can be achieved by choosing the right quantity of water glass content [37].

Another interesting study was reported by Gapsari et al. on iron powder addition to polyester resin to form Timoho fiber/UP composites [11]. They used 30% fiber as reinforcement and 10% iron powder as an additive to the polyester resin. LOI of the samples improved from 21 to 29 by addition of iron powder. Char yield improved from 9 to 28% and combustion speed slowed down from 31 mm/min to 20 mm/min, which clearly implies better fire-retardancy. TG analysis also confirmed thermal stability [11].

Recently, Karuppasamy and Rangarathan used hemp chopped fiber with powdered *Lansium parasiticum* fruit shell to create sustainable polyester-based biocomposites [38]. They used 40% of hemp and 1-3-5% of fruit shell. Probably due to high content of hemp, only biocomposites with 1 and 3% fruit shell passed vertical UL-94 test with a V-1 value, where polyester itself got V-0 [38], which indicates biological additives should be carefully applied with coating or other precautions.

Diharjo et al. studied flammability characteristics of composites reinforced with *Ceiba petandra* fiber (CPF) 30% and supplemented with snail powder (SP) up to 10% [39]. With an increase in SP filler, the composite's flame rate metrics increased, such as the percentage of char yield and the limitation index oxygen (LOI). Also, because the SP filler impeded the flow of oxygen, flammable gas, and products of polyester degradation, the composite's flame rate was decreased [39]. The FS (natural fiber/UPR) composite's flame properties, like char yield and LOI, were improved by adding SP filler. With a char yield percentage of 5.45 (%), the FS10 (natural fiber + snail powder) composite had the highest, while the FS0 (only natural fiber) composite has the lowest, at 1.73 (%). The inclusion of SP filler causes the composite's limiting oxygen index value to significantly increase. The FS10 composite resulted in the largest LOI increase, 19.66. Timoho fiber-reinforced composite with filler iron powder had a somewhat higher LOI value than the composite did. The LOI was found more than 21 which indicates that the flame resistance of the composite was enhanced by the 10% (wt) filler. On the other hand, they reported the maximum flame rate as 30.26 (mm/min) for FS0 composite, where after addition of 10% (wt) SP filler, the flame rate decreased to 21.05 (mm/min). Diharjo et al. suggested that the enhanced barrier qualities of the labyrinth or winding path that prevents flammable oxygen and gas movement and combustible polymer degradation product through SP filler material provide the CPF/SP-reinforced polyester composite flame resistance. Furthermore, SP fillers facilitate the solid charcoal formation process (5.45%) and enhance thermal resistance by accumulating Ca minerals, which in turn reduces the rate at which heat is released from the surface of the composite materials [39].

The work of Diharjo et al. [39] is very recent and has significant quality in terms of its success and suggesting practical solutions as well as utilization of high natural fiber content (30%) is crucial. On the other hand, it is very clear that there is more room in the research field of fire retardancy of natural fiber/unsaturated polyester composite. Because there are numerous types of fire-retardant agents still have not been applied. In the following section, fire retardancy of bionanocomposites of polyester will be discussed.

3.3 *Flammability of Polyester Resin Bionanocomposites*

Making unsaturated polyester resin (UPR) composites that are ecologically benign while maintaining acceptable fire safety, interfacial quality, and impact strength has proven to be a difficult task. Utilization of nanomaterials of course may be an effective way to improve fire retardancy of UPR composites. Unfortunately, it is fact that only a few numbers of studies have been published so far on the investigation of flammability properties of bionanocomposites. Here, almost all the work that can be found through Scopus is mentioned below.

Chu et al. suggested a nanotechnological approach to improve fire resistance and interfacial interaction of ramie fabric and UPR [40]. They assembled the nontoxic two-dimensional (2D) hexagonal boron nitride (h-BN) nanosheets on the surface of ramie fabrics to improve the interfacial performance of fabric-reinforced unsaturated polyester resin composites. Sodium alginate (SA) acts as a green dispersant to disperse the h-BN sheets during the assembly process. The biobased phosphorus-containing toughening agent (PCTA) was then created in order to increase the composite's fire safety and impact strength at the same time. The AF (assembled fabrics)/UPR@PCTA20 composite demonstrated the maximum 41.2% decrease in the value of peak heat release rate (PHRR) and a maximum 17.8% decrease in the value of total heat release (THR) with the application of h-BN nanosheets-assembled fabric and 20 weight percent of PCTA. It also achieved V-0 rating in the vertical burning test. In contrast to RF/UPR, the AF/UPR@PCTA20 composite's limiting oxygen index increased considerably, from 24.0 to 29.5%. Additionally, the AF/UPR@PCTA20 composite demonstrated a noteworthy 68.8% increase in impact strength when compared to the neat RF/UPR composite, suggesting an exceptional toughening effect of PCTA on UPR composites. The authors suggest that in addition to suppressing the live radical reaction in the gas phase, PCTA may facilitate the creation of the whole char layer structure in the condensed phase. Additionally, the 2D h-BN sheets promote the development of char residues. Due to the extended alkyl structure of the bio-based cardanol, the AF/UPR@PCTA20 composite exhibited a noteworthy 68.8% increase in impact strength when compared to the plain RF/UPR composite, suggesting that PCTA has a favorable toughening effect on UPR composites. It was concluded that utilizing biobased components including cardanol, ramie fabric, and nontoxic h-BN sheets, it is anticipated that the research will offer a workable green technique for the future production of environmentally acceptable UPR composites [40] (Fig. 5).

Uz Zaman et al. [41] reported one of the bionanocomposites studies. They focused on the examination the effects of treating jute fiber, adding ZnO nanoparticles to polyester resin, and using chemical-resistant resin on the mechanical characteristics, fire retardancy, and chemical resistance (alkali and acid) of jute/unsaturated polyester composites. Weaved jute fabrics underwent mercerization, scouring, and hybrid fluorocarbon (HFC) treatments. Using a sonicator, zinc oxide (ZnO) nanoparticles were distributed across normal (Nrml) and chemical resistant (CR) unsaturated polyester resins [41]. The fire-retardant effect of ZnO nanoparticles on biocomposites' burning time was measured using ASTM D6413. It was noted that without ZnO, control and fiber treated groups have burning times between 197 and 179 s, but after ZnO nanoparticle addition, the burning time dropped about 145 s, which indicates resistance to flammability and was associated to the large surface area of ZnO nanoparticles [41].

In Table 2, some flammability properties of the above-mentioned articles are presented as a summary of their findings. Ignition time, UL-94 test results or LOI values are presented in the table, as well as type of the natural fiber and additives.

New opportunities for "green" materials have been made possible by the recent success in isolating cellulose nanofibers (CNF) or crystals (CNC) with minimal

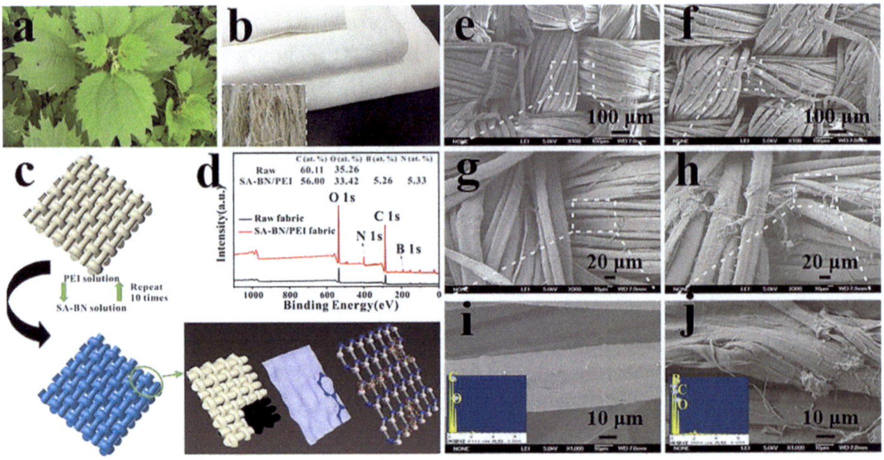

Fig. 5 Images of ramie and ramie textiles (**a**, **b**). **c** Schematic representation of the LbL assembly process using SA-BN anionic dispersion and PEI cationic solution. **d** Raw ramie fabric and constructed fabric XPS results. SEM pictures of unprocessed ramie fabrics with different magnification (**e**, **g**). SEM pictures of treated ramie fabrics with different magnification (**f**, **h**). The EDX images of the raw fabrics and the SA-BN constructed fabrics (**i**, **j**), respectively. Picture adapted from Ref. [40] with permission from ACS group

energy consumption. The main load-bearing component of plant fibers are cellulose nanofibrils, which have a high tensile modulus. The last ten years have seen a steady rise in interest in materials based on nanocellulose, particularly composite materials reinforced with nanocellulose [42–44]. Interestingly there are only a few articles published with nanocellulose reinforced **unsaturated polyester** resin biocomposites and none of them investigating flammability properties of the created bionanocomposite [20, 45–48]. Ansari et al. (2015) reported one of the first studies utilizing nanocellulose to reinforce unsaturated polyester. They noted that robust CNF–UP synergy in the composites, indicating the materials' effective nanostructuring. Both modulus and strength rose with increasing CNF content up to the maximum value (45 vol%), despite the use of a more brittle UP thermoset. Additionally, the T_g increased for the 45 vol% composite from 66 °C for the neat material to 78 °C, indicating favorable secondary CNF–UP interactions through Van der Waals interactions, hydrogen bonding, and other comparable mechanisms. Well-dispersed individual CNFs in a continuous UP matrix were supported by the composites' ductility and the fracture surfaces' microscopy [20]. With this study, effectiveness of CNF/UPR biocomposites clearly provide better reinforcement against plant fibers (about 30 μm in diameter) in terms of strength, ductility, and moisture sensitivity because of the lignin and hemicellulose content of plant fibers as well as the less ideal fiber/matrix interface. Similar T_g increase was also reported by Chirayil et al. [48].

Utilization of nanocellulose in flammability research of the unsaturated polyester bionanocomposites needs to be done immediately with an organized strategy. The flammability properties of nanocellulose unsaturated polyester bionanocomposites

Table 2 Article published until 2024 on flammability properties of cellulosic fibers reinforced unsaturated polyester resin biocomposites and bionanocomposites [10–12, 31–41]

Fiber type	Fiber content (%)	Additive	Ignition time (s)	UL-94	LOI	Reference
Oil palm (7:3) Glass fiber	30	Aluminum trihydrate		–	40	Kumar et al. [31]
Jute	30	–	205 (burning time)	–	–	Rodríguez et al. [33]
Jute	30	–	51	–	–	Manfredi et al. [32]
Kenaf	20	Phenolic resin 40%	–	V-1	41	Marliana et al. [10]
Pine cone fiber	5	Boron oxide 8%	–	–	22	Borazan and Gokdai [12]
Ramie fabric	10	Ammonium polyphosphate (APP) 10%	58	V-0	–	Chu et al. [34]
Ramie fabric	30	Cationic chitosan + anionic pytic acid + cationic melamine (coating)	60	–	34.5	Yu et al. [35]
Kenaf fiber	–	Water glass 20%	–	V-0	38	Salim et al. [37]
Hemp fiber	3	Phytic acid (coating)	33	–	22.1	Zhang et al. [36]
Timoho fiber	30	Iron powder 10%	–	–	28.9	Gapsari et al. [11]
Hemp fiber	40	*Lansium parasiticum* 1%	–	V-1 (vertical)	–	Karuppasamy and Ranganathan [38]
Ceiba petandra fiber	30	Snail powder 10%	–	–	19.6	Diharjo et al. [39]
Ramie fabric	20	Phosphorus-containing toughening agent 20%	–	V-0	29.5	Chu et al. [40]
Jute fiber	–	ZnO nanoparticles	145 (burning time)	–	–	Uz Zaman et al. [41]

can be influenced by various factors, including the type and concentration of nanocellulose, the type of unsaturated polyester resin used, and the processing methods employed. Here are some ways these properties can be managed. (a) Nanocellulose type and concentration: The type and concentration of nanocellulose used in the composite can significantly affect its flammability properties. For instance, incorporating a higher concentration of nanocellulose can enhance the flame retardancy of the composite due to the formation of a protective char layer during combustion. (b) Surface modification: Surface modification of nanocellulose can improve its compatibility with the unsaturated polyester matrix and enhance the dispersion of nanocellulose within the composite. This can lead to improved flame retardancy by promoting the formation of a more uniform char layer during combustion. (c) Addition of flame-retardant additives: Incorporating flame-retardant additives, such as phosphorus-based or nitrogen-based compounds, into the nanocellulose unsaturated polyester bionanocomposites can enhance their flame retardancy by promoting char formation, reducing heat release, and suppressing combustion reactions.

4 Conclusions

In conclusion, the incorporation of cellulosic fibers into unsaturated polyester matrices presents a promising avenue for enhancing the flame-retardant properties of composite materials with or without coating strategies or additives. Through the incorporation of various nanomaterials, bio-based additives, nitrogen or phosphorous containing fire retardants, and reinforcement ability of plant fibers, nanocellulose, lignin, or other natural fibers, researchers have achieved notable improvements in flame retardancy while simultaneously addressing environmental concerns associated with toxic flame retardants.

Studies have demonstrated that the addition of plant fibers can effectively reduce the flammability of unsaturated polyester composites by promoting char formation, improving thermal stability, and inhibiting the release of combustible gases during combustion (UL-94V-0, improved LOI, etc.). Furthermore, the synergistic effects between the polymer matrix and bio-based additives have been shown to enhance the overall flame-retardant performance of the materials.

However, challenges remain in optimizing the flame-retardant properties of unsaturated polyester biocomposites, including achieving a balance between flame retardancy and mechanical properties, as well as addressing concerns regarding processing compatibility and cost-effectiveness. Future research efforts should focus on further elucidating the mechanisms underlying the flame-retardant behavior of biocomposites and exploring novel bio-based additives to continue advancing the development of sustainable and flame-retardant polymer materials.

References

1. Rashid M, Chetehouna K, Cablé A, Gascoin N (2021) Analysing flammability characteristics of green biocomposites: an overview. Fire Technol 57:31–67. https://doi.org/10.1007/s10694-020-01001-0

2. Manu T, Nazmi AR, Shahri B, Emerson N, Huber T (2022) Biocomposites: a review of materials and perception. Mater Today Commun 31:103308. https://doi.org/10.1016/j.mtcomm.2022.103308
3. John MJ, Thomas S (2008) Biofibres and biocomposites. Carbohydr Polym 71:343–364. https://doi.org/10.1016/J.CARBPOL.2007.05.040
4. Bahrami M, Abenojar J, Martínez MÁ (2020) Recent progress in hybrid biocomposites: mechanical properties, water absorption, and flame retardancy. Materials 13:5145. https://doi.org/10.3390/ma13225145
5. Ermeydan MA (2024) New methodologies to improve the interfacial interaction in natural fibre polymer composites. In: Krishnasamy S, Hemath Kumar M, Parameswaranpillai J, Mavinkere Rangappa S, Siengchin S (eds) Interfacial bonding characteristics in natural fiber reinforced polymer composites: fiber-matrix interface in biocomposites. Springer Nature, Singapore, pp 23–45. https://doi.org/10.1007/978-981-99-8327-8_2
6. Prabhakar MN, Naga Kumar C, Dong Woo L, Jung-IL S (2022) Hybrid approach to improve the flame-retardant and thermal properties of sustainable biocomposites used in outdoor engineering applications. Compos Part A Appl Sci Manuf 152:106674. https://doi.org/10.1016/j.compositesa.2021.106674
7. Penczek P, Czub P, Pielichowski J (2005) Unsaturated polyester resins: chemistry and technology. In: Crosslinking in materials science. Springer, Berlin, Heidelberg, pp 1–95. https://doi.org/10.1007/b136243
8. Mehdipour-Ataei S, Aram E (2023) 18—High-end applications of unsaturated polyester composites. In: Thomas S, Chirayil CJ (eds) Applications of unsaturated polyester resins. Elsevier, pp 421–39. https://doi.org/10.1016/B978-0-323-99466-8.00009-5
9. Wang Y, Zhang L, Yang Y, Cai X (2015) The investigation of flammability, thermal stability, heat resistance and mechanical properties of unsaturated polyester resin using AlPi as flame retardant. J Therm Anal Calorim 122:1331–1339. https://doi.org/10.1007/s10973-015-4875-7
10. Marliana MM, Hassan A, Yuziah MYN, Khalil HPSA, Inuwa IM, Syakir MI et al (2016) Flame retardancy, thermal and mechanical properties of Kenaf fiber reinforced unsaturated polyester/phenolic composite. Fibers and Polymers 17:902–909. https://doi.org/10.1007/s12221-016-5888-y
11. Gapsari F, Purnowidodo A, Setyarini PH, Suteja S, Abidin Z, Rangappa SM et al (2022) Flammability and mechanical properties of Timoho fiber-reinforced polyester composite combined with iron powder filler. J Mater Res Technol 21:212–219. https://doi.org/10.1016/j.jmrt.2022.09.025
12. Borazan AA, Gokdai D (2018) Pine cone and boron compounds effect as reinforcement on mechanical and flammability properties of polyester composites. Open Chem 16:427–436. https://doi.org/10.1515/chem-2018-0054
13. Saravanakumar P, Karuppuswamy P, Binoj JS (2024) Agave Tequilana biowaste fiber-treated biocomposites: influence of SiC nanoparticles on mechanical, morphological, tribological and moisture absorption properties. Biomass Convers Biorefin 14:5689–5698. https://doi.org/10.1007/s13399-023-04463-8
14. Anish MC, Pandey KK, Kumar R (2024) Preparation and characterization of unsaturated polyester infused transparent wood composites. Eur J Wood Wood Prod 82:503–513. https://doi.org/10.1007/s00107-023-02023-5
15. Thangapandian P, Seenivasan S, Raj Kumar P, Kailasavalli S, Paulsingarayar S (2024) Effects of chemical treatment on natural (rice straw/neem saw dust) fiber morphology and mechanical properties of its composite material. Mater Lett 364:136288. https://doi.org/10.1016/j.matlet.2024.136288
16. dos Santos AJG, Ribeiro MM, Corrêa A de C, Rodrigues J da S, Silva DS, Junio RFP et al (2024) Morphological, chemical and mechanical properties of hybrid polyester composites reinforced with bamboo fibers and kaolin waste. J Mater Res Technol 30:1–15. https://doi.org/10.1016/j.jmrt.2024.03.003
17. Burgueño R, Quagliata MJ, Mohanty AK, Mehta G, Drzal LT, Misra M (2004) Load-bearing natural fiber composite cellular beams and panels. Compos Part A Appl Sci Manuf 35:645–656. https://doi.org/10.1016/j.compositesa.2004.02.012

18. Mehta G, Drzal LT, Mohanty AK, Misra M (2006) Effect of fiber surface treatment on the properties of biocomposites from nonwoven industrial hemp fiber mats and unsaturated polyester resin. J Appl Polym Sci 99:1055–1068. https://doi.org/10.1002/app.22620
19. Gañán P, Barajas J, Zuluaga R, Castro C, Marín D, Tercjak A et al (2023) The evolution and future trends of unsaturated polyester biocomposites: a bibliometric analysis. Polymers (Basel) 15:2970. https://doi.org/10.3390/polym15132970
20. Ansari F, Skrifvars M, Berglund L (2015) Nanostructured biocomposites based on unsaturated polyester resin and a cellulose nanofiber network. Compos Sci Technol 117:298–306. https://doi.org/10.1016/j.compscitech.2015.07.004
21. Seraji SM, Song P, Varley RJ, Bourbigot S, Voice D, Wang H (2022) Fire-retardant unsaturated polyester thermosets: the state-of-the-art, challenges and opportunities. Chem Eng J 430:132785. https://doi.org/10.1016/j.cej.2021.132785
22. Wu P, Peng Y, Zhang X, Zhang G, Ran J, Xu M (2022) Unsaturated polyester resin modified with a novel reactive flame retardant: effects on thermal stability and flammability. J Polym Eng 42:818–826. https://doi.org/10.1515/polyeng-2021-0317
23. Zhou J, Xu M, Zhang X, Leng Y, He Y, Li B (2019) Preparation of highly efficient flame retardant unsaturated polyester resin by exerting the fire resistant effect in gaseous and condensed phase simultaneously. Polym Adv Technol 30:1684–1695. https://doi.org/10.1002/pat.4599
24. Abdel-Azim AA, Mahmoud BH, Farhat MS (1994) Synthesis of unsaturated polyester resins for large sized composites. Polym Adv Technol 5:269–274. https://doi.org/10.1002/pat.1994.220050505
25. Hai Y, Jiang S, Zhou C, Sun P, Huang Y, Niu S (2020) Fire-safe unsaturated polyester resin nanocomposites based on MAX and MXene: a comparative investigation of their properties and mechanism of fire retardancy. Dalton Trans 49:5803–5814. https://doi.org/10.1039/D0DT00686F
26. Kicko-Walczak E (2003) Study of self-extinguishing unsaturated polyester resins. a brief survey of industrial and new achievements. Int Polym Sci Technol 30:41–46. https://doi.org/10.1177/0307174X0303000311
27. Qiu X, Li Z, Chevali V, Song P, Li X, Zhang Z (2023) Functionalized mesoporous silica fire retardant via hierarchical assembly for improved fire retardancy of unsaturated polyester. ACS Appl Polym Mater 5:2203–2212. https://doi.org/10.1021/acsapm.2c02195
28. Pan LL, Li GY, Su YC, Lian JS (2012) Fire retardant mechanism analysis between ammonium polyphosphate and triphenyl phosphate in unsaturated polyester resin. Polym Degrad Stab 97:1801–1806. https://doi.org/10.1016/j.polymdegradstab.2012.06.002
29. Liu L, Xu Y, Xu M, He Y, Li S, Li B (2020) An efficient synergistic system for simultaneously enhancing the fire retardancy, moisture resistance and electrical insulation performance of unsaturated polyester resins. Mater Des 187:108302. https://doi.org/10.1016/j.matdes.2019.108302
30. Kim NK, Dutta S, Bhattacharyya D (2018) A review of flammability of natural fibre reinforced polymeric composites. Compos Sci Technol 162:64–78. https://doi.org/10.1016/j.compscitech.2018.04.016
31. Kumar RN, Wei LM, Rozman HD, Abusamah A (1997) Fire resistant sheet moulding composites from hybrid reinforcements of oil palm-fibres and glass fibre. Int J Polym Mater Polym Biomater 37:43–52. https://doi.org/10.1080/00914039708031476
32. Manfredi LB, Rodríguez ES, Wladyka-Przybylak M, Vázquez A (2006) Thermal degradation and fire resistance of unsaturated polyester, modified acrylic resins and their composites with natural fibres. Polym Degrad Stab 91:255–261. https://doi.org/10.1016/j.polymdegradstab.2005.05.003
33. Rodríguez E, Petrucci R, Puglia D, Kenny JM, Vázquez A (2005) Characterization of composites based on natural and glass fibers obtained by vacuum infusion. J Compos Mater 39:265–282. https://doi.org/10.1177/0021998305046450
34. Chu F, Yu X, Hou Y, Mu X, Song L, Hu W (2018) A facile strategy to simultaneously improve the mechanical and fire safety properties of ramie fabric-reinforced unsaturated polyester resin composites. Compos Part A Appl Sci Manuf 115:264–273. https://doi.org/10.1016/j.compositesa.2018.10.006

35. Yu X, Pan Y, Wang D, Yuan B, Song L, Hu Y (2017) Fabrication and properties of biobased layer-by-layer coated ramie fabric-reinforced unsaturated polyester resin composites. Ind Eng Chem Res 56:4758–4767. https://doi.org/10.1021/acs.iecr.7b00101
36. Zhang J, Zhang H, Wang X, Zhang M (2022) An environment friendly hemp fiber modified with phytic acid for enhancing fire safety of automobile parts. J Polym Eng 42:535–543. https://doi.org/10.1515/polyeng-2021-0269
37. Salim MS, Ahmad Rasyid MF, Abdullah MA, Mat Taib R, Mohd Ishak ZA (2018) Mechanical, thermal and flammability properties of nonwoven kenaf reinforced acrylic based polyester composites: effect of water glass treatment. IOP Conf Ser Mater Sci Eng 368:012040. https://doi.org/10.1088/1757-899X/368/1/012040
38. Karuppasamy K, Ranganathan B (2024) Development of sustainable polyester biocomposites using Lansium parasiticum fruit shell powder and hemp chopped fibre: mechanical, wear, hydrophobic and flammability behaviour. Biomass Convers Biorefin 14:5745–5758. https://doi.org/10.1007/s13399-023-05199-1
39. Diharjo K, Gapsari F, Andoko A, Wijaya MN, Mavinkere Rangappa S, Siengchin S. Flammability and thermal resistance of Ceiba petandra fiber-reinforced composite with snail powder filler. Polym Compos 45:4947–4960. https://doi.org/10.1002/pc.28100
40. Chu F, Zhang D, Hou Y, Qiu S, Wang J, Hu W et al (2018) Construction of hierarchical natural fabric surface structure based on two-dimensional boron nitride nanosheets and its application for preparing biobased toughened unsaturated polyester resin composites. ACS Appl Mater Interfaces 10:40168–40179. https://doi.org/10.1021/acsami.8b15355
41. Uz Zaman S, Shahid S, Shaker K, Nawab Y, Ahmad S, Umair M et al (2022) Development and characterization of chemical and fire resistant jute/unsaturated polyester composites. J Textile Inst 113:484–493. https://doi.org/10.1080/00405000.2021.1889131
42. Qi Y, Guo Y, Liza AA, Yang G, Sipponen MH, Guo J et al (2023) Nanocellulose: a review on preparation routes and applications in functional materials. Cellulose 30:4115–4147. https://doi.org/10.1007/s10570-023-05169-w
43. Gu H, Gao C, Zhou X, Du A, Naik N, Guo Z (2021) Nanocellulose nanocomposite aerogel towards efficient oil and organic solvent adsorption. Adv Compos Hybrid Mater 4:459–468. https://doi.org/10.1007/s42114-021-00289-y
44. Ghasemlou M, Daver F, Ivanova EP, Habibi Y, Adhikari B (2021) Surface modifications of nanocellulose: From synthesis to high-performance nanocomposites. Prog Polym Sci 119:101418. https://doi.org/10.1016/j.progpolymsci.2021.101418
45. Saravanan KG, Kaliappan S, Natrayan L, Patil PP (2023) Effect of cassava tuber nanocellulose and satin weaved bamboo fiber addition on mechanical, wear, hydrophobic, and thermal behavior of unsaturated polyester resin composites. Biomass Convers Biorefin. https://doi.org/10.1007/s13399-023-04495-0
46. Chirayil CJ, Mathew L, Hassan PA, Mozetic M, Thomas S (2014) Rheological behaviour of nanocellulose reinforced unsaturated polyester nanocomposites. Int J Biol Macromol 69:274–281. https://doi.org/10.1016/j.ijbiomac.2014.05.055
47. Rusmirović JD, Rančić MP, Pavlović VB, Rakić VM, Stevanović S, Djonlagić J et al (2018) Cross-linkable modified nanocellulose/polyester resin-based composites: effect of unsaturated fatty acid nanocellulose modification on material performances. Macromol Mater Eng 303:1700648. https://doi.org/10.1002/mame.201700648
48. Chirayil CJ, Joy J, Mathew L, Koetz J, Thomas S (2014) Nanofibril reinforced unsaturated polyester nanocomposites: morphology, mechanical and barrier properties, viscoelastic behavior and polymer chain confinement. Ind Crops Prod 56:246–254. https://doi.org/10.1016/j.indcrop.2014.03.005

Flammability Performance of the Graphene-Infused Sisal/Epoxy and Hemp/Epoxy Composites

Preet Parashram Harer, K. Balaji, Muthukumar Chandrasekar, S. Jeyanthi, K. Senthilkumar, T. Senthil Muthu Kumar, and R. Sundaramoorthy

Abstract In this experimental work, the flammability performance of sisal/epoxy and hemp/epoxy composites infused with graphene nanocomposites at varying weight percentages was studied. The composite samples were fabricated using the hand lay-up technique, and the samples were subjected to the UL-94 test in both vertical and horizontal directions. The results revealed that sisal/epoxy samples exhibited a V-1 rating, which was attributed to the absence of dripping. However, the hemp/epoxy samples showed dripping, and resulted in a V-2 rating. Further, the burn time ranged from 170 to 210 s for all composite samples. A decrease in burn time was observed for all sisal-based composite samples with varying graphene content, whereas hemp-based samples exhibited a significant reduction in burn time only at 3 wt% of graphene. Based on these observations, sisal-based composite samples can be proposed for fire-sensitive applications due to their superior flammability resistance behavior.

P. P. Harer
Department of Aerospace Engineering, Sanjay Ghodawat University, Atigre, Maharashtra, India

K. Balaji
Department of Aeronautical Engineering, Parul Institute of Engineering and Technology, Parul University, Vadodara, Gujarat, India

M. Chandrasekar (✉) · R. Sundaramoorthy
Department of Aerospace Engineering, SIMCRASH Centre, Hindustan Institute of Technology and Science, Padur, Chennai, Tamil Nadu, India
e-mail: chandrasekar.25j@gmail.com

S. Jeyanthi
School of Mechanical Engineering, Vellore Institute of Technology, Chennai, India

K. Senthilkumar
Department of Mechanical Engineering, PSG Institute of Technology and Applied Research, Coimbatore, Tamil Nadu, India

T. Senthil Muthu Kumar
Department of Mechanical Engineering, Kalasalingam Academy of Research and Education, Krishnankoil, Tamil Nadu, India

C. Muthukumar et al. (eds.), *Flammability Performance of Biocomposites and Bionanocomposites*, Composites Science and Technology,
https://doi.org/10.1007/978-981-95-2389-4_5

Keywords Sisal · Hemp · Epoxy · Graphene · Flammability · UL-94 · Biocomposites

1 Introduction

In recent years, the use of biofiber-reinforced composite materials has been significant advancement. This development has driven by several factors, especially the increasing cost of fossil fuels and the global shift toward environmentally friendly composite materials, due to environmental legislation and the REACH Act (Registration, Evaluation, Authorization and Restriction of Chemical substances) [1, 2]. Agricultural products are increasingly used as sources for producing biocomposites and offer sustainable solutions to traditional materials. By utilizing these agricultural resources for fabricating biocomposites, these materials offer solutions to waste disposal and help reduce environmental pollution caused by the burning of agricultural waste. Furthermore, biocomposites offer an economic solution for many developing countries due to the contribution of environmental benefits as well as economic sustainability [3–5].

Biofiber composites have numerous advantages, such as partially or fully recyclable, renewable, low-cost, and most importantly, environmentally friendly. These characteristics are highly attractive in producing sustainable materials [6–8]. Though biocomposites have many advantages, they also have some inherent disadvantages, including poor fire resistance property and a hydrophilic nature. The hydrophilic nature of biofibers and/or biocomposites will lead to moisture absorption and resulting in affects their durability in various applications [9–11]. However, the poor fire resistance property could be improved by incorporating flame retardant additives or fiber surface treatments. Thus, biocomposites can be made suitable for many applications including fire-prone environments.

The flammability of biocomposites can be assessed using various testing methods. Researchers have reported several widely used techniques in both academic and industry laboratories. These methods can be used for small, medium-sized and full-scale laboratories tests, based on the type of analysis. The techniques including (i) Ohio State University (OSU) heat release test, (ii) underwriters' laboratories 94 (UL94), (iii) limiting oxygen index (LOI), (iv) pyrolysis combustion flow calorimetry (PCFC) and (v) cone calorimetry [1, 12]. These techniques measure various parameters based on the type of polymeric materials including smoke production, product distribution, burn and spread rates, delay timing and ignitability. Each technique provides valuable insights into material behavior during combustion and can be utilized to evaluate the fire performance of biocomposite materials [1].

Flame retardants can be used to give flame retardancy behavior to many materials including thermosets, thermoplastics, coatings, etc. The flame retardants are classified into many categories, such as (i) phosphorus based, (ii) halogen based, (iii) silicon based, (iv) mineral flame based, and (v) nano-metric-based materials.

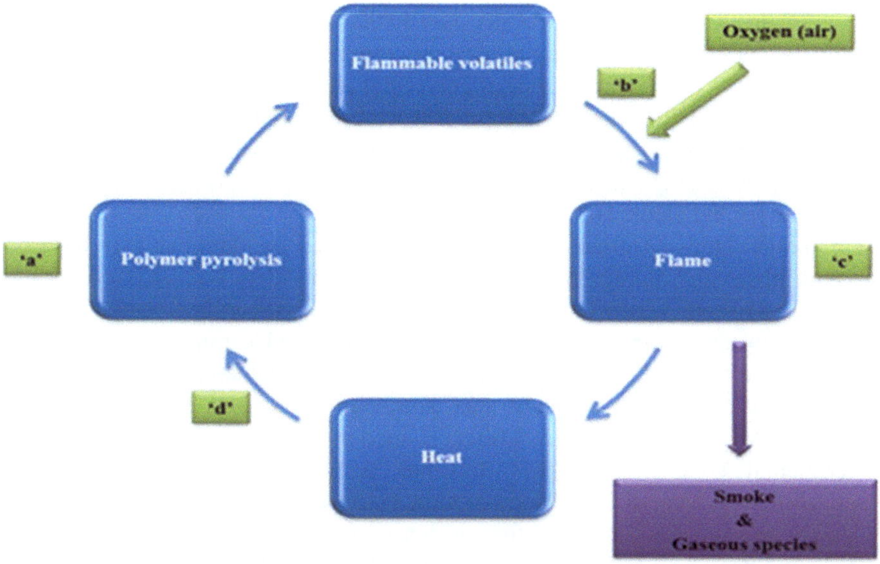

Fig. 1 Explaining the self-sustaining polymer combustion cycle [1]

These flame retardants can be used to reduce, prevent, suppress or stop the combustion process of various materials. Furthermore, they help interrupt the self-sustaining polymer combustion process as shown in Fig. 1; thus, resulting in extinguishing the flame [1].

(Fig. 1 is reused with permission from Elsevier, License number: 5885430263699).

Among the various flame retardants, nanomaterial-based materials (e.g., graphene) were utilized in this experimental work to improve the flammability properties of biocomposites. The biocomposites were fabricated using sisal and hemp fibers with an epoxy matrix. Furthermore, nanomaterial-based materials can be used to improve mechanical and thermal behavior when dispersed within polymeric matrices.

Graphene, a unique 2-D material is known for its mechanical, thermal and electrical properties. In recent times, graphene has been increasingly incorporated into polymer materials with low filler loadings due to its high aspect ratio. Similar to graphene, other nanomaterials such as montmorillonites (MMTs), and carbon nanotubes (CNTs) are also commonly used for flame retardancy purposes in biocomposites. Yuan et al. [13] examined the flammability of polypropylene composites incorporating graphene materials. The study reported that incorporating graphene helped reduce flame intensity and leading to a smaller flame size. Furthermore, lower graphene loadings supported a decrease in heat and smoke release from the composite samples. However, further addition of graphene worsened fire safety. In another work [14], the decomposition behavior of polyimide with the addition of nanographene was studied. The results showed that the rate of decomposition significantly decreased, with increased char retention in the composites. Guobo et al. [15] analyzed the fire retardant behavior of poly(vinyl alcohol) reinforced with graphene nanosheets using cone calorimetry test. It was observed that peak heat release rate was significantly reduced due to the addition of graphene nanosheets. For example, a loading of 3 wt% graphene resulted in a 49% reduction in heat release rate compared to pure poly(vinyl alcohol).

In another study, Amreen Taj et al. [16] examined the flammability properties of kenaf fiber/nano-aluminum oxide and nano-graphene/epoxy matrix composites using UL-94 (both vertical and horizontal tests). The researchers varied the fiber loading between 50 and 60%, with nano-aluminum oxide ranged from 0 to 6% and graphene from 0 to 3% and epoxy matrix was maintained at 40%. The results reported that composites with 6% nano-aluminum oxide and 3% nano-graphene exhibited a V-2 rating in the UL-94 test (i.e., moderate flame retardancy). In another experimental work [17], flax/polypropylene composites were analyzed for their flammability performance by incorporating graphene nanoparticles. The addition of graphene supported the formation of denser and rigid char layers and resulted in reducing peak heat release rate and total heat release. Additionally, the mechanical performance of the composites improved with inclusion of 1 wt% graphene nanoplatelets.

Based on the literature studies, combining sisal, hemp and graphene nanoparticles is limited. Therefore, in this present work, the flammability of the biocomposites fabricated from sisal, hemp and an epoxy matrix were examined using UL-94 test (vertical and horizontal). Based on the results observed from these combinations indicate that the developed biocomposite materials may be suitable for fire-sensitive applications.

2 Materials and Method

2.1 Materials

The sisal and hemp fabric of 200GSM and individual ply thickness of 0.3 mm used as reinforcement was purchased from Gogreen Products, Chennai. The LY556 epoxy and hardener used as matrix was purchased from Herenba Instruments, Chennai. The graphene nanoparticles used were procured from a local supplier in Chennai. The specifications of the constituents are provided in Tables 1 and 2, respectively.

Table 1 Fiber reinforcement and matrix specifications [18]

Properties	Sisal	Hemp	Epoxy
Density (g/cm^3)	1.3–1.5	1.4	1.15–1.20
Tensile strength (MPa)	507–855	550–1110	3.77
Young's modulus (GPa)	9.4–28	30–70	–
Cellulose (%)	78	70–74	–
Hemicellulose (%)	10	15–20	–
Lignin (%)	8	3.5–5.7	
Pectin (%)	–	0.8	–
Wax (%)	2	1.2–6.2	–
Ash (%)	1	0.8	–

Table 2 Specifications of graphene nanoparticle

Properties	Value
Diameter (nm)	5–10
Length (μm)	5–10
Density (g/ cm^3)	0.15
Surface area (m^2/g)	200–240

2.2 *Fabrication Method*

The composites were fabricated through the hand layup technique. The graphene nanofillers were gradually infused in the epoxy resin through the mechanical stirrer operated at 500 rpm for 30 min. The sisal and hemp fabric layers were placed in a mold of dimension 300 × 300 × 3 mm^3 and applied with the nanoparticle infused resin (different wt% as shown in Table 3) and hardener. The fabric and resin mixture were allowed to cure in the room temperature and removed from the mould after 24 h. The composites without nanofillers were also fabricated and tested for comparison.

Table 3 Composites investigated in this study and their notation [16]

Notation	Hemp (wt%)	Sisal (wt%)	Graphene (wt%)	Epoxy (wt%)
H0	40	–	0	60
H1	40	–	1	59
H2	40	–	2	58
H3	40	–	3	57
S0	–	40	0	60
S0.5	–	40	0.5	59.5
S1	–	40	1	59
S2	–	40	2	58

2.3 UL-94 Flammability Test

UL-94 vertical and horizontal burn tests were performed on the specimens of dimension 127 × 13 × 3 mm^3 as per the ASTM D 3801 and ASTM D635, respectively. In the vertical burn test, all the prepared samples were subjected to ignition twice for ten seconds and allowed the flame to extinguish to study their flammability. The specimen was ignited at the bottom for 10 s and burnt upward. The material undergoing Ul-94 vertical burn test and horizontal burn test is classified as shown in Table 1 (Table 4).

Table 4 Inferences from the UL-94 horizontal and vertical burn test

Test	Characteristic	Description
Horizontal burn test	HB	• The free end of the specimen is ignited with flame for 30 s • Burning rate not exceeding 40 mm per minute, If thickness of tested material is between 3 and 13 mm • Burning rate should not exceed 75 mm per minute. for material with a thickness of less than 3 mm, • If material ceases to burn before the 100 mm mark independent of thickness
Vertical burn test	V-0	• If the after-flame time for the individual specimen during the first or second impingements does not exceed 10 s, the total after-flame time for any number of specimens, which includes adding time required for the first impingement and second impingement, does not exceed 50 s • Cotton does not ignite or no dripping
	V-1	• If the after-flame time for the individual specimen during the first or second impingements does not exceed 30 s, the total after-flame time for any number of specimens, which includes adding time required for the first and second impingements does not exceed 250 s • Cotton does not ignite or no dripping
	V-2	• If the after-flame time for the individual specimen during the first or second impingements does not exceed 30 s, the total after-flame time for any number of specimens, which includes adding time required for the first and second impingements does not exceed 250 s • Cotton is ignited causing dripping
	Burning	• If the entire sample is consumed

Table 5 Flammability characteristics from the UL-94 VB test

Composites	Burn time (seconds)	Burn length (mm)	Dripping	Rating
H0	190	38	Yes	V-2
H1	194	39		
H2	170	43		
H3	209	48		
S0	183	49	No	V-1
S0.5	202	44		
S1	188	48		
S2	196	47		

3 Results and Discussion

3.1 *Vertical Burn Characteristics*

Table 5 depicts the burn characteristics from the UL-94 VB test which shows that the burn time was recorded between 170 and 210 s for all the composites. Though there was marginal difference in burn time between the hemp and sisal-based nanocomposites, dripping occurred in the former over the latter. Thus, the hemp-based nanocomposites exhibited superior vertical burn characteristics with V-1 rating over the sisal-based nanocomposites which displayed V-2 rating. Researchers reported that thermal degradation of natural fibers is influenced by their chemical compositions such as cellulose, hemicellulose, lignin, etc. For example, researchers Chapple et al. [19], Ryszard and Maria [20] reported that a higher cellulose content may support in improving flammability behavior, while a higher lignin content could reduce flammability property. Thus, by comparing the lignin content of hemp and sisal fibers, sisal fibers have a higher percentage, such as 8% [18], whereas the hemp fibers have a lignin content of 3.5–5.7% [18]. This higher lignin content in sisal fibers could increase the production of flammable volatiles due to decomposition, contributing to higher susceptibility to ignition and resulting in a V-2 rating.

In order to analyze the influence of nanofiller proportion on vertical burn characteristics, burn time was plotted corresponding to the filler weight for the nanocomposites as illustrated in Fig. 2. The general observation was that adding graphene filler above 1 wt% led to prolonged burning before self-extinguishing as indicated by the higher burning time.

3.2 *Horizontal Burn Characteristics*

The composite specimens post UL-94 HB test is shown in Fig. 3 and the corresponding burn time, burn length and burn rate are illustrated in Table 6. A char

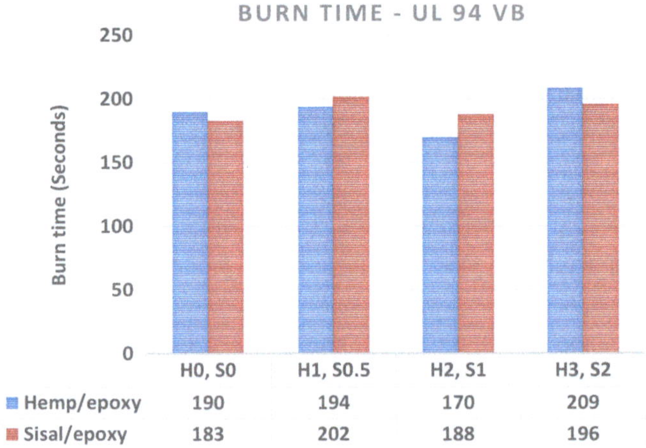

Fig. 2 Burn time of the sisal and hemp-based nanocomposites from the UL-94 VB test

layer with blackish appearance can be observed on the burnt specimens. The char formation is a typical feature of the polymer composite [21].

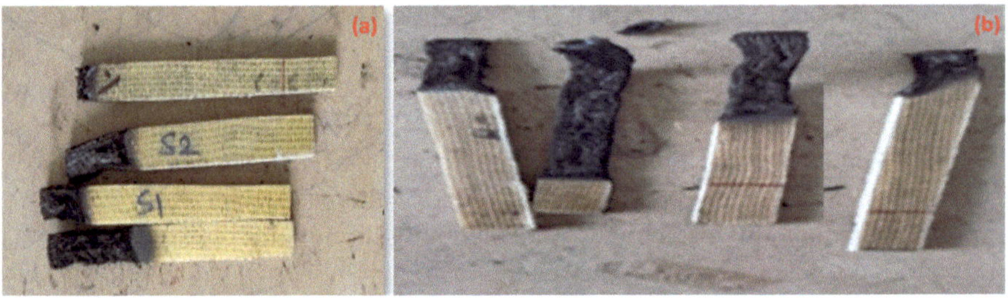

Fig. 3 Composites after Ul-94 HB test **a** S0–S3 and **b** H0–H3

Table 6 Flammability characteristics from the UL-94 HB test

Composites	Burn time (seconds)	Burn length (mm)	Burn rate (mm/min)	Rating
H0	309	55	10.680	HB
H1	644	95	8.851	
H2	372	75	12.097	
H3	202	62	18.416	
S0	358	65	10.894	
S0.5	106	20	11.321	
S1	237	85	21.519	
S2	193	45	13.990	

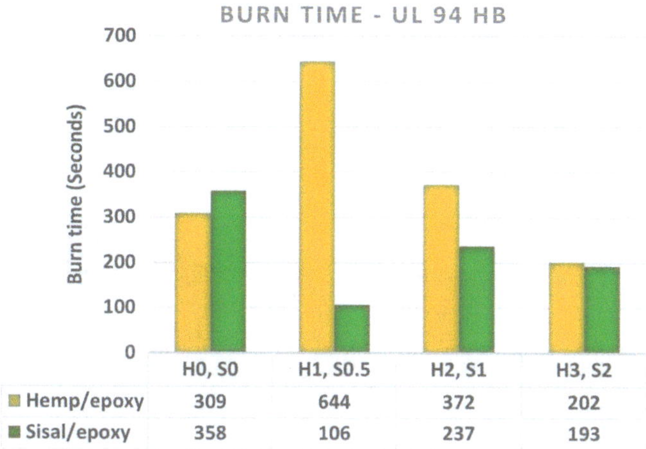

Fig. 4 Burn time of the sisal and hemp-based nanocomposites from the UL-94 HB test

Figure 4 provides the comparison for the vertical burn time between sisal and hemp-based nanocomposites. Incorporation of nanofillers between 0.5 and 2 wt% in the sisal/epoxy composite reduced the burn time significantly while the burn time reduced only with the addition of 3 wt% graphene in case of the hemp/epoxy composite. The major conclusion from the vertical burn test was that the sisal and hemp-based nanocomposites with highest proportion of nanofiller displayed least burning time. Similar to this observation, Bodduru et al. [22] reported that the burning rate of 11.22 mm/min was reduced to 7.6 mm/min in glass/sisal/epoxy composite samples due to the addition of graphene nanoparticles.

The burn rate computed based on the burn length and burn time is depicted in Fig. 5. In general, burn rate increased with the addition of the filler. It can be observed that the hemp-based nanocomposites outperformed the sisal-based nanocomposites with the lower burn rate for the former over the latter. The burn rate was maximum for the composites infused with highest proportion of graphene nanofiller. Similar observation was reported recently [16], where the flame propagation ceased, resulting in a zero burning rate due to the addition of nano-graphene up to 3% in kenaf/epoxy composite samples. This work highlighted the effectiveness of graphene in enhancing flame retardancy properties.

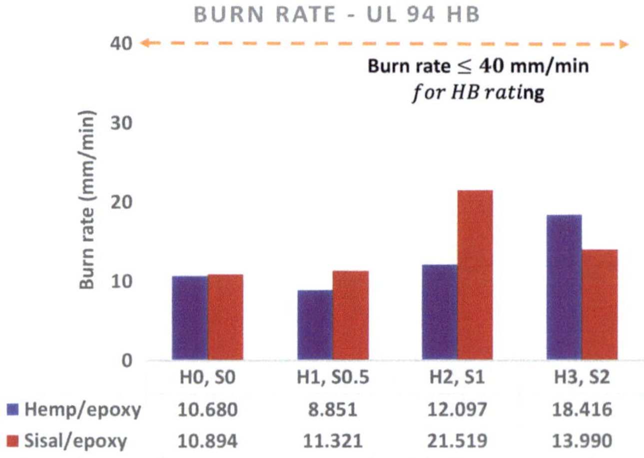

Fig. 5 Burn rate of the sisal and hemp-based nanocomposites from the UL-94 HB test

4 Conclusion

The sisal and hemp-based nanocomposites with the varying weight proportions of graphene as the nanofiller was fabricated. The flammability performance of these composites was then assessed through the UL-94 horizontal and vertical burn test. In case of the vertical burn test, burn time increased with the addition of nanofiller while only a marginal difference in the burning time was observed. However, sisal-based nanocomposites did not cause dripping leading to V-1 rating over V-2 rating for the hemp-based nanocomposites. All the nanocomposites subjected to the horizontal burn test exhibited HB rating. Furthermore, addition of nanofiller at all proportions reduced the burn time for the sisal-based nanocomposites whereas the burn time reduced substantially, only with the addition of 3 wt% graphene. In general, the burn rate increased with the filler addition for nanocomposites.

References

1. Mngomezulu ME, John MJ, Jacobs V, Luyt AS (2014) Review on flammability of biofibres and biocomposites. Carbohydr Polym 111:149–182
2. John MJ (2019) Flammability performance of biocomposites. In: Green composites for automotive applications. Elsevier, pp 43–58
3. Muthukumar C, Naveen J, Senthil Muthu Kumar T et al (2021) Fire retardant properties of bio-phenolic hybrid composites. Phenolic Polym Based Compos Mater, pp 111–121
4. Thirukumaran M, Senthilkumar K, Selvabharathi R (2023) 15 Effect of carbon nanotubes, aluminum hydroxide, and zinc borate on the mechanical and fire properties of epoxy nanocomposite. Nanocomposite Nanohybrid Mater: Process Appl 17:297

5. Krishnasamy S, Sasikumar R, Swaminathan G et al (2024) Hybrid glass/Kevlar fiber reinforced phenolic matrix composites: thermal degradation and flammability studies. Polym Compos. https://doi.org/10.1002/pc.29416
6. Jeyaguru S, Thiagamani SMK, Krishnasamy S et al (2024) Experimental investigation on the thermal characteristics of Kevlar/hemp intraply hybrid composites: Influence of various weaving designs. Ind Crops Prod 221:119280
7. Periasamy D, Prakalathan K, Manoharan B et al (2024) Examining the performance of eco-friendly flax fiber composites: mechanical, thermal and durability properties. Biomass Convers Biorefin, pp 1–15
8. Senthilkumar K, Aravind D, Pulikkalparambil H et al (2024) Mechanical performance of biofibers and biocomposites. In: Biocomposites-bio-based fibers and polymers from renewable resources. Elsevier, pp 143–163
9. Krishnasamy S, Ungtrakul T, Chandrasekar M et al (2024) Analysis of the thermal properties in short sansevieria cylindrica fibre/PLA composites processed by twin screw extruder followed by hot press molding technique. Mater Res Express 11:035506
10. Krishnasamy S, Hemath Kumar M, Parameswaranpillai J et al (2024) Interfacial bonding characteristics in natural fiber reinforced polymer composites. Springer
11. Thiagamani SMK, Krishnasamy S, Muthukumar C et al (2019) Investigation into mechanical, absorption and swelling behaviour of hemp/sisal fibre reinforced bioepoxy hybrid composites: effects of stacking sequences. Int J Biol Macromol 140:637–646. https://doi.org/10.1016/j.ijbiomac.2019.08.166
12. Kim NK, Dutta S, Bhattacharyya D (2018) A review of flammability of natural fibre reinforced polymeric composites. Compos Sci Technol 162:64–78
13. Yuan B, Fan A, Yang M et al (2017) The effects of graphene on the flammability and fire behavior of intumescent flame retardant polypropylene composites at different flame scenarios. Polym Degrad Stab 143:42–56
14. Akinyi C, Longun J, Chen S, Iroh JO (2021) Decomposition and flammability of polyimide graphene composites. Minerals 11:168
15. Huang G, Gao J, Wang X et al (2012) How can graphene reduce the flammability of polymer nanocomposites? Mater Lett 66:187–189
16. Taj A, Swamy RP, Naik K, Bharath KN (2023) Effect of nano-filler aluminum oxide and graphene on flammability properties of kenaf epoxy composites. J Inst Eng (India): Series D 104:143–154
17. Ali I, Kim NK, Bhattacharyya D (2021) Effects of graphene nanoplatelets on mechanical and fire performance of flax polypropylene composites with intumescent flame retardant. Molecules 26:4094
18. Manaia JP, Manaia AT, Rodriges L (2019) Industrial hemp fibers: an overview. Fibers 7:106
19. Chapple S, Anandjiwala R (2010) Flammability of natural fiber-reinforced composites and strategies for fire retardancy: a review. J Thermoplast Compos Mater 23:871–893. https://doi.org/10.1177/0892705709356338
20. Kozlowski R, Przybylak MW (2001) Natural polymers, wood and lignocellulosic materials. In: Fire retardant materials. Elsevier, pp 293–317
21. Sinha Ray S, Kuruma M (2020) Melt-dripping and char formation, pp 69–82
22. Kamesh B, Gude R, Maddamsetty A et al (2024) Investigation of the mechanical, absorption, flammability and swelling properties of graphene filled sisal/glass fiber reinforced polymer hybrid nanocomposites. Cogent Eng 11:2342433

Flame Retardancy Characteristics of the Polypropylene-Based Biocomposites and Bionanocomposites

Lin Feng Ng, Mohd Yazid Yahya, Chandrasekar Muthukumar, Jyotishkumar Parameswaranpillai, Sanjay Mavinkere Rangappa, and Suchart Siengchin

Abstract Composite materials have received significant attention from researchers and engineers over the past decades. With several fascinating merits offered by composite materials, the steady growth of such materials in various engineering sectors has been witnessed. However, one of the critical weaknesses of most composite materials is associated with their poor fire retardancy properties, which lead to material decomposition and the release of toxic substances upon burning.

L. F. Ng (✉) · M. Y. Yahya
Centre for Advanced Composite Materials (CACM), Universiti Teknologi Malaysia, 81310 Johor Bahru, Johor, Malaysia
e-mail: linfeng@utm.my; linfeng@utm.my

Faculty of Mechanical Engineering, Universiti Teknologi Malaysia, 81310 Johor Bahru, Johor, Malaysia

M. Y. Yahya
e-mail: yazidyahya@utm.my

C. Muthukumar
SIMCRASH Centre, Department of Aerospace Engineering, Hindustan Institute of Technology & Science, Kelambakkam, Chennai 603103, India
e-mail: chandrasm@hindustanuniv.ac.in

J. Parameswaranpillai
Department of Science and AU—Sophisticated Testing and Instrumentation Center, Alliance University, Chandapura 562106, Bengaluru, India
e-mail: jyotishkumar.p@alliance.edu.in

S. Mavinkere Rangappa · S. Siengchin
Natural Composites Research Group Lab, Department of Materials and Production Engineering, The Sirindhorn International Thai-German School of Engineering (TGGS), King Mongkut's University of Technology North Bangkok (KMUTNB), Bangkok, Thailand
e-mail: mavinkere.r.s@op.kmutnb.ac.th

S. Siengchin
e-mail: suchart.s@tggs.kmutnb.ac.th

C. Muthukumar et al. (eds.), *Flammability Performance of Biocomposites and Bionanocomposites*, Composites Science and Technology,
https://doi.org/10.1007/978-981-95-2389-4_6

When dealing with biocomposites incorporated with cellulosic fibers, it is particularly vital to understand their fire retardancy potential as these materials are generally highly flammable. Since biocomposite materials have been employed in various engineering applications, it is essential to improve their fire retardancy to lessen the risk of catastrophic failure. This chapter intends to provide a comprehensive understanding of the combustion mechanism and the techniques for enhancing the flame retardancy of composite materials. In addition, this chapter also aims to delve deeper into the flammability of cellulosic fibers, flame retardancy of biocomposites and eco-friendly flame retardants that have immense potential to develop flame-retardant biocomposites. This chapter is expected to enhance the understanding of the readers by providing a comprehensive overview of the fire retardancy of biocomposites.

Keywords Composite materials · Biocomposites · Bionanocomposites · Polypropylene · Fire retardancy

1 Introduction

The continuous exploitation of non-renewable natural resources such as fossil fuels has reached an alarming level. Certainly, unconstrained exploitation of non-renewable natural resources may negatively impact the future existence of human beings. The emphasis on developing products based on non-renewable natural resources has led to the ecological systems being plundered by human beings. At this moment, fossil fuel has been utilized as the primary energy source in several engineering sectors. In the transportation sector, fossil fuel has been used as an energy source to move vehicles. It has been identified that the amount of fossil fuel required to move the vehicles is highly dependent on the vehicle weight. In the construction sector, fossil fuel has also been used to power heavy machinery and equipment. Due to the high dependency on fossil fuels in the construction sector, this sector is considered one of the significant contributors to greenhouse gas emissions. In order to mitigate the dependency on fossil fuels, weight reduction of structural components is an effective method to reduce the energy required to operate heavy machinery. From this standpoint, partial substitution of metal alloys with lightweight composite materials has been the contemporary trend in innumerable engineering sectors to achieve weight reduction. With years of investigation on composite materials, the findings revealed that these materials manifest some fascinating merits compared to metal alloys. In addition to lightweight characteristics, composite materials possess a high strength-to-weight ratio, superior fatigue properties and good corrosion resistance. Furthermore, composite materials also offer flexibility in material design, allowing engineers and scientists to tailor their material properties based on the criteria of particular applications [1]. Since composite materials are formed by combining two or more solid phases, their miscellaneous properties can be efficiently designed through appropriate material selection. However, it is worth mentioning that the compatibility between constituents in composite materials plays a significant role in determining the resulting properties, and thus, their compatibility should be taken into consideration during the material design stage.

When scrutinizing the commercially available composite materials, it can be seen that most composite materials are mainly based on synthetic fibers such as carbon, glass and aramid fibers [2, 3]. The driving factors behind the broad utilization of synthetic fibers are outstanding mechanical properties and affordable cost. Among the three types of synthetic fibers, glass fiber stands out to be the most attractive reinforcement for composite materials due to its balance in mechanical properties and cost. Nonetheless, the broad applications of these synthetic fibers have led to severe environmental issues due to their non-biodegradable characteristic. The anthropogenic pollution caused by the waste accumulation has shown a significant adverse effect on the biodiversity. Besides, high energy consumption for producing synthetic fibers escalates greenhouse gas emissions, consequently leading to global warming. Today, the ever-growing issues related to environmental degradation have spurred the interest in using green materials from natural resources in a wide variety of applications. The gradual shift from synthetic fiber-based composites to biocomposites has been witnessed by incorporating natural fibers into the materials. Certain natural fibers show high feasibility to be potential reinforcement owing to their high economic value and decent mechanical properties. Flax, hemp, kenaf, and jute fibers have been extensively explored in innumerable research studies, and they have shown a very high potential to be used as reinforcement for composite materials. Recently, natural fibers from secondary fiber crops such as pineapple leaf, banana and sisal fibers have started to gain attention from researchers and scientists. With the waste-to-wealth policy, the research studies focusing on secondary fiber crops tremendously increase, intending to develop value-added products.

Fire accidents can happen instantaneously and catastrophically, and they might pose a severe threat to human life. In addition, burning polymeric materials also manifests a serious issue related to environmental and ecological concerns. The globe has recently seen an increase in the frequency of fire incidents, causing the loss of innumerable human lives and property [4]. In the construction sector, fire outbreak is commonly due to the combustible polymeric composite materials used as building components [5]. Endeavor has been given to supersede metal alloys with polymer composites for various engineering applications. However, the flammability of polymer composites, especially thermoplastic-based composites, has been one of the shortcomings that limits the expansion of their applications [6]. Therefore, it is necessary to look into the fire retardancy of composite materials and enhance their fire resistance since these materials, particularly natural fiber-reinforced composites, have a relatively high thermal and fire sensitivity. When referring to the outcomes from years of experimental investigations, fire retardancy of composite materials is affected by various factors, such as fiber and matrix types, fiber content and amount of filler in the composite materials. Based on the above argument, it is possible to enhance fire retardancy by adequately designing the composite materials for applications where fire retardancy is crucial.

2 Overview of the Biocomposites and Bionanocomposites

Biocomposites refer to the composite materials consisting of different phases of green materials. On the other hand, bionanocomposites refer to green composite materials having at least one solid phase has a dimension in nanoscale. The incorporation of nanoscale additive, such as nanoclay, nanotube and graphite, is a common way to improve mechanical properties and flame retardancy of nanocomposites. They are generally considered composite materials sans any harmful effect on the ecosystem, environment and human health. Using biocomposites instead of conventional composite materials is vital to securing the future of human beings and the biodiversity of the world. Today, the rising pollution level has reached a worrying state, severely affecting biodiversity and Mother Earth. Undoubtedly, air pollution has been recognized as one of the scourges in this modern era, showing an inferior impact on climate change and individual health [7]. It has been estimated that outdoor air pollution has resulted in 4.2 million death cases annually [8]. Even though air pollution manifests a profound inferior impact on the world, soil and water pollution should not be neglected as they could also lead to environmental devastation. In this context, waste accumulation is a colossal challenge nowadays because of the extensive use of non-biodegradable materials. In the realm of composite materials, researchers have attempted to gradually reduce the reliance on synthetic fibers as it has been found that the non-biodegradable characteristic of these fibers has led to waste accumulation, which pollutes the environment and may jeopardize the existence of wildlife.

Burning fossil fuels has been identified as the largest source of pollution, leading to global warming and accelerating sea level rise. It is worth mentioning that reducing a collective carbon footprint from humankind and conserving natural resources to tackle climate change will help safeguard Mother Earth to remain habitable for future generations. In an effort to reduce the burning of fossil fuel to generate the energy required for various sectors, reducing the weight of structures is one of the viable options toward the goal. The goal of structural weight reduction can be achieved by superseding metal alloys with lightweight composite materials. It can be noticed that conventional composite materials are dominated by synthetic fibers embedded in the epoxy polymer. Although structural weight reduction can be realized using these composite materials, they are considered non-environmental-friendly materials due to their non-biodegradable characteristic. Because of this reason, there is a shift from the focus on synthetic to natural fibers for developing biocomposite materials.

The initial idea of incorporating fibers in the polymer is to reduce the use of petroleum-based polymers. However, it was found that the inclusion of fibers may also uplift the mechanical properties of the materials. It means that the presence of fibers in composite materials can achieve two aims at once. With the tremendous increase in interest in natural fibers, there has been a sharp rise in embedding natural fibers into polymer matrices to form biocomposite materials. As mentioned previously, substituting metal alloys with composite materials can reduce the overall structural weight, thus undermining energy consumption and mitigating environmental pollution. From this standpoint, it should be emphasized that the replacement of the most extensively used glass fibers with natural fibers can further reduce the overall structural weight as the density of natural fibers is approximately half of the glass fibers.

2.1 *Biofibers*

Biofibers are generally referred to as natural fibers that do not threaten the environment, biodiversity and human health. When it comes to biofibers, they should be biodegradable, renewable, carbon dioxide neutral, and, most importantly, non-toxic. High production cost, inferior environmental impact and difficulty in recycling are those rational factors that galvanize a switch focus from synthetic to natural fibers. Figure 1 shows the classification of natural fibers based on their origins. Natural fibers can be categorized into plant, animal, and mineral fibers. Each category of natural fibers has distinctive chemical constituents which make their mechanical properties vary. Cellulose is the principal element in all plant fibers, whereas the primary components in animal and mineral fibers are protein and natural rock substances, respectively. The scientific communities today are more interested in plant fibers due to their high availability and superior mechanical performance. The mechanical properties of plant fibers are governed by their physical, chemical and morphological properties, such as fiber orientation, cellulose content and fiber aspect ratio. Figure 2 shows the fiber structure with three main chemical constituents. Among the three main chemical constituents of plant fibers, cellulose is the main constituent that provides plant fibers with strength and stiffness. In order to obtain the optimum fiber strength and stiffness, it is essential to remove hemicellulose and lignin from the plant fibers as they do not contribute to fiber strength and stiffness. By removing unnecessary chemical constituents from the fibers, more reactive functional groups of cellulose can be bonded with the polymer matrices, which in turn, improve the mechanical strength and stiffness of biocomposites. In addition, purging unnecessary chemical constituents may also offer plant fibers more flexibility to realign microfibrils toward the loading direction, which is particularly beneficial to the mechanical properties when subjected to external loads. Regarding moisture uptake, hemicellulose plays a significant role in absorbing water for plant fibers. In this case, it is necessary to eliminate hemicellulose from plant fibers to avoid high moisture uptake as it has a significant negative impact on the mechanical properties of plant fiber-reinforced composites.

Although plant fibers have various fascinating characteristics, they are not demerit-free natural resources. The commonly cited weaknesses of plant fibers include low mechanical strength, high moisture sensitivity, batch-to-batch variation, poor compatibility with non-polar polymer matrices and low thermal resistance. In order to resolve the shortcomings of plant fiber-reinforced composites, several effective ways have been developed and widely studied by researchers worldwide.

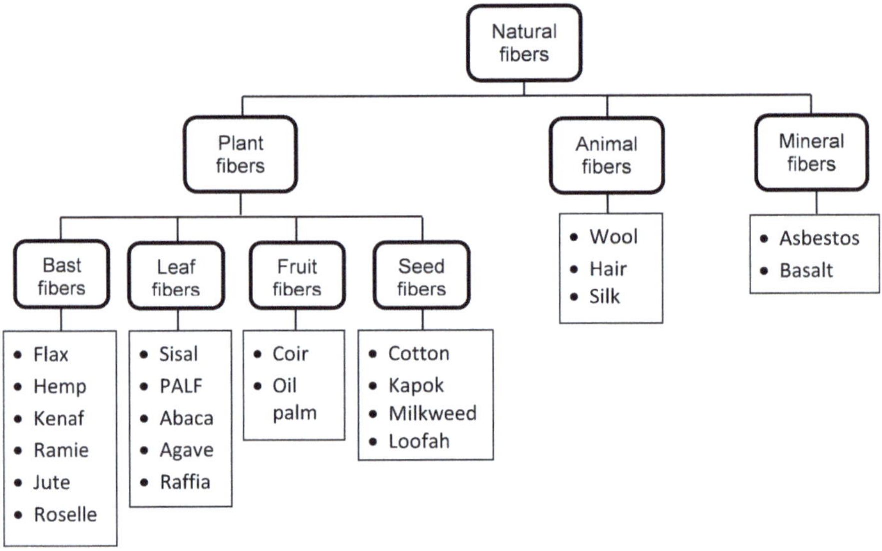

Fig. 1 Classifications of natural fibers based on their origins

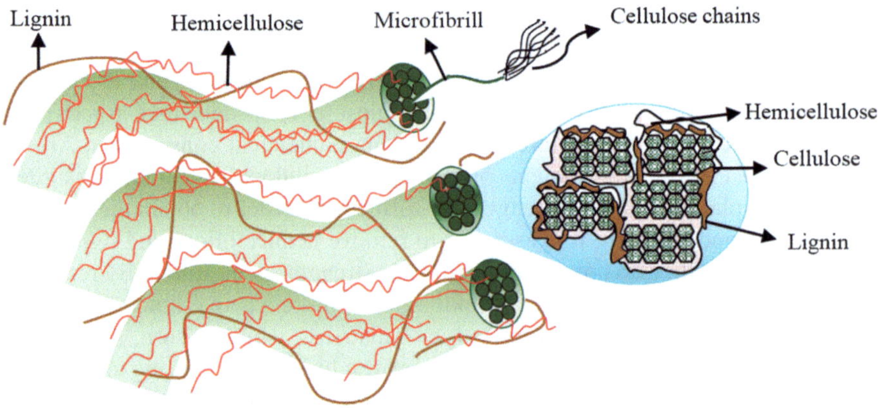

Fig. 2 Plant fiber structure with main chemical constituents [10]

Chemical treatments, hybridization and the addition of nanofillers have been found to be effective ways to resolve the demerits of plant fiber-reinforced composites. In addition, the physical properties of plant fibers, such as fiber length, diameter and surface roughness, can also impact the mechanical properties of their associated biocomposites. From Fig. 1, it can be seen that plant fibers can be sub-categorized into bast, leaf, fruit and seed fibers. Nonetheless, bast fibers are the most promising plant fibers as they possess the greatest mechanical properties, even higher than most animal fibers except silk. In comparison with leaf and fruit fibers, bast fibers generally have a higher fiber aspect ratio which is defined as the ratio of length to diameter. A higher degree of fiber aspect ratio allows an effective load transfer from the polymer matrix to the fiber, thereby improving the mechanical properties of the biocomposites.

Flame retardancy of plant fibers has received considerable attention nowadays as many engineering applications require materials to have excellent thermal and flame stability. Unfortunately, plant fibers are highly sensitive to flame and tend to degrade and decompose when exposed to high-temperature environments. The weak thermal stability and flame retardancy of plant fibers are ascribed to their chemical constituents: cellulose, hemicellulose and lignin which consist of dense carbohydrates moieties [9]. In fact, the thermal degradation of plant fibers is primarily governed by lignin and cellulose contents. When plant fibers are subjected to elevated temperatures, their thermal degradation can result in the formation of a significant amount of dense carbonaceous char owing to the cellulose degradation triggered by lignin.

2.2 *Biopolymers*

Petroleum-based polymers mostly dominate the contemporary matrix phase utilized to form composite materials. Petroleum-based bisphenol A-type epoxy resin represents the most commonly used thermoset polymer matrix in multiple engineering applications. In the meantime, the adoption of thermoplastic polypropylene (PP) polymer has been on an upward trend due to its low-cost characteristics. Thermoplastic polymers are receiving growing attention mainly because they are recyclable and possess higher ductility than thermoset polymers. For instance, PP-based composites can be recycled several times without apparently degrading their mechanical properties. Due to their high ductility, thermoplastic polymers can easily be formed into three-dimensional complex shapes, which is especially desirable for applications requiring excellent formability. Nevertheless, the mechanical strength and stiffness of thermoplastic polymers are generally lower than those of thermoset polymers.

Among all the thermoplastic polymers, PP has found a wide range of applications in various engineering sectors such as food packaging, automotive, textile, etc. [11]. In addition to low-cost characteristic, the prevalent use of PP polymer is also due to its unique characteristics such as excellent durability, ease of processing, good chemical resistance and decent mechanical performance. An added advantage of using PP as a polymer matrix for natural fiber-reinforced composites is that it provides lower fatigue sensitivity than epoxy-based composites [12]. Even though epoxy-based composites have greater fatigue strength inherited from their tensile strength, their fatigue strength degradation rate is higher than PP-based composites. Apart from the virtues mentioned above, the recyclability of PP-based composites is particularly promising as this feature may alleviate the environmental burden due to the alarming level of waste accumulation on the earth. The remarkable recyclability of PP-based composites allows such materials to be recycled without prominently deteriorating their mechanical properties after their service life. Ngaowthong et al. [13] reported that the tensile properties and thermal stability of sisal fiber-reinforced PP composites were retained after the repeated recycling process. Lila et al. [14] obtained similar findings in which recycled PP/natural fiber-based composites successfully preserved their inherent characteristics during recycling. Due to increasing interest in PP as a potential matrix to form composite materials, it has been increasingly used to merge with innumerable plant fibers to develop green composite materials.

Although PP is recyclable, it is not considered a biopolymer as it is partly biodegradable, implying that certain measure has to be taken to modify PP characteristic. PP may take several decades to degrade naturally in the environment. Difficulty in the biodegradation of PP is principally due to its hydrophobic nature and absence of functional groups as well as high molecular weight that inhibit the attachment of microorganisms on the PP surface for the biodegradation process. Accordingly, a dedicated method via the prodegradant technology is indispensable to turn partly degradable into fully biodegradable PP. In this technique, pro-oxidants such as cobalt stearate (Co^{2+}), magnesium stearate (Mn^{2+}) and iron stearate (Fe^{3+}) are added to PP to initiate thermo-oxidation and photo-oxidation [15]. These two processes eventually change the PP surface nature from hydrophobic to hydrophilic and speed up the breakdown of long-chain molecules. Once the hydrophilic functional groups are formed, PP will absorb moisture from the surroundings. The moisture content of PP will then attract microorganisms to attach to its surface, degrading PP biotically by consuming its carbon backbone fragment to form water, carbon dioxide and biomass which are the final products of the oxo-biodegradation process.

3 Combustion Mechanism of Composite Materials

Composite materials are one of the most extensively used materials for structural applications thanks to their high specific mechanical properties and lightweight characteristics. In fact, fiber-reinforced composites have become an indispensable material for various structural applications today [16]. These materials have been employed as fire-retardant materials since a few decades ago. The flame retardancy and flammability of composite materials are different from their individual components because these properties are also affected by factors such as fiber-matrix compatibility and structure of the composites in addition to fiber and matrix types. Typically, organic fibers such as glass, carbon and ceramic fibers are used as reinforcement in fire retardant applications as they exhibit remarkable fire resistance and do not burn when exposed to a high-temperature environment. Conversely, when exposed to fire, organic fibers tend to decompose into some flammable and non-flammable volatile substances, solid carbonaceous char and smoke. Following the contemporary trend of developing green and sustainable composite materials based on natural fibers, diving deeper into the fire retardancy of these emerging natural fiber-reinforced composites is essential. The design goal for natural fiber-reinforced composites should include a wide range of desirable properties, including fire retardancy. Notwithstanding the commonly cited virtues of natural fibers, a critical shortcoming of natural fiber-based composites for engineering applications is their poor fire resistance. Thus, searching for ways to improve their flame retardancy is essential before further expanding their possible applications.

In order to improve the flame retardancy of fiber-reinforced composites, an understanding of the combustion mechanism of composites is necessary. Figure 3 shows the process flow of the composite combustion and flame retardancy approach. The combustion mechanism of composite materials is not complicated as it only involves two major processes: pyrolysis and combustion. Heat, oxygen and combusting materials are the three essential components to initiate combustion. When heat from either flame or radiation is applied to materials, the temperature increases and eventually reaches pyrolysis temperature. At this pyrolysis temperature, char, flammable and non-flammable gases and liquid condensate will be produced. When heat is continuously applied to the materials after pyrolysis, the temperature will increase further and eventually attain the combustion temperature where heat, light and smoke are yielded. The continuation of the heat supply helps keep the combustion of composite materials. As visible in Fig. 3, enhancing the flame retardancy of composite materials according to each combustion stage is possible. The flame retardancy of composite materials is often unveiled by determining time-to-ignition (TTI), limiting oxygen index (LOI), peak heat release rate (PHRR), total heat release (THR) and flame spread rate. However, the parameters, including total smoke production (TSP), smoke density and smoke toxicity, also need to be identified because they could have an inferior effect on the health of living organisms.

4 Techniques of Improving Flame Retardancy of Composite Materials

Composites have been one of the well-known materials that come under the spotlight nowadays owing to their fascinating attributes. Nonetheless, composite materials possess poor flame retardancy that restricts the expansion of their applications [18]. Flame retardation is thus pivotal for the materials to burn at a slower rate and difficult to ignite. Flame retardancy of composite materials is vital to safeguard human lives and avoid anthropogenic pollution due to composite burning. To that end, several techniques to improve the fire resistance of composite materials have been developed. These techniques tend to disrupt and inhibit the combustion cycle during burning to attain a desirable level of flame retardancy. When looking into the fire retardancy of composite materials, its mechanisms can be grouped into three major categories: heat sink mechanism, condense phase mechanism and gas phase mechanism. All three flame-retardant mechanisms are summarized in Fig. 3. Figure 4 illustrates the flame-retardant mechanism of polymeric materials during combustion. As visible in Figs. 3 and 4, the flame-retardant mechanism is complicated, and its

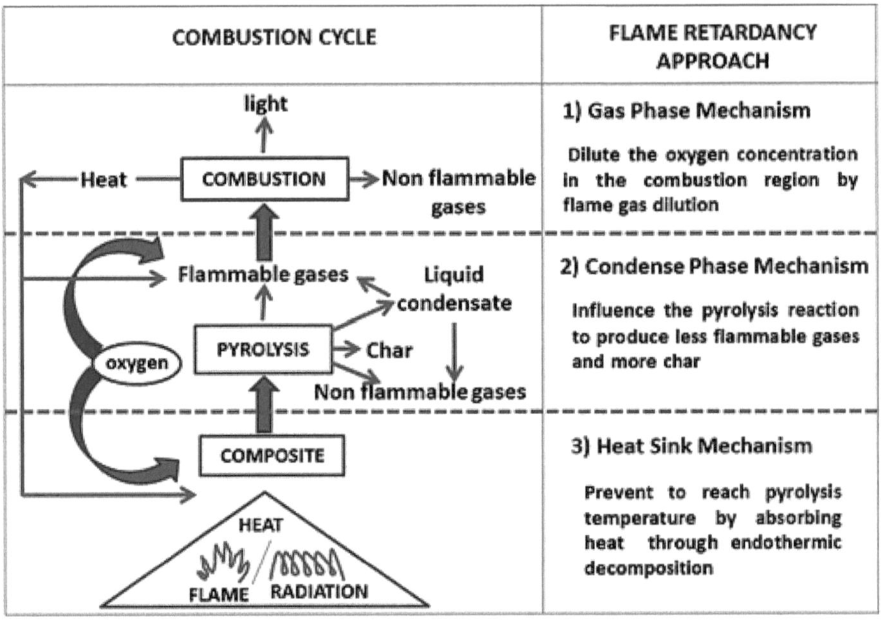

Fig. 3 Process flow of composite combustion and flame retardancy approach (Reprinted with permission obtained from [17])

basic principle is to disrupt the three basic components required for combustion. Each mechanism has its own role in improving the flame retardancy of composite materials. The first mechanism is represented by the heat sink mechanism in which the endothermic decomposition of flame-retardants cools the substrate materials to below combustible level and prevents the heat from reaching pyrolysis temperature when heat is applied to the composite materials. The condense phase mechanism refers to the decomposition of the flame retardants that affect the pyrolysis reaction by producing less flammable gas and more char as a protective layer to prevent the composites from further decomposing. The third mechanism is the gas phase mechanism in which certain flame retardants may decompose during burning to produce water and non-combustible gases, diluting the oxygen concentration and thus preventing the composites from further burning. When it comes to the flame retardancy of polymer composites, it can be enhanced either by improving the flame retardancy of each composite constituent, including matrix and reinforcing materials or by providing a coating as a protective flame-retardant layer around composite materials.

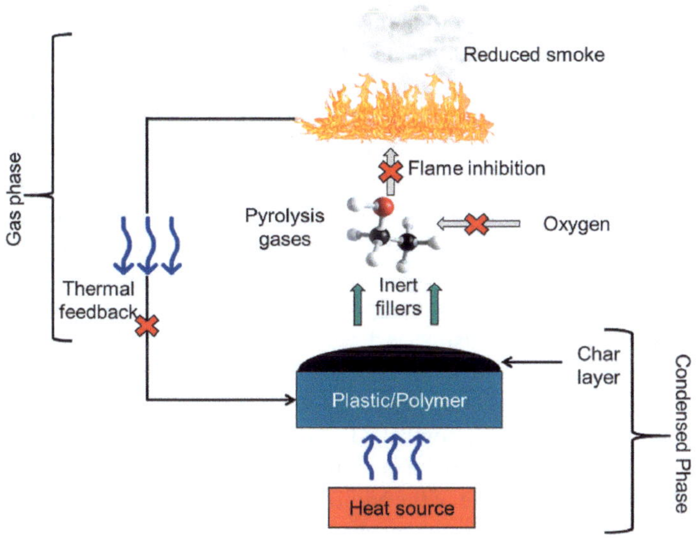

Fig. 4 Illustration of flame retardation mechanism of polymeric materials (Reprinted with permission obtained from [19])

4.1 Modification of Polymer Matrix

Given the excellent thermal resistance and non-flammable properties of conventional synthetic fibers, the flammability of the conventional composites primarily depends on the polymer matrices. Improving the flame retardancy of conventional composites can be realized by modification of the polymer matrix. It is worth mentioning that modification of polymer matrix is not limited to traditional composites only. It is also applicable for emerging composite materials based on natural fibers to enhance their flame retardancy. Several types of flame retardant can be incorporated into polymers to protect the materials from severe fire-related damage. Flame retardants uplift the flame retardancy of polymer matrix either by chemical or physical modification. In terms of chemical modification, the retardants help retard the radical formation, whereas the physical modification involves the mechanism, as shown in Fig. 3. As of today, there are over 175 compounds that have been identified to have flame-retardant properties [20]. Specifically, flame retardants can be grouped into four categories: inorganic, halogenated organic, nitrogen-containing and phosphorus-containing compounds. At present, halogenated organic flame retardants represent the most extensively used chemicals to prevent combustion and inhibit the spread of the fire after ignition. The rationales behind the widespread use of halogenated organic flame retardants are low cost, high miscibility and processability, high efficiency in capturing the free radicals and low reduction in mechanical properties of the materials. However, halogenated organic flame retardants pose a severe threat to the environment and human beings, and therefore, their usage has been limited to a certain extent. In this regard, looking into these flame retardants is pertinent because they show potential hazards to human health and the environment.

Flame retardants can serve as an additive (physically added into polymer matrix) or coreactants (covalently bound to polymer matrix) to suppress, inhibit and retard the combustion of composites, thereby improving their flame retardancy. Additives, such as metal hydroxides, are generally applied for thermoplastic polymers, while the coreactants are suitable for thermoset polymers [19]. Undoubtedly, one way to improve the fire performance of polymer matrices is by incorporating flame-retardant additives. This is considered the most straightforward yet effective method of reducing the flammability of the composites. An effective flame retardant typically encompasses all three mechanisms, as discussed previously, to inhibit the composite decomposition, production of flammable gases and flame spread although one of the mechanisms typically has dominant effect. However, the amount of additives in the composites should be well-controlled as it may adversely affect the mechanical properties of the materials due to the incompatibility of the polymer matrices and the additives. Additionally, the size of the additives, their dispersion and chemical nature are also the key factors affecting the mechanical properties of the composites. In fact, the mechanical properties of composites are typically compromised when the additives are added to improve their flame retardancy [21]. Apart from the mechanical properties, the viscosity of the polymers may increase after the addition of the additives, causing difficulty in processing the polymer melt to form composites [22]. Thus, it is essential to consider all these factors when developing flame-retardant composite materials, as they may significantly impact the functional properties of composites. In order to expand the future applications of composite materials, it is paramount to develop flame-retardant polymer matrix that allow composites to improve flame retardancy, thermal stability and mechanical properties simultaneously.

Despite significant improvement in the flame retardancy of composites, additive flame retardants always come with poor dispersion and compatibility with polymer matrices. These are the key factors rendering the deterioration in the mechanical properties of composites after incorporating additive-type flame retardants. In comparison with additive-type retardants, reactive-type flame retardants provide an additional advantage that is especially attractive to researchers and industry professionals. The main virtue of reactive-type flame retardants is that the negative impact on the mechanical properties of the composites is relatively low. As the reactive flame retardants are directly introduced into the molecular chain of the polymer via covalent bonding, there is no dispersion issue for composites incorporated with reactive flame retardants. In spite of the promising characteristics of reactive-type flame retardants, research studies still focus more on additive-type flame retardants to enhance the flame retardancy of polymeric materials [23]. This scenario is mainly ascribed to the low-cost characteristic of additive-type flame retardants and process simplicity to develop flame-retardant composite materials.

4.2 Developing Flame-Retardant Fiber via Surface Treatment

Due to the interfacial region in fiber-reinforced composites, the burning behavior of the composites becomes more complicated. It should be emphasized that the burning behavior of the polymer around the interfacial region is more intensive because of the wicking/candlewick effect. Because of the wicking effect that is driven by the interfacial tension, the burning rate in the polymer matrix and the interfacial region is apparently different. The polymer melts may spread and flow rapidly along the fiber surface toward the combustion zone at the interfacial area. This phenomenon will accelerate the material decomposition rate and further aggravate the combustion, especially for PP/natural fiber-based composites which are very sensitive to fire. For that reason, an effort has been made to chemically or physically modify the fiber surface to enhance interfacial flame retardation and inhibit the intense combustion of composites resulting from the wicking effect without deteriorating their mechanical properties. This approach is particularly beneficial for natural fiber-based composites since natural fibers are flammable, and the reduction in the flammability of natural fibers may eventually diminish the wicking effect and thus improve the flame retardancy of the materials.

Fiber surface treatments have been a common technique to strengthen the fiber-matrix adhesion in natural fiber-based composites so as to maximize the stress-transferring efficiency. However, it is also possible to boost the flame retardancy of the fibers through chemical surface treatment. Various well-established flame-retardant treatment techniques have been employed to treat the fiber surfaces, including covalent surface modification, plasma-aided surface modification, sol–gel treatment, layer-by-layer assembly and polyelectrolyte complexes. Each surface treatment technique has its own merits and demerits. Many flame-retardant fibers will have higher surface roughness and lower surface energy after the treatment. The surface modification allows the fiber can react with the polymer matrix to promote the formation and growth of char along the fiber surfaces during combustion, forming fibers with rougher and more inert surfaces.

4.3 The Flame-Retardant Coating on the Composite Surface

The flame-retardant coating is a versatile approach that prevents the composite materials from burning, slows the flame propagation, and lowers flame intensity and the amount of smoke produced. Instead of suppressing the fire spread and combustible gas transfer, this approach can even hinder the fire ignition on the composite surface when the materials are exposed to a high-temperature environment. Composite coating is undoubtedly an effective way to protect composite materials from fire damage. An effective flame-retardant coating provides a thermal-protective shield against fire ignition and spread, resulting in a lower rate of flame propagation and heat conduction through the material thickness as well as a longer time-to-ignition. One of the most significant advantages of flame-retardant coating is that it does not alter the fiber-matrix interaction, and therefore, it does not adversely affect the mechanical properties of the composites. The desirable coating should have certain characteristics, such as high durability, low thermal conductivity, non-flammability, inexpensive, lightweight, excellent wear resistance and good compatibility with the polymer surface [24, 25]. As of today, there are plenty of flame-retardant coatings, but none have all the attributes required for ideal coating.

Several techniques have been employed to apply coatings on the composite surface, including brushing, spraying, plasma, resin transfer molding, etc. Resin containing additive- or reactive-type flame retardant is required when applying coatings on the surfaces. These flame retardants are incorporated into coating formulation to hinder or suppress the combustion of the materials. It is worth emphasizing that the coating formulation can be tailored by selecting specific flame retardants to obtain desired properties for intended applications. Generally, flame-retardant coatings can be classified into two groups, which are intumescent and non-intumescent coatings. Intumescent coatings are typically thinner than non-intumescent coatings. Because of their outstanding flame-protective characteristics, intumescent coatings are regarded to be more promising than non-intumescent coatings. The intumescent coatings contain char-forming agents, ammonium polyphosphate-pentaerythritol-melamine, which aim to promote char formation as protective layers on the substrate surfaces. When intumescent coatings are exposed to heat or flame, they set off a chain reaction of physical and chemical reactions that expand the multicellular char layer and serve as a thermal barrier, effectively shielding the elements from a sharp rise in temperature and keeping the structures from collapsing in extreme flame situations. This intumesced char can grow up to 50 times thicker than the original coating thickness when exposed to a high-temperature environment, thus providing a robust shield protecting the composite from burning. Furthermore, Intumescent fire-retardant coatings have a fire-resistance limit of up to 3 h which is far greater than conventional thin and ultra-thin fire-retardant coatings [26]. Despite promising and excellent prospects of intumescent coatings, their durability is always a concern that needs to be carefully addressed.

5 Flammability of Cellulosic Fibers and Flame Retardancy of Biocomposites

A gradual transition from synthetic fiber-based to natural fiber-based composites has been witnessed over the past few years. Natural fibers, especially cellulosic fibers, are an essential component in biocomposites. They are considered the prior choice to substitute synthetic fibers in fiber-reinforced composites [27]. Incorporating these cellulosic fibers into composite materials aims to reduce the dependence on synthetic fibers, primarily because of environmental concerns [28]. However, natural fibers are highly flammable and have poor thermal stability. Therefore, it is necessary to delve deeper into the flammability of natural fibers before they can be applied in those applications where flammability is a concern. In comparison with protein fibers, cellulosic fibers have poor flame retardancy, making them vulnerable to combustion. Cellulose, which is a major component of cellulosic fibers, is thermally decomposed by a series of stages: (i) desorption of water content, (ii) cross-linking of cellulose chains with the evolution of water to form dehydrocellulose, (iii) Formation of char and volatiles owing to the decomposition of the dehydrocellulose, (iv) synthesis of levoglucosan, and (v) Formation of flammable and non-flammable char, volatiles, gases and tar [29, 30]. Once the cellulose of the fibers has thermally degraded, the mechanical strength of the fibers and their associated composites will collapse. This is due to the fact that cellulose is the primary chemical component that provides strength and rigidity to the fibers [31].

Cellulose commences to decompose at a temperature of 260–350 °C, whereas hemicellulose decomposes between 200 and 260 °C. Lignin has a wide range of degradation temperature, 160–400 °C, due to different bonding conditions [32]. However, different cellulosic fibers might have a subtle variation in the degradation temperature. This can be attributed to the different chemical compositions of the cellulosic fibers. When looking into the flammability and flame retardancy of natural fibers, the lignin content of the fibers has a decisive impact as it is the primary component that yields the char layer. Meanwhile, cellulose content could increase the flammability of the fibers. Hence, cellulosic fibers with low cellulose content and high lignin content generally have lower flammability. In addition to chemical composition, crystallinity level also profoundly affects the flammability of the fibers. A high level of crystallinity may yield a higher amount of levoglucosan, leading to increased flammability. Meanwhile, high crystallinity levels may increase the fire ignition temperature as more energy is required to break the crystalline structure. Undeniably, the degree of polymerization and microfibrillar angle could also affect the flammability of cellulosic fibers. High degree of polymerization and highly oriented microfibrils weaken the pyrolysis process, decreasing the flammability. From here, it can be identified that cellulosic fibers with low crystallinity, high degree of polymerization and highly oriented microfibrils have high potential to be utilized as reinforcement for flame-retardant composite materials. The hybridization concept that combines two different cellulosic fibers to obtain a desired level of flame retardancy is also a feasible approach.

A remarkable flame retardancy of biocomposites is particularly the main focus of the researchers as natural fibers and certain types of the polymer matrix can be ignited easily under the ignition temperatures with the supply of oxygen and heat source. Correspondingly, the number of research studies focusing on the flame retardancy of natural fiber-based composites is on an upward trend. Nonetheless, most of the research studies reported a decline in the mechanical properties after improving the flame retardancy of the bio-composites. Therefore, it is paramount to look for an effective technique that can improve flame retardancy without compromising the mechanical properties of biocomposites. Biocomposites is a term that refers to composite materials consisting of biofibers and biopolymers. Toward the sustainable development goal for biocomposites, the flame retardants incorporated in the biocomposites should be environmental friendly and non-toxic. More specifically, all constituents of biocomposites should not pose any potential hazards to the health of living organisms and the environment. Halogenated flame retardants have often been used to enhance the flame retardancy of composites, but their toxicity and corrosive gas emission during combustion have banned the usage of such flame retardants [33]. Alternatively, non-halogen flame retardants such as silicon compounds, minerals, phosphorus- and nitrogen-containing compounds and metallic hydroxide have been the main focus in recent years.

6 Eco-friendly Flame Retardants

Halogen-containing flame retardants have been well-known materials used to inhibit and suppress the flame ignition and combustion of polymeric materials. Halogenated flame retardants, such as chlorine- and bromine-based compounds, have long been the main focus of researchers and industry professionals because they are inexpensive, easily processed and only have a minor effect on the mechanical properties of composite materials. Additionally, they are able to effectively capture free radicals that are essential elements to sustain combustion. Even though halogenated flame retardants exhibit several attractive features, they present some critical demerits related to the environment and health of living organisms under fire conditions. These flame retardants emit toxic gases and tend to decompose into more noxious substances upon combustion. Due to environmental and health concerns, innumerable research studies have been initiated on non-halogenated and eco-friendly flame retardants. Eco-friendly flame retardants refer to those retardants which are halogen-free and do not lead to any harmful effects on Mother Earth and the health of living organisms. Metal hydroxides, intumescent flame retardants and silicon-containing flame retardants represent eco-friendly flame retardants that are in the spotlight today.

6.1 Metal Hydroxide

When mentioning halogen-free flame retardants, metal hydroxides, including magnesium hydroxide ($Mg(OH)_2$) and aluminum trihydrate ($Al(OH)_3$) are among the environmental-friendly inorganic flame retardants showing a high flame-retardant performance by decomposing endothermically to metal oxides and releasing a sufficient amount of water to dilute the basic components required for combustion and cool polymer matrix. Moreover, the endothermic decomposition of the retardant can help absorb heat, thereby lower the temperature of the polymers. It is worth mentioning that the metal oxides formed after the decomposition have high specific surface energy and remarkable absorption capacity. These attributes allow the metal oxides to cover up the surfaces of polymers, forming a protective char layer to inhibit smoke and the spread of fire [34]. The endothermic reaction of magnesium hydroxide and aluminum trihydrate can be represented by Eqs. (1) and (2), respectively.

$$Mg(OH)_2 \rightarrow MgO + H_2O \tag{1}$$

$$2Al(OH)_3 \rightarrow Al_2O_3 + 3H_2O \tag{2}$$

Magnesium hydroxide remains stable below the decomposition temperature of around 350 °C, thus allowing it to be used as flame retardant for most polymer matrices with a processing temperature far below the decomposition temperature of magnesium hydroxide. In comparison with magnesium hydroxide, the decomposition temperature of aluminum trihydrate is relatively low, in which aluminum trihydrate starts decomposing at around 220 °C. Hence, the processing temperature of composite materials should be below such temperature. In spite of the high efficiency of metal hydroxide in suppressing fire ignition and combustion, a significant amount of these retardants ($\geq$ 50%) is generally required to attain the acceptable flame retardancy of the composite materials. In this case, the physical and mechanical properties of the composite materials will be subjected to a significant drop due to the polarity difference and incompatibility between the retardants and the polymers. Zadeh et al. [35] studied the effect of magnesium hydroxide on the flame retardancy and mechanical properties of date palm fiber-reinforced composites based on a polymer blend comprising polypropylene/low-density polyethylene/high-density polyethylene. They reported that increasing the magnesium hydroxide content up to 40 wt% in the composites improved flame retardancy but deteriorated the mechanical strength. Xiao and Kibble [36] obtained similar findings in which adding magnesium hydroxide and aluminum trihydrate up to 75 wt% resulted in enhanced flame retardancy of styrene-ethylene/butylene-styrene block copolymer-based composites, but the mechanical strength of the materials diminished.

A combination of metal hydroxide with other flame retardants is a common approach to uplift the flame retardancy of composites to a greater extent. In this context, zinc borate is one of the flame retardants that has been broadly used as a synergistic agent to combine with metal hydroxide to further improve the flame retardancy of composite materials. Zinc borate in commerce has a chemical formula of $2ZnO:3B_2O_3:3.5H_2O$, and it is a stable substance which is insoluble in water. Therefore, zinc borate does not leach from the composite materials due to its excellent stability and insolubility in water. It may decompose endothermically at a temperature above 290 °C, releasing water and forming a glassy protective B_2O_3 char layer on the polymer surface that serves as a barrier that limits the heat and oxygen transfer from reaching the combustion zone. The water released after its decomposition may also dilute the combustible gases and suppress the formation of smoke. Research studies have reported that adding zinc borate could undeniably improve the flame retardancy of composite materials by enhancing the char formation and quality that provides better protection to the materials [37–39].

6.2 Intumescent Flame Retardants

Intumescent flame retardant system has been regarded as one of the most promising environmental-friendly approaches that can be used to enhance the flame retardancy of composite materials. It is a new generation of halogen-free retardants consisting of three major components: acid source, carbonizing agent and blowing agent. Ammonium polyphosphate (APP), melamine phosphate (MPP), and pyrophosphate (PyMP) are the three primary acid sources, whereas pentaerythrithol, phenol–formaldehyde resins, starch are examples of carbonizing agent of intumescent flame retardants. Meanwhile, melamine, guanidine, urea, chlorinated paraffin represent the blowing agent. The mechanism of intumescent flame retardants is that the intumescent additives in the composites decompose and form a carbonaceous char layer, enfolding polymer surfaces upon combustion. At high temperatures, the acid source decomposes, producing inorganic strong acid, which might increase the dehydration of the carbonizing agent, resulting in a char formation [40]. This carbonaceous char layer plays a significant role in blocking the diffusion of combustible volatile gases, oxygen and heat between the flame and the composites, thus inhibiting the combustion process [23]. Moreover, intumescent flame retardants can also result in a radical trapping effect, further inhibiting the spread of fire and enhancing flame retardancy and thermal stability of composites. Intumescent flame retardants have gained popularity due to their superior properties, including low toxicity, low smoke, low corrosion, high efficiency and no molten drops due to combustion. Today, intumescent flame retardants have been increasingly adopted to improve the flame retardancy of natural fiber-reinforced composites.

A few research studies reported a positive impact on the flame retardancy of composite materials through the encapsulation of intumescent flame retardants in the polymer matrix. Liu et al. [41] successfully improved the flame retardancy of thermoplastic polyurethane composites by adopting dimelamine pyrophosphate (DMPY) as an intumescent flame retardant. Even though the composites experienced a drop in the mechanical properties, it is still in an acceptable range. Figure 5 illustrates the combustion mechanism of the composites with and without intumescent flame retardant. As visible in Fig. 5, the char layer helps prevent the heat and oxygen from transferring from the combustion zone to the interior polymer or vice versa. Moreover, it has also been reported that intumescent flame retardant based on APP is tremendously efficient in escalating the fire retardant performance of natural fiber-reinforced composites [32]. Specifically, phosphorus ester is formed as a result of the phosphorylation of hydroxyl functional groups of natural fibers by phosphoric acid in APP. This process eventually catalyzes the dehydration and results in char and water formation.

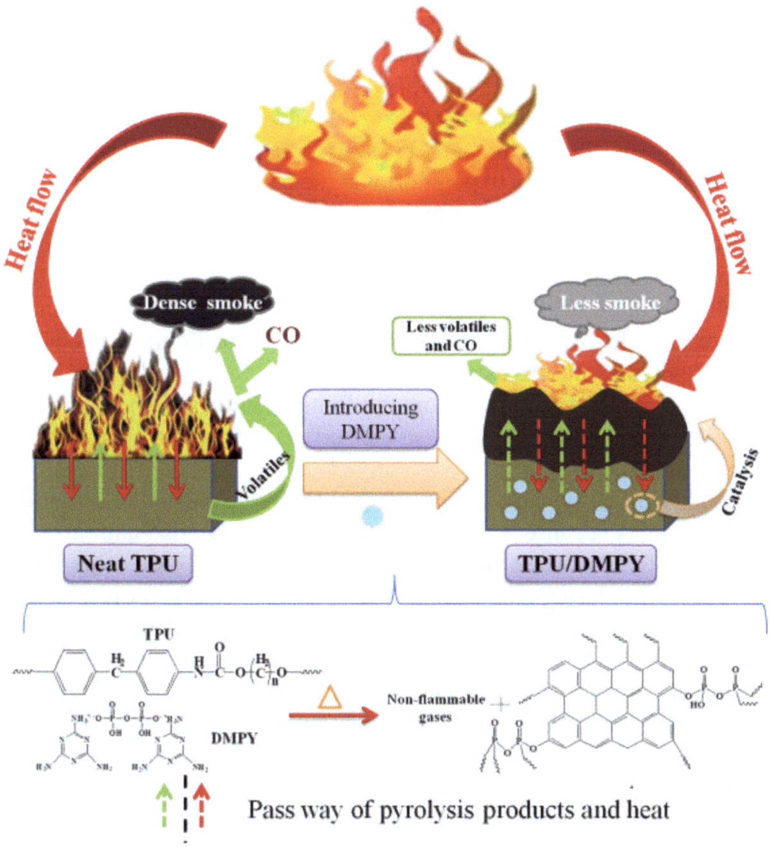

Fig. 5 Difference between the combustion mechanism of composites with and without intumescent flame retardant (Reprinted with permission obtained from [41])

6.3 *Silicon-Containing Flame Retardants*

Silicon-containing flame retardants can be divided into organic and inorganic silicon flame retardants. Both are eco-friendly additives since they present no harmful effects to the environment. Furthermore, they are very efficient in inhibiting and suppressing fire propagation and improving the flame retardancy of composite materials. Inorganic silicon-containing flame retardants exhibit several attractive properties such as low smoke, low burning value, non-toxic, slow fire spread, etc. [42]. Silica (SiO_2) and montmorillonite are the most commonly used inorganic silicon-containing flame retardants to develop flame-retardant nanocomposites. Silica comes in various varieties, including silica gel, fused silica and fumed silica. Small particle size, large specific surface area, and excellent thermal stability are the advantages offered by silica. Velmurugan et al. [43] revealed positive results when utilizing silica as a flame retardant for composite materials. The results showed significant improvement in the flame retardancy and mechanical properties of jute fiber composites. Montmorillonite is a natural clay mineral with a layer structure comprising two tetrahedral sheets and one octahedral central sheet. It can also be encapsulated in nanocomposite materials to enhance their flame retardancy and thermal stability while augmenting mechanical properties. However, altering the functional group of the polymer matrix is necessary to convert its non-polar nature to polar prior to the incorporation of montmorillonite. This is due to the fact that weak compatibility between the non-polar polymer matrix and the polar surface of montmorillonite may weaken the mechanical properties of the composite materials. Aside from the weak compatibility, hydrophilic surface of montmorillonite often leads to agglomeration of nanoclay, causing the composites to have poor particle dispersion. Alternatively, the compatibility between montmorillonite and polymer matrix can be improved by developing organic montmorillonite. The hydrophilic surface of montmorillonite can be modified by introducing organic cation as a modifier to diminish the surface energy and change hydrophilic to hydrophobic surface.

On the other hand, organic silicon is often the prior choice as co-additives for composite materials because it is highly efficient, a good smoke inhibitor, and environmental friendly. As flame retardant, organic silicon acts by migrating toward the material surface during combustion. The protective char layer is then formed to block oxygen transfer into the composites and prevent the volatile substances from escaping. Organic silicon can be in additive-type or reactive-type, inferring that organic silicon can be directly added to the polymer matrix without forming chemical bonds or copolymerizing with flame-retardant units and grafting or crosslinking with macromolecules [44, 45]. However, more research studies focused on additive-type organic silicon flame retardants because of their simplicity in the composite preparation process. Organic silicon has indeed shown a very high performance as a flame retardant that can be used to solve the health and safety concerns raised by halogenated flame retardants.

7 Conclusion

Biocomposites consisting of natural fibers are emerging as a viable option to supersede non-environmental friendly synthetic fiber-reinforced composites. Nevertheless, natural fibers, especially cellulosic fibers, possess weak thermal stability and are susceptible to fire and heat. Thus, it is pivotal to dive deeper into the flammability of cellulosic fibers to gain a better understanding of the topic. Flame retardancy is the key specification for many engineering sectors, such as building, automotive and electrical equipment. To that end, several approaches have arisen, aiming to uplift the flame retardancy and fire resistance of the composite materials. Even though innumerable flame retardants are available for composite materials, some are considered non-environmental friendly or even toxic, particularly halogenated flame retardants, which might endanger the health of living organisms and result in devastating damage to the environment. Therefore, when selecting flame retardants for composite materials, priority should always be given to bio-based flame retardants. As of today, a number of non-halogenated flame retardants have been explored and identified to be effective in enhancing the flame retardancy of composites. These green flame retardants include metal hydroxide, intumescent flame retardant and silicon-containing nanoparticles. Although there are several options to develop sustainable flame-retardant composites, these flame retardants still pose a challenge that needs to be carefully addressed. For instance, adding metal hydroxide may improve the flame retardancy but deteriorate the mechanical properties of the composites. By eliminating the doubts or concerns about these green flame retardants, it is anticipated that flame-retardant biocomposites with remarkable functional properties can be developed.

References

1. Feng NL, Malingam SD, Subramaniam K et al (2019) The influence of fibre stacking configurations on the indentation behaviour of pineapple leaf/glass fibre reinforced hybrid composites. Def S T Tech Bull 12:113–123
2. Feng NL, Malingam SD, Subramaniam K et al (2020) The investigation of the tensile and quasi-static indentation properties of pineapple leaf/kevlar fibre reinforced hybrid composites. Def S T Tech Bull 13:117–129
3. Ng LF, Yahya MY, Leong HY et al (2023) Evaluation of physical and mechanical properties of pineapple leaf and kenaf fabrics as potential reinforcement in bio-composites. Biomass Convers Biorefinery. https://doi.org/10.1007/s13399-023-04525-x
4. Shen J, Liang J, Lin X et al (2022) The flame-retardant mechanisms and preparation of polymer composites and their potential application in construction engineering. Polymers (Basel) 14:82. https://doi.org/10.3390/polym14010082
5. Tan S, Moinuddin K, Joseph P (2023) The ignition frequency of structural fires in Australia from 2012 to 2019. Fire 6:35. https://doi.org/10.3390/fire6010035
6. Szolnoki B, Bocz K, Sóti PL et al (2015) Development of natural fibre reinforced flame retarded epoxy resin composites. Polym Degrad Stab 119:68–76. https://doi.org/10.1016/j.polymdegradstab.2015.04.028
7. Manisalidis I, Stavropoulou E, Stavropoulos A, Bezirtzoglou E (2020) Environmental and health impacts of air pollution: a review. Front Public Heal 8:14. https://doi.org/10.3389/fpubh.2020.00014
8. Shaddick G, Thomas ML, Mudu P et al (2020) Half the world's population are exposed to increasing air pollution. NPJ Clim Atmos Sci 3:23. https://doi.org/10.1038/s41612-020-0124-2
9. Prabhakar MN, Naga Kumar C, Dong Woo L, Jung-IL S (2022) Hybrid approach to improve the flame-retardant and thermal properties of sustainable biocomposites used in outdoor engineering applications. Compos Part A Appl Sci Manuf 152:106674. https://doi.org/10.1016/j.compositesa.2021.106674
10. Lee HV, Hamid SBA, Zain SK (2014) Conversion of lignocellulosic biomass to nanocellulose: structure and chemical process. Sci World J 2014:631013. https://doi.org/10.1155/2014/631013
11. Gao Y, Wang J, Liu X et al (2023) Fabrication of durable and non-leaching triclosan-based antibacterial polypropylene. Eur Polym J 187:111892. https://doi.org/10.1016/j.eurpolymj.2023.111892
12. Sivakumar D, Ng LF, Lau SM, Lim KT (2018) Fatigue life behaviour of glass/kenaf woven-ply polymer hybrid biocomposites. J Polym Environ 26:499–507. https://doi.org/10.1007/s10924-017-0970-0

13. Ngaowthong C, Borůvka M, Běhálek L et al (2019) Recycling of sisal fiber reinforced polypropylene and polylactic acid composites: thermo-mechanical properties, morphology, and water absorption behavior. Waste Manag 97:71–81. https://doi.org/10.1016/j.wasman.2019.07.038
14. Lila MK, Singhal A, Banwait SS, Singh I (2018) A recyclability study of bagasse fiber reinforced polypropylene composites. Polym Degrad Stab 152:272–279. https://doi.org/10.1016/j.polymdegradstab.2018.05.001
15. Hayoune F, Chelouche S, Trache D et al (2020) Thermal decomposition kinetics and lifetime prediction of a PP/PLA blend supplemented with iron stearate during artificial aging. Thermochim Acta 690:178700. https://doi.org/10.1016/j.tca.2020.178700
16. Ng LF, Yahya MY, Muthukumar C (2022) Mechanical characterization and water absorption behaviors of pineapple leaf/glass fiber-reinforced polypropylene hybrid composites. Polym Compos 43:203–214. https://doi.org/10.1002/pc.26367
17. Bar M, Alagirusamy R, Das A (2015) Flame retardant polymer composites. Fibers Polym 16:705–717. https://doi.org/10.1007/s12221-015-0705-6
18. Gunes OC, Gomek R, Tamar A et al (2018) Comparative study on flame retardancy, thermal, and mechanical properties of glass fiber reinforced polyester composites with ammonium polyphosphate, expandable graphite, and aluminum tri-hydroxide. Arab J Sci Eng 43:6211–6218. https://doi.org/10.1007/s13369-018-3397-6
19. Mensah RA, Shanmugam V, Narayanan S et al (2022) A review of sustainable and environment-friendly flame retardants used in plastics. Polym Test 108:107511. https://doi.org/10.1016/j.polymertesting.2022.107511
20. Yogui GT, Sericano JL (2009) Polybrominated diphenyl ether flame retardants in the U.S. marine environment: a review. Environ Int 35:655–666. https://doi.org/10.1016/j.envint.2008.11.001
21. Liu T, Peng J, Liu J et al (2021) Fully recyclable, flame-retardant and high-performance carbon fiber composites based on vanillin-terminated cyclophosphazene polyimine thermosets. Compos Part B Eng 224:109188. https://doi.org/10.1016/j.compositesb.2021.109188
22. Shi XH, Li XL, Li YM et al (2022) Flame-retardant strategy and mechanism of fiber reinforced polymeric composite: a review. Compos Part B Eng 233:109663. https://doi.org/10.1016/j.compositesb.2022.109663
23. Li Y, Qi L, Liu Y et al (2022) Recent advances in halogen-free flame retardants for polyolefin cable sheath materials. Polymers (Basel) 14:2876. https://doi.org/10.3390/polym14142876
24. Abdeen DH, El Hachach M, Koc M, Atieh MA (2019) A review on the corrosion behaviour of nanocoatings on metallic substrates. Materials (Basel) 12:210. https://doi.org/10.3390/ma12020210
25. Awang M, Khalili AA, Pedapati SR (2020) A review: thin protective coating for wear protection in high-temperature application. Metals (Basel) 10:42. https://doi.org/10.3390/met10010042
26. Zhou G, Li S, Zhang X et al (2022) Synthesis and properties of a fire-retardant coating based on intercalated expandable graphite-modified cellulose for steel structures. J Build Eng 51:104270. https://doi.org/10.1016/j.jobe.2022.104270
27. Feng NL, Malingam SD, Ishak NM, Subramaniam K (2021) Novel sandwich structure of composite-metal laminates based on cellulosic woven pineapple leaf fibre. J Sandw Struct Mater 23:3450–3465. https://doi.org/10.1177/1099636220931479
28. Dhar Malingam S, Feng NL, Kamarolzaman AA et al (2021) Mechanical characterization of woven kenaf fabric as reinforcement for composite materials. J Nat Fibers 18:653–663. https://doi.org/10.1080/15440478.2019.1642827
29. Horrocks AR (1983) An introduction to the burning behaviour of cellulosic fibres. J Soc Dye Colour 99:191–197. https://doi.org/10.1111/j.1478-4408.1983.tb03686.x
30. Das O, Sarmah AK (2015) Mechanism of waste biomass pyrolysis: effect of physical and chemical pre-treatments. Sci Total Environ 537:323–334. https://doi.org/10.1016/j.scitotenv.2015.07.076
31. Feng NL, Malingam SD, Ping CW (2021) Mechanical characterization of kenaf/PALF reinforced composite-metal laminates: effects of hybridization and weaving architectures. J Reinf Plast Compos 40:193–205. https://doi.org/10.1177/0731684420956719

32. Kim NK, Dutta S, Bhattacharyya D (2018) A review of flammability of natural fibre reinforced polymeric composites. Compos Sci Technol 162:64–78. https://doi.org/10.1016/j.compscitech.2018.04.016
33. Schirp A, Su S (2016) Effectiveness of pre-treated wood particles and halogen-free flame retardants used in wood-plastic composites. Polym Degrad Stab 126:81–92. https://doi.org/10.1016/j.polymdegradstab.2016.01.016
34. Feng C, Liang M, Jiang J et al (2016) Flame retardancy and thermal degradation behavior of efficient intumescent flame retardant LDPE composite containing 4A zeotile. J Anal Appl Pyrolysis 118:9–19. https://doi.org/10.1016/j.jaap.2015.12.008
35. Zadeh KM, Ponnamma D, Al Ali Al-Maadeed M (2017) Date palm fibre filled recycled ternary polymer blend composites with enhanced flame retardancy. Polym Test 61:341–348. https://doi.org/10.1016/j.polymertesting.2017.05.006
36. Xiao WD, Kibble KA (2008) Comparison of aluminium hydroxide and magnesium hydroxide as flame retardants in SEBS-based composites. Polym Polym Compos 16:415–422. https://doi.org/10.1177/096739110801600702
37. Cheng X, Wu J, Yao C, Yang G (2019) Flame-retardant mechanism of zinc borate and magnesium hydroxide in aluminum hypophosphite-based combination for TPE-S composites. J Fire Sci 37:273–300. https://doi.org/10.1177/0734904119851270
38. Sabet M, Hassan A, Ratnam CT (2013) Effect of zinc borate on flammability/thermal properties of ethylene vinyl acetate filled with metal hydroxides. J Reinf Plast Compos 32:1122–1128. https://doi.org/10.1177/0731684413494942
39. Orhan T, Isitman NA, Hacaloglu J, Kaynak C (2011) Thermal degradation mechanisms of aluminium phosphinate, melamine polyphosphate and zinc borate in poly(methyl methacrylate). Polym Degrad Stab 96:1780–1787. https://doi.org/10.1016/j.polymdegradstab.2011.07.019
40. Yiga VA, Lubwama M, Pagel S et al (2021) Flame retardancy and thermal stability of agricultural residue fiber-reinforced polylactic acid: a review. Polym Compos 42:15–44. https://doi.org/10.1002/pc.25835
41. Liu L, Xu Y, He Y et al (2019) An effective mono-component intumescent flame retardant for the enhancement of water resistance and fire safety of thermoplastic polyurethane composites. Polym Degrad Stab 167:146–156. https://doi.org/10.1016/j.polymdegradstab.2019.07.006
42. Petersen H, Jakubowicz I, Enebro J, Yarahmadi N (2015) Organic modification of montmorillonite for application in plasticized PVC nanocomposites. Appl Clay Sci 107:78–84. https://doi.org/10.1016/j.clay.2015.01.006
43. Canıtez H, Yüksel Yılmaz AN, Çelik Bedeloğlu A (2023) The effect of silicon dioxide and zinc borate on the flame retardancy, thermal and mechanical properties of jute/epoxy hybrid composite. J Nat Fibers 20:4875–4888. https://doi.org/10.1080/15440478.2022.2131683
44. Liu L, Zhang W, Yang R (2020) Flame retardant epoxy composites with epoxy-containing polyhedral oligomeric silsesquioxanes. Polym Adv Technol 31:2058–2074. https://doi.org/10.1002/pat.4929
45. Nguyen VH, Manh VuC, Thi HV (2021) MWCNT grafted with POSS-based novel flame retardant-filled epoxy resin nanocomposite: fabrication, mechanical properties, and flammability. Compos Interfaces 28:843–858. https://doi.org/10.1080/09276440.2020.1816068

Flammability Performance of the HDPE and LDPE Biocomposites and Bionanocomposites

Konstantinos G. Gatos

Abstract HDPE and LDPE composites are considered in various applications in everyday life. The need for a sustainable future suggests a low environmental impact of the composites in use throughout their life cycle. In this view, the reduction in carbon dioxide emissions can be approached by exploiting biocomposites and bionanocomposites. Consequently, natural fibers play a key role in such a product design. Current standards in critical sectors such as construction and transportation require, among others, high levels of fire resistance. Therefore, polyethylene biocomposites should also successfully meet the relevant stringent criteria. In this book chapter, several strategies which are advancing the flammability performance of the HDPE and LDPE biocomposites and bionanocomposites are discussed.

Keywords HDPE · LDPE · Flame retardants · Biocomposites · Bionanocomposites · Fire resistance · Wood fibers · Non-wood natural fibers · Green composites · Halogen-free flame retardants

1 Introduction to Flammability Performance of HDPE and LDPE Biocomposites and Bionanocomposites

Within the group of thermoplastics, polyethylene holds a major position finding a plethora of applications [1]. Due to its versatility in structure, there are various forms of polyethylene; however, the high-density polyethylene (HDPE) and the low-density polyethylene (LDPE) belong to its main representatives [2]. The quest for a sustainable future has directed scientists and engineers to find ways to reduce the environmental impact of polymeric products and components of synthetic origin [3]. The petroleum-derived composite materials although yield a high performance, raise strong concerns regarding their overall carbon footprint.

K. G. Gatos (✉)
TERNA MAG S.A., 11526 Athens, Greece

C. Muthukumar et al. (eds.), *Flammability Performance of Biocomposites and Bionanocomposites*, Composites Science and Technology,
https://doi.org/10.1007/978-981-95-2389-4_7

HDPE and LDPE serve in many cases as a matrix in polymeric composites; thus, solutions to lowering their carbon footprint are directed to their biocomposite versions. In this view, the matrix can be bio-based wherein the starting monomers are originated by biological resources such as sugar cane and starch crops [4]. However, bio-based polyethylene is not readily biodegradable with unclear end-of-life management indicating that the use of renewable resources alone does not imply substantial environmental benefits [5]. Besides, natural fibers are widely accepted in the community as sustainable materials with prominent renewability and biodegradability. Nevertheless, there are drawbacks in using natural fibers in the polymer matrices. For example, it has been presented that they yield reduced impact strength and creep behavior compared to their synthetic versions [6]. During the previous decades, significant efforts have been taken place to improve the interfacial adhesion of the natural fibers with the matrix, as well as, to reduce the overall moisture absorption [7]. On the one hand, the initial chemical treatment serves in surface cleaning and induces hydrophobicity to the natural fibers [8], and on the other hand, any additional surface modification is advancing further the bonding of the fiber with the matrix [9].

Although it was evident that natural fibers were readily flammable compared to their synthetic ones, it has not been raised any great concern in practice since their applications were governing sectors such as packaging, outdoor constructions, sports, and leisure products. By introducing biocomposites and bionanocomposites in areas like transportation and aviation, there was a need for improving their flame-retardant performance, too [10]. A typical fire scenario involves the stages of its ignition, growth, full development, and final decay [11]. The mechanism involved in the combustion of HDPE and LDPE biocomposites relates to that of a typical polymer combustion. Specifically, as the temperature increases inducing fire ignition, heat is released stimulating decomposition of the biocomposite in presence of air, wherein the flammable volatile species assist the fire evolvement. To extinguish the fire, the sequence of heat, oxygen, and fuel (reported also as the fire triangle or the combustion triangle) should be disrupted. In general, strategies for combating fire in the case of the HDPE and the LDPE biocomposites are combining the tactics that have been presented for improving the flammability performance of the matrix [12] and that of the natural fiber [13]. The analysis in the current study for the flammability performance of the HDPE and the LDPE biocomposites and bionanocomposites is focused on the two major categories of natural fibers, namely the wood and the non-wood (plant) natural fibers. The first are available from pulp mills with rather uniform performance, showing processability relevant to short fibers and do not consume food corps, while the latter present high cellulose content and crystallinity, providing processability relevant to long fibers and adequate mechanical properties [14]. The categories of animal [15] and mineral natural fibers [16] are excluded from current analysis.

2 Flammability Performance of Wood Fiber/HDPE and LDPE Polymer Biocomposites

The wood fiber/HDPE and LDPE composites represent an important class of biocomposites finding plenty of applications in markets such as constructions, civil, marine, and automotive ones [17]. The low melting temperature of the matrix, which is below the degradation temperature of the wood fibers, facilitates said wood plastic composite (WPC) fabrication [18].

Stark et al. [19] compounded an HDPE of a melt flow index of 5 g/10 min with a 40-mesh wood flour (WF) in a twin-screw co-rotating extruder at a temperature of about 200 °C. To improve mixing, a lubricant blend of complex modified fatty acid ester at 5 wt.% was used. The flammability performance of the compound was improved when 60 wt.% of WF was added to the mixture. More specifically, the limited oxygen index (LOI) of the plain HDPE was increased from 19.3 to 20.2 for the respective biocomposite. At the same time, the peak heat release rate (HRR) and the total heat release (THR) of the biocomposite were significantly diminished compared to those of the matrix. On the downside, the ignition time (IT) of the HDPE was reduced for the respective WPC from 82.2 s to 24.9 s, respectively.

To improve the flammability performance of said biocomposite, various flame-retardant additives were premixed with HDPE prior to WPC extrusion. These additives relate to decabromodiphenyl oxide (BR), antimony trioxide (AT), magnesium hydroxide (MH), zinc borate (ZB), melamine phosphate (MP), and ammonium polyphosphate (AP). Figure 1 depicts the HRR curves of said additives at 10 wt.% compounded with 50 wt.% of WF in the relevant WPC versus the WPC with 60 wt.% WF (WF-60). It appears that all additives can yield a lower HRR curve compared to that of WF-60.

For the best overall fire performance, it has been found that the addition of 10 wt.% of ammonium polyphosphate (AP) increased LOI value to around 27, while induced the lowest HRR and mass loss rate (MLR) values among the examined additives, even though its ignition time was further reduced. However, the flexural modulus of elasticity (MOE) of this compound presented the lowest values highlighting the plasticizing side effect of AP. Another flame-retardant additive that worked quite well was the synthetic magnesium hydroxide (MH). It was found that even at 10 wt.% of MH addition, there was a moderate improvement in LOI, HRR, THR, and MLR values, while the mechanical properties of the WPC were not adversely affected. Moreover, MH yielded the highest improvement in the observed ignition time. In general, MH is expected to effectively deploy its flame-retardant action when it is mixed at a much higher percentage in the recipe [20].

The incorporation of the flame-retardant (FR) additive can be performed also in a two-stage processing technique. More specifically, the FR substance can be added at the pretreatment of the wood fibers as well as during the melt processing of the latter. In the example of ammonium phosphate, Schrimp et al. [21] treated at first place thermomechanical pulp from spruce wood chips with an aqueous solution of a short chain ammonium polyphosphate (APP) at 45 wt.% solid content. This initial

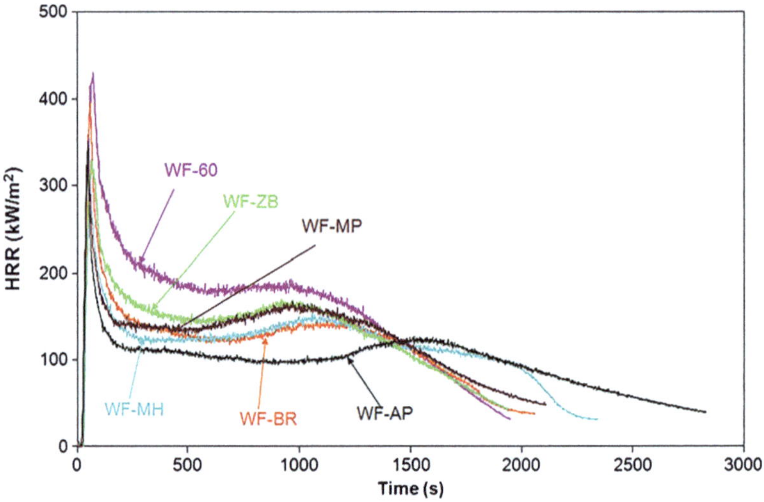

Fig. 1 Representative HRR curves for 60% wood flour-polyethylene composites and for 50% wood flour-polyethylene composites with 10% fire retardant "Reprinted from Stark et al. [19], with permission from Elsevier"

processing step was conducted continuously over several days, and the final APP content in the dried fibers, after their cutting and sieving at 2.6 mm, was calculated at maximum 20 wt.%. Following, the pulp at 50 wt.% was compounded with an HDPE of a melt flow index 2.1 g/10 min in an internal mixer. For improving interfacial adhesion, a maleic anhydrite grafted LLDPE (with 1.5 wt.% maleic content) was in action at 2 wt.%. The processability was favored using an oxidized HDPE wax at 1 wt.%. For advancing flame retardancy, as a second step, APP in powder form at 12 wt.% was mixed directly during kneading (replacing part of the matrix material). Finally, the batches were milled into granules to produce samples for LOI testing through injection molding. It was found that the LOI value was increased from 19.2 for the untreated WPC to 29.9 for the two-stage treated WPC. For comparison purposes, the sample with the untreated fibers, however, with the conventional addition of 12 wt.% APP during melt mixing yielded 26.1 as a LOI value. Nevertheless, a moderate swelling along with an increase in water uptake and decrease in tensile strength was also apparent in case of the two-stage technique.

García et al. [22] presented a recipe with self-extinguished behavior under EN ISO 11925-2 test in the example of HDPE with a melt flow rate of 23 g/10 min along with 45.5 wt.% wood fibers of 300–500 μm in length. These components were compounded in a co-rotating double screw extruder at 180 °C. During extrusion, a maleic anhydrite polyethylene at 2.7 wt.% was used for improving the interfacial bonding of the matrix with the fiber. For enhancing processing, 7.3 wt.% of micro-talc was in action. The testing samples were produced by performing injection molding at 195 °C. It was presented that the biocomposite with 9.1 wt.% of ammonium polyphosphate or encapsulated ammonium polyphosphate maintained its integrity during the fire testing and the fire was effectively extinguished when the flame was removed.

The positive effect of ammonium polyphosphate on improving the flame resistance of the biocomposite has been presented in case of LDPE, too. Altun et al. [23] compounded LDPE of a melt flow index of 2.5 g/10 min with pine wood flour of a sizing under 20-mesh. The mixing took place in a counter rotating twin-screw extruder at 190 °C, followed by compression molding of the extrudate at the same temperature. The initial biocomposite at a wood-flour percentage of 30 wt.% yielded a LOI value of 17.5. At the same time, the UL94 testing of a 3.2 mm sample thickness presented a full burning up to the clamp. Following, by substituting 30 wt.% of the LDPE with AP had increased the LOI value at 24.2, without however improving the most demanding UL94 test result. Besides, an enhancement in overall fire properties was reported when red phosphorous (RP) was mixed mainly due to its gas phase activity. Adding 3 wt.% of RP and deducting the same amount of AP resulted in a LOI value increase up to 26.3 achieving in parallel a UL94 rating enhancement of V0. Nonetheless, the initial deterioration of the tensile properties of the biocomposite due to AP addition was not altered further by the incorporation of RP.

3 Flammability Performance of Non-wood Natural Fiber/ HDPE and LDPE Polymer Biocomposites

The non-wood natural fibers present a fast growth requiring some months to a few years as compared to the softwood and the hardwood, which need about twenty and thirty years to reach maturity, respectively [24]. Apart from the conventional plant fibers such as jute, hemp, flax, bamboo, sisal, ramie, coir, kenaf, there are also the agricultural based fiber residues such as rice husk, wheat straw, bagasse, corn stalk, coconut stem, which may also find applications in relevant HDPE and LDPE biocomposites.

The selection of the most suitable non-wood fiber for each application is based on mechanical, physical, chemical, as well as economical evaluation criteria [25]. The flammability performance of the non-wood natural fibers has been shown to present some differences among those of bast and leaf origin, wherein the former is presenting lower heat release rate than the latter under cone calorimetry testing [26]. Besides, processing techniques such as extrusion, injection molding, or hot-pressing affect the flame performance, too [27]. Nonetheless, in practice the fire properties of the HPDE and LDPE biocomposite require further improvement in most cases.

Wang et al. [28] mixed HDPE of a melt flow index of 18 g/10 min with hemp fibers having an average diameter of 22 μm. The compounding of the HDPE with hemp fiber sections of 3–6 mm length was carried out in an internal mixer at 180 °C followed by hot-pressing at 180 °C to produce the plates for cutting the samples. The addition of 15 wt.% hemp fibers alone not only did not improve the fire properties of the biocomposite but also reduced the time to ignition compared to the neat matrix

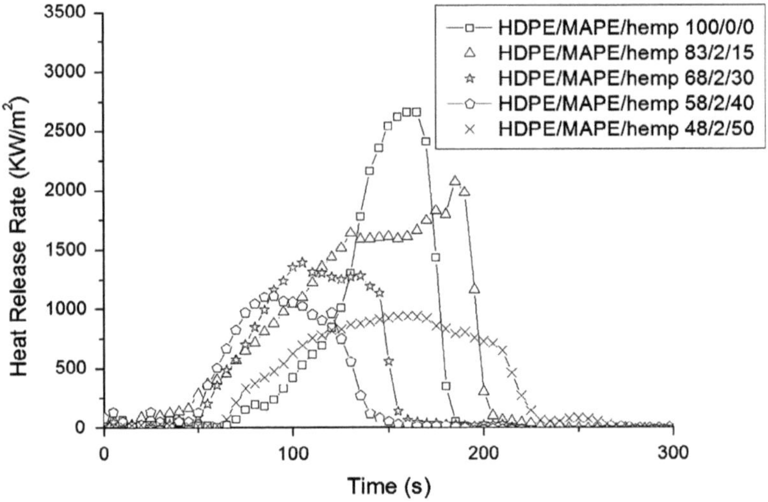

Fig. 2 Effect of hemp content on the heat release rate of HDPE/MAPE/hemp composites "Reprinted from Wang et al. [28], with permission from Elsevier"

as this was evidenced under cone calorimetry. The situation was ameliorated when 2 wt.% of maleic anhydrite grafted polyethylene (MAPE) was compounded with the HDPE prior to the addition of the 15 wt.% hemp fibers in the internal mixer. The peak heat release rate was reduced by 25%; however, the time to ignition was not increased. The positive effect of the maleic anhydrite compatibilizer was further investigated at higher fiber fractions. As shown in Fig. 2, increasing the addition of the hemp fibers to 30 wt.%, 40 wt.%, and 50 wt.% reduced the peak of the heat release rate of the biocomposite by 48%, 58%, and 65%, respectively.

The positive effect in flame retardancy of the high filler loading parameter has been revealed clearly in a rice husk biocomposite. Zhao et al. [29] mixed HDPE of a melt flow index value of 20 g/10 min with maleic anhydrite linear low-density polyethylene in a twin-screw extruder at 175 °C along with rice husk particles of 60 mesh size. The samples for fire testing were finally produced by compression molding at 170 °C. The addition of 70 wt.% rice husk (RH-70) required an increased amount of compatibilizer up to 10 wt.% for favoring the filler dispersion and the filler-matrix adhesion. The total heat evolved under cone calorimetry was reduced from 88.6 MJ/m^2 for the neat matrix to 68.4 MJ/m^2 for the biocomposite. At the same time, the fire growth rate index was diminished from 2.550 kW/m^2 s to 1.135 kW/m^2 s, respectively. Besides, the time to ignition was reduced from 76 s for the plain HDPE to 33 s for the 70 wt.% biocomposite in consistency with prior art due to the lower thermal stability of the natural filler compared to the HDPE. Evidence for the enhanced performance of RH-70 is depicted in Fig. 3 wherein the role of the inherent silica that rice husk contains apart from cellulose, hemicellulose, and lignin was responsible for the creation of a silica-ash layer after combustion. The thickness of the char layer in case of RH-70 was more efficient compared to the RH-20 or the

Fig. 3 Digital optical photographs for the combusted residues of HDPE/RH ecocomposites after cone calorimetry testing: **a** RH-20, **b** RH-50, **c** and **d** RH-70 "Reprinted from Zhao et al. [29], with permission from Elsevier"

RH-50 compounds acting as a thermal shield, which restricted also effectively the access of oxygen to the unburned compound.

In another example of exploiting agricultural waste, Zhang et al. [30] mixed HDPE with biochar, which was produced by rice husk burned at 600 °C under nitrogen atmosphere. The size of the biochar after sieving was below 150 μm and the matrix possessed a melt flow index value of 0.95 g/10 min. Biocomposites with biochar content up to 70 wt.% were produced in a twin-screw extruder at 180 °C, while the samples for testing were prepared after the injection molding of the batches at the same temperature. The best flame-retardant performance of the biocomposites was delivered by the 70 wt.% biochar-filled compound, increasing the limited oxygen index value to 25.1 from the value of 18.2 that was measured for the neat matrix. However, the higher overall mechanical performance was yielded by the 50 wt.% biochar-filled compound. This latter sample presented a 22.4 limited oxygen index value. The subsequent property compromise should be considered for relevant applications bearing in mind that biochar is originating abundantly by agricultural waste after pyrolysis and it is an economic way of producing biocomposites without the absolute need of compatibilizers.

4 Flammability Performance of Hybrid Filler/HDPE and LDPE Polymer Biocomposites and Bionanocomposites

The combination of various types of fillers either of synthetic or of natural origin has already been in action for improving the properties of the biocomposites [31]. This concept of hybridization has accomplished a great impact on the fire properties of the respective compounds [32]. The inherent drawback of natural fiber components and specifically of the lignin, the hemicellulose, and the cellulose to present an early decomposition at the temperature ranges of 160–400 °C, 200–260 °C, and 260–350 °C, respectively, can be compensated with the appropriate combination of other fillers, which possess a better thermal stability [33]. The range of fillers that are potentially available for wood and non-wood biocomposites and bionanocomposites for such purposes is rather wide [34].

Regarding nanoparticles, Pan et al. [35] extruded at 140 °C an HDPE of a melt flow index of 0.95 g/10 min with 30 wt.% of natural fibers screened below 80mesh, along with various percentages of ammonium polyphosphate (APP) and nano silicon dioxide (nano-SiO_2) particles of 15 nm in diameter. The compounds were molded at 150°C to produce samples for cone calorimetry and limited oxygen index testing. As shown in Table 1, adding 14 wt.% of APP or nano-silica, separately, it rather restricted flammability. However, the best fire performance of the bionanocomposites was presented combining 8 wt.% APP with 6 wt.% of nano-SiO_2 in view of the average heat release rate (HRR), the peak HRR, the total heat release (THR), the mass loss rate (MLR), and the limited oxygen index (LOI). The improvement was not revealed for the time to ignition (TTI) and the total smoke release (TSR), where the plain matrix demonstrated better compared to the bionanocomposites. The observed property improvement is attributed to the synergistic role of nano-silica to interact with APP forming silicon pyrophosphate and a more stable char layer. Kumar et al. [36] presented in case of bamboo fibers and HDPE matrix that the synergy of nano silica with APP can be promoted when the natural fibers are treated with borax and boric acid. In said compounds, the heat release rate as well as the total smoke production was both improved compared to the plain biocomposite. Besides, Zhou et al. [37] observed an increase in smoke production rate and total smoke production in HDPE/poplar wood fiber when nano silica was in action, which was eliminated when micro silica had been used in the recipe. The difference in the performance was related to the cohesion of the char layer, which was better in the latter compound.

In an example of synthetic fiber usage, Lu et al. [38] combined cotton stalks of an average size between 100 and 80 mesh with Kevlar fibers of about 10 μm in diameter, which possess inherently high thermal stability. The hybrid filler was examined for its fire resistance in an HDPE matrix with a melt flow index of 0.05 g/10 min. For favoring compatibilization and compounding, a maleic anhydride grafted polyethylene along with a polyethylene wax, respectively, each at 5 wt.% was used. The biocomposite was melt mixed in a twin-screw extruder followed by injection molding, while both processes were operated at the temperature of 180 °C. The

Table 1 Cone calorimeter and LOI results of WPCs based on HDPE and 30 wt% of wood fiber "Reprinted from Pan et al. [35], with permission from Elsevier"

Sample code	Average HRR (kW/m^2)	Peak HRR (kW/m^2)	THR (MJ/m^2)	TSR (m^2/m^2)	TTI (s)	MLR (1000 mg/s)	LOI (%)
HDPE	145 (34.1)	677 (48.5)	89 (5.0)	911 (43)	54 (2.8)	46 (5.1)	18.2 (0.15)
WPC	165 (5.8)	471 (11)	117 (1.2)	1478 (52)	23 (2.1)	52 (1.5)	19.6 (0.15)
WPC/APP 14%	148 (5.7)	355 (15)	109 (2.7)	2590 (424)	33 (2.8)	52 (3.2)	22.9 (0.21)
WPC/nano-SiO_2 14%	115 (9.9)	337 (24)	102 (1.3)	1510 (51)	37 (1.4)	38 (4.1)	21.2 (0.24)
WPC/APP 12%/nano-SiO_2 2%	117 (2.4)	297 (1.6)	101 (6.2)	2070 (16)	36 (4.2)	42 (1.0)	23.4 (0.15)
WPC/APP 10%/nano-SiO_2 4%	131 (17.8)	356 (46.7)	101 (4.1)	1949 (107)	37 (2.1)	45 (4.0)	24.1 (0.24)
WPC/APP 8%/nano-SiO_2 6%	95 (6.0)	262 (2)	96 (4.9)	1786 (22)	41 (7.1)	35 (1.4)	23.3 (0.21)

Values in parentheses are the standard deviations of the two measurements

addition of Kevlar fiber content delayed the time to ignition and improved the overall fire resistance as well as the mechanical properties. The self-extinguishing behavior was obtained for the biocomposite at the combination of 30 wt.% cotton stalks with 20 wt.% of Kevlar fibers. At the same filler fraction, the biocomposite yielded high tensile, impact, and flexural strength. A further increase in Kevlar content reaching a combination of 10 wt.% cotton stalks with 40 wt.% of Kevlar fibers deteriorated the flame retardance of the biocomposite along with the mechanical properties. The self-extinguishing performance was attributed to the optimum filler dispersion at the specific filler ratio, which created a dense internal structure along with the appropriate amount of natural material in the HDPE matrix.

Using a solvent-assisted technique, Fei et al. [39] prepared HDPE bionanocomposites reinforced with 55 wt.% bamboo fibers of 100-150μm in length along with 4 wt.% of titanium oxide nanotubes and 7 wt.% of poly(ethylene-co-vinyl acetate). The samples for testing were compounded in an internal mixer at 170 °C for 15 min followed by hot-pressing at the same temperature. The addition of nanotubes improved the fire resistance of the conventional HDPE/bamboo biocomposite as this was evidenced under cone calorimetry testing. The performance was further improved when the nanotubes were doped with cerium at 10.1 wt.% content. It was found that the time to ignition was increased from 29 s for the conventional biocomposites to 36 s and 41 s for the bionanocomposites and cerium doped bionanocomposites, respectively. The improved performance was attributed to the absorbance of the combustion process intermediates by the high surface area and the tubular structure of the nanotubes, while cerium likely restrained the flammable gases to the flame zone assisting the barrier action.

A material that has been proven beneficial in enhancing the fire performance of polar and apolar polymers relates to layered silicate [40]. Kord et al. [41] compounded HDPE presenting a melt flow index of 20 g/10 min with flour from bagasse stalks having a size of 60 mesh and montmorillonite modified with bis-2-hydroxyethyl tallow quaternary ammonium. For improving filler dispersion, a polyethylene grafted maleic anhydrite was in action at an amount of 2 phr. Initially, 50 wt.% HDPE was melt-mixed with the compatibilizer and the organoclay in an internal mixer at 180 °C followed by the addition of 50 wt.% of the bagasse flour. The obtained batch was grinded, and injection molded at 190 °C for producing the samples for testing. On the one hand, the fire behavior of the biocomposite was continuously improving by increasing the amount of the organoclay; however on the other hand, the optimum mechanical performance was delivered at 2 phr organoclay. Overall, the hybrid compound with 6phr of organoclay significantly improved the limited oxygen value of the initial biocomposite from 22.4% to 34.1%. Additionally, the hybrid biocomposite performed better under cone calorimetry increasing in parallel the time to ignition from 34 to 59 s compared to the HDPE/bagasse biocomposite. The described enhanced performance was attributed to the nanostructures, which were obtained due to intercalation of the organoclay by the matrix, as this was presented by XRD spectra.

The potential of the layered silicates in a polymer matrix can be fully deployed when the nanosized in thickness layers of the mineral are dispersed in the continuous medium resulting in the so-called exfoliated nanocomposites [42]. In this view, Lee et al. [43] mixed HDPE with a melt flow index of 17 g/10 min with 30 wt.% of softwood fibers having an average particle size of 75–125 μm. The hybridization of the biocomposite progressed with a clay modified with dimethyl dehydrogenated tallow alkyl ammonium, while for the compatibilization it was exploited an HDPE-grafted maleic anhydride. In a first approach, all components were mixed in a counter rotating twin-screw extruder at 160 °C followed by injection molding for sample production. In a second approach, the maleic anhydride was initially melt-blended at 90/10 ratio with the organoclay at 170 °C following by its mixing with the HDPE and the wood fibers in a counter rotating twin-screw extruder at 160 °C before their injection molding for obtaining the samples. The first and the second methods were specified as direct and masterbatch mixing process, respectively. As shown in Fig. 4, the transmission electron microscopy revealed that the masterbatch technique delivered exfoliated silicate layers in the matrix at 3 wt.% or at 5 wt.% organoclay content, whereas in the direct mixing, there were apparent intercalated bundles of silicate nanolayers at the amount of 5 wt.% organoclay. This difference in the morphology of the compounds affected their fire performance as it was evaluated through ASTM D635 testing. Both bionanocomposites achieved lower burning rates than the conventional biocomposite; however, the exfoliated hybrid bionanocomposite presented an even lower burning rate as compared to its intercalated version at all examined organoclay filler fractions. The enhanced performance is attributed to the insulating silicate char, which is built up on the surface of the compound during

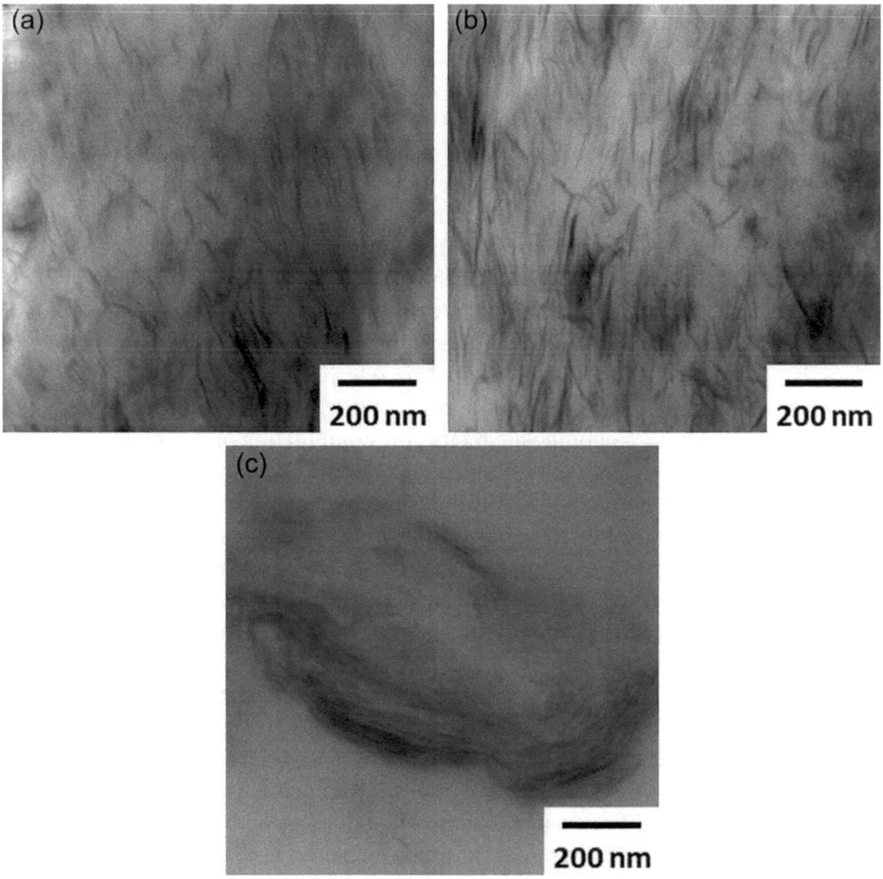

Fig. 4 TEM micrographs of **a** 3% clay/wood fiber (30%)/HDPE nanocomposites melt-blended by masterbatch method, **b** 5% clay/wood fiber (30%)/HDPE nanocomposites melt-blended by masterbatch method, and **c** 5% clay/wood fiber (30%)/HDPE nanocomposites by direct melt blending "Reprinted from Lee et al. [43], with permission from John Wiley and Sons"

burning along with the reduced dripping due to the increased melt viscosity induced by the dispersed clay platelets.

5 Trends on Flame Retardance of HDPE and LDPE Polymer Biocomposites and Bionanocomposites

The value of blending polymers in enhancing the desired properties of the matrix is a popular way in compounding [44]. In case of improving the flame retardancy, incorporation in the polyethylene biocomposite recipe a less flammable polymer can act beneficial. In such an example, HDPE with a melt flow index of 9 g/10 min was mixed with polycarbonate (PC) with a melt flow index of 10 g/min [45]. In general, PC is considered as a less flame-retardant polymer [46]. The HDPE/PC blend was melt-mixed in a counter rotating twin-screw extruder at 220 °C with 30 wt.% wood flour (WF) of 20–80 mesh, which has been treated previously with boric acid. The samples for fire testing were prepared in injection molding at the same temperature range. For improving the compatibility of the biocomposite, a maleic modified polyethylene at 4 wt.% was exploited. The addition of PC at 28 wt.% significantly reduced the heat release rate and the total heat release of the HDPE/WF biocomposite; however, the smoke production rate and the total smoke production were undesirably increased.

The improvement of the matrix-filler interface in polymer biocomposites is normally a requirement for property upgrade in relevant applications [47], and the use of coupling agents is of paramount importance [48]. Apart from the single selection of a coupling agent in a biocomposite recipe, the combination of compatibilizers may act favorable in the biocomposites performance. In the example of HDPE reinforced with rice husk, a maleic anhydride grafted polyethylene has been combined with γ-aminopropyltriethoxysilane providing improved thermal stability in the biocomposite [49]. In another study, cellulose fibers, which were produced by rice straw, had been treated subsequently with phosphoric acid and ammonia solutions to achieve doping of phosphorus and nitrogen elements on the surface of the said fibers. The P,N-doped fibrils were melt mixed with an HDPE of a melt flow index of 0.95 g/10 min on an open mill at 170 °C followed by compression molding at the same temperature. The incorporation of 7 wt.% of P,N-doped cellulose in the HDPE matrix yielded a LOI value of 25.7%, while the peak heat release rate was reduced by 72.9% compared to that of the neat matrix. Nevertheless, the total smoke release was increased compared to that of the plain HDPE. It was speculated that during combustion, the P,N-doped fibrils decompose to create pyro/polyphosphoric acid. On the one hand, this effect may catalyze the dehydration and esterification of cellulose, and on the other hand, it creates a char layer on the surface of the biocomposite, which promotes flame retardancy.

In many real-life applications, the relevant products in use require a low weight and increased specific properties. Trying to create biocomposites and bionanocomposites of a lower weight, the researchers are investigating various strategies to improve the foaming methods involved. In such an example, an LPDE with a melt flow index of 0.8 g/10 min was melt-mixed in an internal mixer with oligomeric-polyethylene grafted lignin (oPE-g-lignin) in the presence of azodicarbonamide and dicumyl peroxide followed by compression molding [50]. The foamed biocomposite, which was developed, had presented a uniform cell structure at 20 wt.% oPE-g-lignin. In another example, HDPE of a melt flow index of 17 g/10 min was compounded with 30 wt.% wood fibers along with clay modified with dimethyl dihydrogenated tallow alkyl ammonium. For compatibilization purposes, a maleic anhydrite grafted HDPE was used for dispersing initially the organoclay by exploiting the masterbatch process, while the blowing agent in the recipe was nitrogen. It was presented that the exfoliated organoclay assisted the cell uniformity even at 5 wt.% [51]. Such structures may yield flame retardancy, too.

In most cases, the flame retardancy of a product is achieved by the improvement of the relevant properties of the bulk. However, its fire resistance might further be successfully enhanced by applying flame-retardant surface treatments [52]. In a characteristic example of surface treatment, an HDPE mixed with 60 wt.% of poplar wood flour of 80 mesh particle size was further processed applying the layer-by-layer technique [53]. The manufacturing of the WPC had been favored by the addition of 3 wt.% of maleic anhydrite grafted polyethylene along with 1 wt.% of polyethylene wax. For the surface treatment of the WPC, an anionic colloidal montmorillonite (MMT) water solution and a cationic protonated polyethyleneimine (PEI) water solution with graphitized multi-walled carbon nanotubes (GMWCNT) were prepared. In both water solutions, the filler percentage was set at about 1 wt.% and the final average coating thickness after drying was 27 ± 1 μm. For the appropriate adhesion of the layers, the surface of the WPC was polished beforehand, washed with ethanol, and treated with oxygen plasma. When the surface treatment involved subsequent spraying applications of PEI and MMT solutions, the coated WPC achieved a reduction in heat release rate and smoke production rate as well as a delay of the time to ignition from 49 to 249 s for the plain versus the PEI/MMT coated WPC, respectively. Besides, when the PEI layer was substituted by the PEI reinforced with GMWCNT, the PEI-GMWCNT/MMT coated WPC maintained the previously superior performance under cone calorimetry and at the same time yielded a self-extinguishing behavior when it was exposed to a flame (Fig. 5). It is worth noticing that the limited oxygen index (LOI) value of the PEI-GMWCNT/MMT coated WPC was significantly increased reaching the value of 39%, compared to the values of 23% and 26% for the uncoated WPC and PEI/MMT coated WPC, respectively. The superior fire resistance of the PEI-GMWCNT/MMT coating was attributed to the dense structure of its char formation after combustion creating a brick-mortar-rebar analogous.

The construction sector is a promising market for applying biocomposites and especially in the segment of wood plastic composites [54]. Current efforts for exploiting the non-wood natural fiber/HDPE polymer biocomposites in this market relate to the enhancement of their flame retardancy [55]. Relevant suggestions include the development of appropriate flame-retardant recipes or surface treatments, as well as co-extruded core–shell structures wherein the shell would provide among other properties, superior fire resistance [56]. Besides, in the production of complex structures where conventional manufacturing techniques find limitations, the additive manufacturing may provide competent solutions [57]. Biocomposites and bionanocomposites would be considered in 3D printing for high-tech market sectors by the time that relevant processing issues are resolved [58].

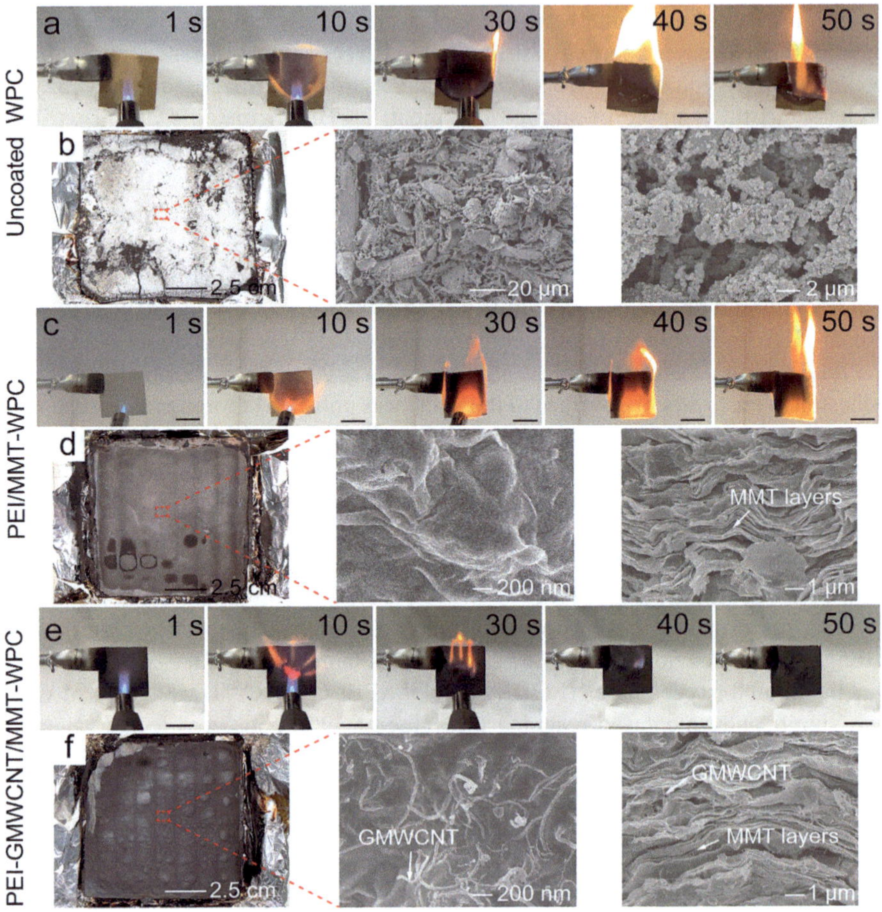

Fig. 5 Flammability tests and morphologies of the char residues. The photos taken at different burning times **a** uncoated WPC, **c** PEI/MMT-WPC, and **e** PEI-GMWCNT/MMT-WPC with respect to time in 50 s, respectively. The scale bar is 3 cm in **a, c,** and **e**. The photographs and morphologies of **b** uncoated WPC, **d** PEI/MMT-WPC, and **f** PEI-GMWCNT/MMT-WPC after burning, respectively. WPCs are coated a layer with 80 BLs "Reprinted from Zhou et al. [53], with permission from Elsevier"

6 Conclusions

The commitment of the society to reduce the overall carbon footprint of the products in use boosts the development of the HDPE and the LDPE biocomposites in various applications. Sectors such as construction and transportation nowadays require fire resistant components for advancing safety issues, and a plethora of methodologies have been developed therein. Notably, the flammability performance of the HDPE and the LDPE biocomposites and bionanocomposites is improved when the wood or the non-wood natural fiber dispersion is favored by compatibilizers. Flame-retardant additives are imperative for yielding enhanced fire protection; however, a challenge is posed to avoid side effects and deterioration of other crucial properties. In most cases, the enhancement of the fire behavior of the biocomposites under cone calorimetry is not depicted also in the time to ignition due to the natural origin of the reinforcement. Nevertheless, additional available techniques such as hybridization or surface treatment may further improve the flame-retardant properties of the respective compounds and customized designs can be proposed for each application.

References

1. Spalding MA, Chatterjee AM (eds) (2017) Handbook of industrial polyethylene and technology. Scrivener Publishing, Beverly
2. Peacock AJ (2000) Handbook of polyethylene, structures, properties, and applications. Marcel Dekker, New York
3. Yashas Gowda TG, Sanjay MR et al (2019) Natural fibers as sustainable and renewable resource for development of eco-friendly composites: a comprehensive review. Front Mater Sci 6:226. https://doi.org/10.3389/fmats.2019.00226
4. Siracusa V, Blanco I (2020) Bio-polyethylene (Bio-PE), bio-polypropylene (Bio-PP) and bio-poly(ethylene terephthalate) (Bio-PET): recent developments in bio-based polymers analogous to petroleum-derived ones for packaging and engineering applications. Polymers 12:1641. https://doi.org/10.3390/polym12081641
5. Rosenboom J-G, Langer R et al (2022) Bioplastics for a circular economy. Nat Rev Mater 7:117–137
6. Faruk O, Bledzki AK et al (2012) Biocomposites reinforced with natural fibers: 2000–2010. Prog Polym Sci 37:1552–1596
7. Zafeiropoulos NE (ed) (2011) Interface engineering of natural fibre composites for maximum performance. Woodhead Publishing, Cambridge
8. John MJ, Francis B et al (2008) Effect of chemical modification on properties of hybrid fiber biocomposites. Compos A: Appl Sci 39:352–363
9. Sanjay MR, Siengchin S (2019) A comprehensive review of techniques for natural fibers as reinforcement in composites: preparation, processing and characterization. Carbohydr Polym 207:108–121
10. Gatos KG (2024) Flame retardant properties of biocomposites for aircraft applications. In: Muthukumar C, Thiagamani SMK, Krishnasamy S, Parameswaranpillai J, Siengchin S (eds) Biocomposites for industrial applications, 1st edn. Woodhead Publishing, Elsevier, Cambridge
11. Kim Y, Lee S et al (2021) Fire-safe polymer composites: flame-retardant effect of nanofillers. Polymers 13:540. https://doi.org/10.3390/polym13040540
12. Ghomi ER, Khosravi F et al (2020) The flame retardancy of polyethylene composites: from fundamental concepts to nanocomposites. Molecules 25:5157. https://doi.org/10.3390/molecules25215157
13. Mngomezulu ME, John MJ et al (2014) Review on flammability of biofibres and biocomposites. Carbohydr Polym 111:149–182
14. Madsen B, Gamstedt EK (2013) Wood versus plant fibers: similarities and differences in composite applications. Adv Mater Sci Eng. https://doi.org/10.1155/2013/564346
15. Galaska ML, Sqrow LD (2019) Flammability characteristics of animal fibers: single breed wools, alpaca/wool, and llama/wool blends. Fibers 7:3. https://doi.org/10.3390/fib7010003
16. Wang S, Zhong J (2020) Mechanical properties, flame retardancy, and thermal stability of basalt fiber reinforced polypropylene composites. Polym Compos 41:4181–4191. https://doi.org/10.1002/pc.25702

17. Ramesh M, Rajeshkumar L et al (2022) A critical review on wood-based polymer composites: processing, properties, and prospects. Polymers 14:589. https://doi.org/10.3390/polym14030589
18. Mokhena TC, Sadiku ER et al (2021) Mechanical properties of fire retardant wood-plastic composites: a review. EXPRESS Polym Lett 15(8):744–780
19. Stark NM, White RH et al (2010) Evaluation of various fire retardants for use in wood flourepolyethylene composites. Polym Degrad Stab 95:1903–1910
20. Haveriku S, Meucci M et al (2021) Optimization of the mechanical properties of polyolefin composites loaded with mineral fillers for flame retardant cables. Micro 1:102–119
21. Schirp A, Hellmann A (2019) Fire retardancy improvement of high-density polyethylene composites based on thermomechanical pulp treated with ammonium polyphosphate. Polym Compos 40:2410–2423. https://doi.org/10.1002/pc.25106
22. García M, Hidalgo J et al (2009) Wood–plastics composites with better fire retardancy and durability performance. Compos A: Appl Sci 40:1772–1776
23. Altun Y, Doğan M et al (2016) The effect of red phosphorous on the fire properties of intumescent pine wood flour—LDPE composites. Fire Mater 40:697–703
24. Paridah MT, Juliana AH et al (2015) Non-wood based composites. Curr Forestry Rep 1:221–238
25. AL-Oqla FM, Sapuan MS et al (2015) Selecting natural fibers for bio-based materials with conflicting criteria. Am J Appl Sci 12(1):64–71
26. Kozłowski R, Władyka-Przybylak M (2008) Flammability and fire resistance of composites reinforced by natural fibers. Polym Adv Technol 19:446–453
27. Wang C, Cai L et al (2019) Thermal and flammable properties of bamboo pulp fiber/high-density polyethylene composites: influence of preparation technology, nano calcium carbonate and fiber content. Renew Energ 134:436–445
28. Wang K, Addiego F et al (2014) Dynamic behavior and flame retardancy of HDPE/hemp short fiber composites: effect of coupling agent and fiber loading. Compos Struct 113:74–82
29. Zhao Q, Zhang B et al (2009) Flame retardancy of rice husk-filled high-density polyethylene ecocomposites. Comp Sci Technol 69:2675–2681
30. Zhang Q, Zhang D et al (2020) Biochar filled high-density polyethylene composites with excellent properties: towards maximizing the utilization of agricultural wastes. Ind Crops Prod 146:112185. https://doi.org/10.1016/j.indcrop.2020.112185
31. Atmakuri A, Palevicius A et al (2020) Review of hybrid fiber based composites with nano particles—material properties and applications. Polymers 12:2088. https://doi.org/10.3390/polym12092088
32. Mochane MJ, Mokhena TC et al (2019) Recent progress on natural fiber hybrid composites for advanced applications: a review. EXPRESS Polym Lett 13(2):159–198
33. Kovačević Z, Grgac SF et al (2021) Progress in biodegradable flame retardant nano-biocomposites. Polymers 13:741. https://doi.org/10.3390/polym13050741
34. Gatos KG, Leong YW (2016) Classification of nanomaterials and nanocomposites. In: Parameswaranpillai J, Yu Y, Hammed N, Kurian Th (eds) Nanocomposite materials: synthesis, properties and applications, 1st edn. CRC Press, Taylor & Francis, Boca Raton
35. Pan M, Mei C et al (2014) Synergistic effect of nano silicon dioxide and ammonium polyphosphate on flame retardancy of wood fiber–polyethylene composites. Compos Part A Appl Sci 66:128–134
36. Kumar R, Chauhan S (2022) Effect of ammonium polyphosphate as synergist with nano silica dioxide on flammability of boron compound pretreated bamboo flour-HDPE composite. Fire Saf J 133:103647. https://doi.org/10.1016/j.firesaf.2022.103647
37. Zhou H, Hao X et al (2020) Comparative study on the effects of silica size and dispersion mode on the fire retardancy of extruded wood fiber/HDPE composites. Polym Compos 41:4920–4932
38. Lu W, Yu W et al (2021) Kevlar fibers reinforced straw wastes-polyethylene composites: combining toughness, strength and self-extinguishing capabilities. Compos Part B Eng 223:109117. https://doi.org/10.1016/j.compositesb.2021.109117
39. Fei P, Guo Z et al (2019) The enhancement of the flame retardance of bamboo fibre/HDPE composites: cerium doped $H_2Ti_2O_5 \cdot H_2O$ nanotubes effects. Constr Build Matter 201:728–735

40. Gatos KG, Karger-Kocsis J (2011) Rubber-clay nanocomposites based on apolar diene rubber. In: Galiberti M (ed) Rubber-clay nanocomposites: science, technology, and applications, 1st edn. Wiley, Hoboken
41. Kord B, Ravanfar P et al (2017) Influence of organically modified nanoclay on thermal and combustion properties of bagasse reinforced HDPE nanocomposites. J Polym Environ 25:1198–1207
42. Gatos KG, Karger-Kocsis J (2010) Rubber/clay nanocomposites: Preparation, properties and applications. In: Thomas S, Stephen R (eds) Rubber nanocomposites: preparation, properties and applications, 1st edn. Wiley, Singapore
43. Lee YH, Kuboki T et al (2010) The effects of clay dispersion on the mechanical, physical, and flame-retarding properties of wood fiber/polyethylene/clay nanocomposites. J Appl Polym Sci 118:452–461
44. Gatos KG (2014) Natural rubber based polar synthetic rubber blends. In: Thomas S, Chan CH, Pothen LA, Rajisha KR, Maria HJ (eds) Natural rubber materials, vol 1: blends and IPNs, 1st edn. The Royal Society of Chemistry, Cambridge
45. Zhang J, Koubaa A (2021) Flammability, thermal stability, and mechanical properties of wood flour/ polycarbonate/polyethylene bio-based composites. Ind Crops Prod 169:113638. https://doi.org/10.1016/j.indcrop.2021.113638
46. Zhang J, Hua Y et al (2023) Constructing flame retardant silica nanoparticles through styrene maleic anhydride copolymer grafting for PC/ABS composites. Compos Part A Appl Sci Manuf 175:107825. https://doi.org/10.1016/j.compositesa.2023.107825
47. Zafeiropoulos NE, Baillie CA et al (2002) Engineering and characterisation of the interface in flax fibre/polypropylene composite materials. Part II. The effect of surface treatments on the interface. Compos Part A Appl Sci Manuf 33:1185–1190
48. Stark NM, Rowlands RE (2003) Effects of wood fiber characteristics on mechanical properties of wood/polypropylene composites. Wood Fiber Sci 35(2):167–174
49. Sun J, Pang Y (2019) Improvement of rice husk/HDPE bio-composites interfacial properties by silane coupling agent and compatibilizer complementary modification. Polymers 11:1928. https://doi.org/10.3390/polym11121928
50. Hong S-W, Hwang S-H (2022) Construction, physical properties and foaming behavior of high-content lignin reinforced low-density polyethylene biocomposites. Polymers 14:2688. https://doi.org/10.3390/polym14132688
51. Lee YH, Kuboki T et al (2011) The effects of nanoclay on the extrusion foaming of wood fiber/polyethylene nanocomposites. Polym Eng Sci 51(5):1014–1022. https://doi.org/10.1002/pen.21739
52. Lazar ST, Kolibaba TJ et al (2020) Flame-retardant surface treatments. Nat Rev Mater 5:259–275
53. Zhou X, Fu Q et al (2021) Efficient flame-retardant hybrid coatings on wood plastic composites by layer-by-layer assembly. J Clean Prod 321:128949. https://doi.org/10.1016/j.jclepro.2021.128949
54. Oliver-Ortega H, Chamorro-Trenado MÀ et al (2018) Macro and micromechanical preliminary assessment of the tensile strength of particulate rapeseed sawdust reinforced polypropylene copolymer biocomposites for its use as building material. Constr Build Mater 168:422–430
55. Rocha JS, Escócio VA et al (2021) Thermal and flammability properties of polyethylene composites with fibers to replace natural wood. J Reinf Plast Compos 40(19–20):726–740
56. Wu Q, Chi K et al (2014) Mechanical, thermal expansion, and flammability properties of co-extruded wood polymer composites with basalt fiber reinforced shells. Mater Des 60:334–342
57. Banks JD, Emami A (2022) Carbon-based piezoresistive polymer nanocomposites by extrusion additive manufacturing: process, material design, and current progress. 3D Print Addit Manuf. https://doi.org/10.1089/3dp.2022.0153
58. Tarrés Q, Melbø JK (2018) Bio-polyethylene reinforced with thermomechanical pulp fibers: mechanical and micromechanical characterization and its application in 3D-printing by fused deposition modelling. Compos B Eng 153:70–77

Flammability Characteristics of Additively Manufactured ASA/Nylon Polymer Blends: Influence of Blend Ratios

Eekshit, Muthukumar Chandrasekar, Wadie Sultan, Lin Feng Ng, S. Jeyanthi, V. Parthasarathy, K. Senthilkumar, and T. Senthil Muthu Kumar

Abstract Polymer blend consists of two or more polymers blended together to form a material that combines the advantage of the individual base materials. In this work, the polymer blend coupons based on the nylon filament and Acrylic Styrene Acrylonitrile (ASA) were developed. The blends with various proportions of ASA and nylon created from the extrusion process and fused deposition modeling were subjected to UL-94 flammability test (vertical burn and horizontal burn). According to the results from this study, blending nylon with ASA improved the flame retardancy. The blends, regardless of the proportion of nylon, exhibited superior flame retardancy marked by a lower flame propagation rate and higher burn time than ASA. Increasing

---

Eekshit · M. Chandrasekar (✉)
Department of Aerospace Engineering, SIMCRASH Centre, Hindustan Institute of Technology and Science, Padur, Chennai, Tamil Nadu, India
e-mail: chandrasekar.25j@gmail.com

W. Sultan
Department of Chemistry, Al-Quds University, East Jerusalem, Palestine

L. F. Ng
Centre for Advanced Composite Materials (CACM), Universiti Teknologi Malaysia, Johor Bahru, Johor, Malaysia

Faculty of Mechanical Engineering, Universiti Teknologi Malaysia, Johor Bahru, Johor, Malaysia

S. Jeyanthi
School of Mechanical Engineering, Vellore Institute of Technology, Chennai, India

V. Parthasarathy
Department of Physics, Rajalakshmi Institute of Technology, Chennai 600124, Tamil Nadu, India

K. Senthilkumar
Department of Mechanical Engineering, PSG Institute of Technology and Applied Research, Coimbatore, Tamil Nadu, India

T. Senthil Muthu Kumar
Department of Mechanical Engineering, Kalasalingam Academy of Research and Education, Krishnankoil, Tamil Nadu, India

C. Muthukumar et al. (eds.), *Flammability Performance of Biocomposites and Bionanocomposites*, Composites Science and Technology,
https://doi.org/10.1007/978-981-95-2389-4_8

the nylon content also increased the elapsed burn time for the blend in the case of the vertical and horizontal burn test.

Keywords Thermoplastics · Polymer blend · ASA: Nylon UL-94 test · Fused deposition modeling

1 Introduction

Developing new materials and replacing conventional materials in engineering applications have been the prime focus of researchers and industries. Specifically, the development of new polymers and their application on a global scale has reached new heights. Thermoplastics such as Polyethylene (PE), polypropylene (PP), polystyrene (PS), polyurethane (PU), polyvinyl chloride (PVC), polyethylene terephthalate (PET), and polyamide (PA) are widely used plastics as they are lightweight, can be recycled and provide superior mechanical, impact and electrical properties, which makes them suitable for automotive and aircraft interiors [15, 19, 12]. According to recent studies, the demand for plastic materials is estimated at 624.8 billion in 2023 and is expected to grow 4.2% annually from 2024 to 2030. The main reason behind this drastic growth is the increased substitution of conventional materials such as metals, glass, natural rubber, and wood with thermoplastic polymer materials [29].

Despite several advantages and a diversity of applications, the thermal degradation temperature of thermoplastic materials is low. Also, due to the presence of hydrocarbons and volatile contents, they can melt quickly and flow when heated, enhancing the spread of flames. Thermoplastic polymers burn rapidly, producing toxic gases, smoke, and heat [21]. The rapid burning of thermoplastic and softening under fire is dangerous to humans which can lead to fire fatalities [16]. Recen studies have reported that than 40% of loss of life is caused by fire and smoke accidents in aircraft [8, 20]. The flammability of thermoplastic polymers varies depending on the chemical composition, the degree of polymerization, molecular weight, and the concentration of additives [4]. For example, due to its chemical composition, the thermal degradation temperature of polypropylene ranges between 150 and 190 °C, which can easily be flammable. However, polypropylene has a wide range of applications in textiles, automotive, furniture, and electronic casings [12]. As most thermoplastic materials are growing as potential candidates in various sectors, it is necessary to improve their flammability characteristics as they are concerned with the safety of the passengers. Subsequently, automotive manufacturing industries are also showing great interest in enhancing the flame retardancy of polymers [25].

Therefore, based on the requirement, flammability techniques can be classified into large-scale, official, and laboratory testing techniques. Among them, official techniques are preferred for industrial testing, whereas large-scale techniques are preferred for real-time conditions. Finally, laboratory techniques are selected for academic and research& development sectors for developing new products [19]. Firstly, official techniques include three primary testing methods, namely Cone

Calorimetry, Radial Panel, and Smoke chamber testing methods. Significant flammability properties such as sample weight and loss, heat release rate (HRR), smoke density, ease of ignition, and ease of extinguishing can be determined using these methods [1]. All these techniques include high-performance equipment and involve skilled persons to perform the testing as they are highly effective in determining the flammable characteristics on a large scale and in actual time conditions. Generally, official testing techniques in automotive manufacturing industries, such as seats, cushions, panels, etc., are preferred in manufacturing industries. Next is the laboratory testing methods. These methods include Underwrite Laboratory test (UL-94 which included vertical and horizontal burn test), and Limited Oxygen Index (LOI) testing methods, generally chosen for product development by researchers [7].

The primary mechanism behind the flame retardancy mechanism is to slow down the spread of the flame, which eventually slows down the flammability of the materials. So, when the burn is put off, the material should be capable of self-extinguishing from the burning process. Hence, heat, oxygen, and fuel are some of the external parameters that must be shattered from the burning process [23]. Researchers have found that reinforcement content, surface treatment, polymer types, and types of flame-retardant material are some of the factors that highly influence flammability. Among these factors, the influence of flame-retardant fillers and the type of polymer play a crucial role in improving the flame-retardant characteristics for developing a synergic composite for futuristic applications [10, 30]. Researchers have experimentally reported that incorporating fillers, modifying the matrix material, changing the polymer composition, or the processing process can effectively enhance the flame-retardant characteristics of the material [17]. It has been experimentally proven that all these techniques can drastically lower the oxygen diffusion rate and fuel amount. Flame-retardant fillers such as minerals, halogenated, phosphorous, silicon, nitrogen, carbon nanotubes, nanoclays, and metallic oxides are generally preferred to reduce flammable properties [27].

Apart from these techniques, researchers have recently revealed that blending a polymer mixture by physical means provides a significant route to enhance the flame-retardant features of the material effectively and to create a novel route for fabricating new products for futuristic applications [5, 6, 13]. Blends such as rubber-toughened plastics and acrylonitrile butadiene styrene (ABS) have also achieved excellent outcomes and attracted industries in this manner [28]. According to reports, developing polymer blends with another polymer can also be effective, cheaper, and time-saving [35]. Also, the selection of specific polymers can result in blends that can create building blocks, improving mechanical, thermal, and electrical resistance [4]. Melt blending technique was to prepare the polymer blends made of polylactic acid (PLA) and polybutylene adipate-co-terephthalate (PET) along with the graphene nanoplatelets. The graphene infused polymer blends resulted in a V-0 rating in the UL-94 test [31]. The flame-retardant characteristics of thermoplastic PU and PVC by adding nanoclays was investigated. Incorporating organo-nanoclay with thermoplastic polyurethane enormously improved the flame retardancy properties of nanocomposites [2]. The influence of graphene platelets on the flammable properties of polyethylene terephthalate/polypropylene nanocomposites were examined and adding exfoliated graphene nanoplatelets with the polymer blend was found to significantly enhance the flame-retardant characteristics of the nanocomposites. In this study, the nanocomposites have shown reduced heat release rate, mass loss rate, and delayed ignition time due to the addition of graphene platelets with polypropylene [31]. In a recent study, adding 5% nano clay with a polymer blend of PU/PP and bromine increased the limited oxygen index to 32%, which was 18% in the case of a neat PP/PU blend [15]. A comparative analysis of CNT and nano clay on the flammability behavior of the polyamide/SEBS blends was done recently. The results showed that CNT performed better than nanoclay in flame retardancy. The research showed that by doping nano clay with polyamide and styrene-ethylene/butylene-styrene graphitized with maleic anhydride (SEBS) blend, the nanocomposite exhibited a V-2 rating in UL-94 testing. In contrast, CNT exhibited no dripping with a V-0 rating [22].

On the other hand, the development of three-dimensional (3D) printing materials is attracting researchers and becoming a revolutionary technique in additive manufacturing due to its unique structure and versatile properties [32, 34]. Fused deposition modeling (FDM) has become the most commonly used 3D printing technique for thermoplastics due to its simplicity, low cost, and eco-friendly features in various product development and manufacturing processes, including aerospace, medical appliances, household appliances, and automobiles [9, 33]. Polymer filament blend consisting of two or more polymers can be drawn by extrusion and is used to make 3D printed parts. In this work, Nylon and Acrylic Styrene Acrylonitrile (ASA) filaments were blended to form the filament. The blended filament was then used to print coupons, and their flammability characteristics were examined.

2 Experimental Procedure

2.1 Materials

3D Printing Filaments such as ePA Nylon and eASA sold by eSun™ were used in this research work and their specifications are given in Table 1

Table 1 Material specifications

Parameter	ASA	Nylon
Color	Black	White
Filament diameter (mm)	1.75	1.75
Density (g/cm^3)	1	1.12
Tensile strength (MPa)	50	57
Elongation at break (%)	30	196

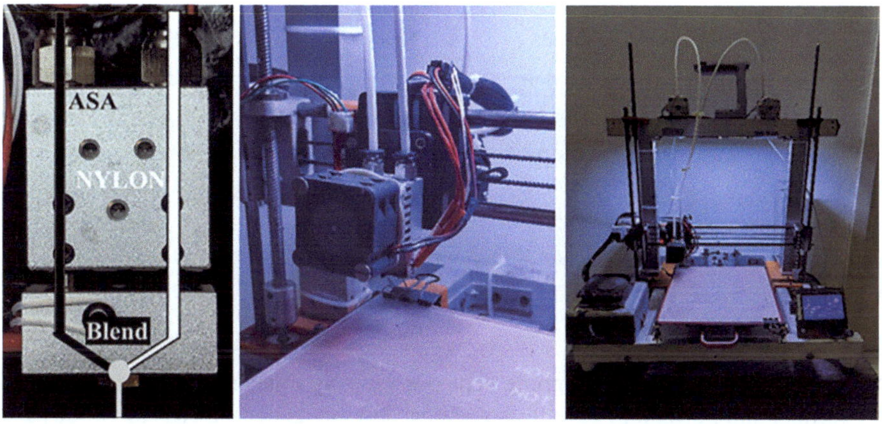

Fig. 1 Extruder setup for making polymer blend filament and 3D printer

Table 2 Printing parameters

Parameter	Description
Layer height	0.2 mm
Speed	45 mm/s
Nozzle temperature of nylon	260 °C
Nozzle temperature of ASA	270 °C
Nozzle temperature of blend	275 °C
Bed temperature	110 °C
Nozzle size	0.4 mm
Filament diameter	1.75 mm
Infill	100% (solid lines)

2.2 *Fabrication Method*

The fabrication process and the printing parameters associated with it are provided in Fig. 1 and Table 2, respectively. The blend filament was created through an extruder, shown in Fig. 1a, and printed as coupons for testing. The formulation examined in this work is provided in Table 3.

2.3 *UL-94 Flammability Test*

The flammability of plastic materials can be assessed by performing the UL94 flammability test. It measures the ability of the plastic material to extinguish the

Table 3 Formulation of the ASA/Nylon blend

Code	Nylon (wt%)	ASA (wt%)
Nylon	–	100
ASA	100	–
N3A7	30	70
N4A6	40	60
N5A5	50	50
N6A4	60	40
N7A3	70	30

fire upon ignition based on dripping behavior, burning time, and burning speed. UL-94 vertical and horizontal burn tests were carried out as per the ASTM D 3801 and ASTM D635, respectively. Specimens of dimension $127 \times 13 \times 3$ mm^3 were used for both tests.

In the vertical burn test, all the prepared samples were subjected to ignition twice for ten seconds and allowed the flame to extinguish to study their flammability. The specimen was ignited at the bottom for 10 s and burnt upward, as shown in Fig. 2. The time required for the flame to self-extinguish after removal of the burner is recorded and also checked for the dripping of the cotton placed underneath the specimen. If the after-flame time for the individual specimen during the first or second impingements does not exceed 10 s, the total after-flame time for any number of specimens, which includes adding time required for the first impingement and second impingement, does not exceed 50 s and does not ignite the cotton, is classified as V-0. If the after-flame time for the individual specimen during the first or second impingements does not exceed 30 s, the total after-flame time for any number of specimens, which includes adding time required for the first and second impingements does not exceed 250 s and does not ignite the cotton, is classified as V-1. If the after-flame time for the individual specimen during the first or second impingements does not exceed 30 s, the total after-flame time for any number of specimens, which includes adding time required for the first and second impingements does not exceed 250 s, and the cotton is ignited causing dripping, it is classified as V-2. If the entire sample is consumed, the material is classified as 'burning'.

In the horizontal burn test, the specimen is mounted horizontally as a cantilever beam. The free end of the specimen is ignited with flame for 30 s. If thickness of the tested material is in the range of 3–13 mm and burning rate does not exceed 40 mm/min, it is classified as an HB material. If the thickness of the tested material is less than 3 mm, the burning rate should not exceed 75 mm/min. The material will also be rated as HB if it ceases to burn before the 100 mm mark independent of thickness. The flammability of plastic materials has been rated in different categories, such as HB, V-2, V-1, and V-0. HB is the lowest UL94 rating [26, 34].

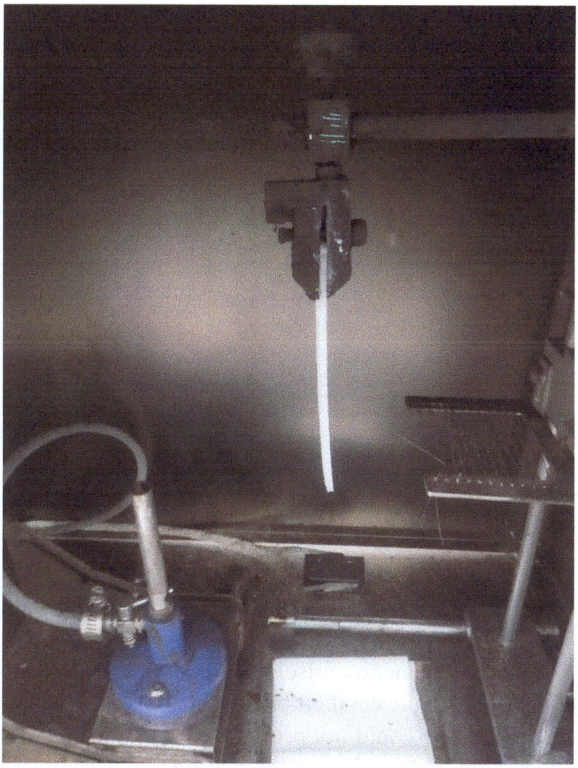

Fig. 2 UL-94 vertical burn test setup

3 Results and Discussion

The specimens before and after the flammability test are shown in Fig. 3a–c. The general observation from burnt specimens was that the burnt length of the UL-94 HB (Fig. 3b) is lower in magnitude than that of the UL-94 VB (Fig. 3c). This implies that the flame was extinguished in a shorter time during the horizontal burn.

3.1 Ul94-HB

Table 4 presents the results from UL-94 HB test. Among the materials investigated, nylon possessed the least self-extinguishing time, followed by ASA. In the case of blends, the elapsed time increases with the increase in nylon content. The general observation was that all the 3D-printed Nylon/ASA blends exhibited a lower flame propagation rate accompanied by moderate dripping compared to that of the pristine nylon and ASA. It can be noticed from Table 4 that nylon possess a burn rate above

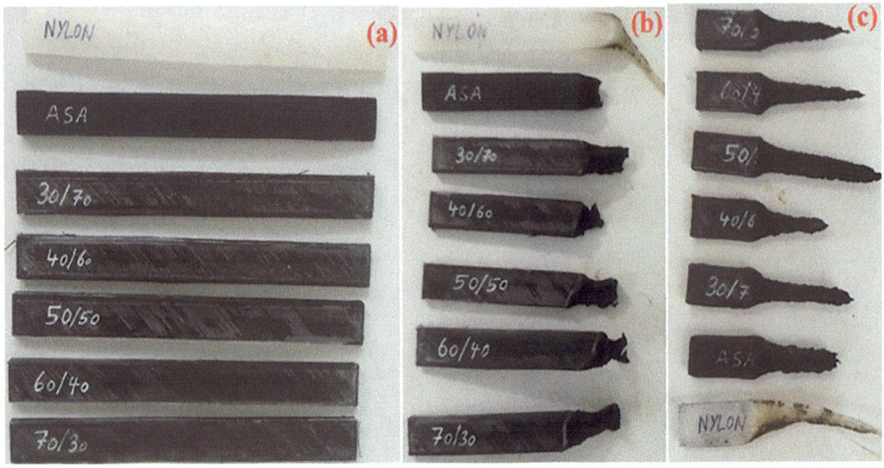

Fig. 3 ASA/Nylon blend specimens **a** Before the flammability test, **b** After UL-94 HB test and **c** After UL-94 VB test

40 mm/min and cannot be classified under the HB rating while adding ASA to nylon helps to slow down burning such that polymer blends exhibit HB rating.

The homogeneous miscibility between nylon and ASA led to the formation of complex structures in their blend state, resulting in a low fire propagation rate. Therefore, the flame could not propagate easily through the Nylon/ASA blends. It can be understood that the flame propagation rate was lower for nylon/ASA blends than for nylon and ASA owing to their strong adhesion. All the Nylon/ASA blends exhibited good flammability compared to pristine nylon and ASA due to their synergistic effects in both UL-94HB and UL-94VB (Table 5). Among the materials investigated, only nylon displayed V-0 rating. Polymeric materials with a V-0 rating possess good flame retardancy [24]. The Nylon/ASA blends showed an improvement in flammability from V-2 to V-1 category with moderate dripping. As the flame was extinguished, the mass and heat transfer processes were hindered due to the formation of a dense char layer on the samples. Interestingly, the phenol–formaldehyde (PF)

Table 4 Ignition time and burn rate from UL-94 HB test

Code	Elapsed time (seconds)	Burnt length (mm)	Burn rate (mm/min)	Rating
Nylon	23	63	164.348	–
ASA	105	48	27.429	HB
N3A7	191	60	18.848	
N4A6	174	45	15.517	
N5A5	187	55	17.647	
N6A4	178	53	17.865	
N7A3	231	60	15.584	

composites with woven fabric reinforcement showed a higher flammability rate of 0.8518 (UL-94HB) and 1.186 (UL-94VB) compared to that of Nylon/ASA blends (0.284 and 0.293) [3].

The elapsed time from the UL-94HB and UL-94VB tests is shown in Fig. 4. It can be seen that nylon self-extinguished at a far lower elapsed time than the ASA and other blends. Increasing the proportion of nylon in the blend led to an increase in burning time, and the vertical burn time was much higher than the horizontal burn time for configurations of N4A6, N5A5 and N6A4, respectively. In the case of the horizontal burn test, the elapsed time increased gradually with the increase in the proportion of nylon in the blend.

Table 5 UL-94 VB rating

Code	Elapsed time (seconds)	Burnt length (mm)	Dripping	Rating
Nylon	12	50	No	V-0
ASA	120	70	Yes	V-2
N3A7	150	53	Yes	V-2
N4A6	290	72	Yes	V-2
N5A5	217	52	Yes	V-2
N6A4	236	60	Yes	V-2
N7A3	175	65	Yes	V-2

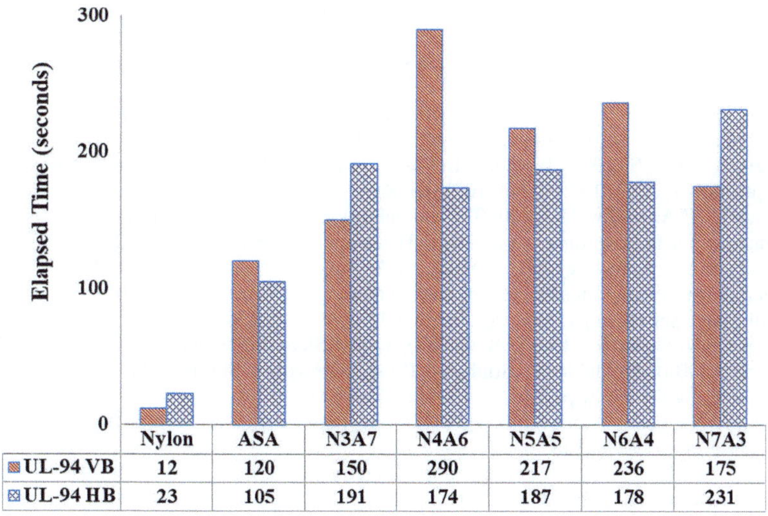

Fig. 4 Comparison of the elapsed time for UL-94 HB and UL-94 VB

4 Conclusion

This work investigated the flammability aspects of the ASA/Nylon blends through the UL-94 horizontal and vertical burn tests. The blends were created through an extrusion and further fabricated through fused deposition modeling. Nylon and ASA exhibited V-0 rating with no dripping and V-2 rating with extensive dripping from the UL-94 VB test. However, the blends exhibited V-1 rating with moderate dripping as compared to the V-2 rating for the ASA. The burn rate was not reported for the horizontal burn test as the flame did not reach a length of 100 mm before self-extinguishing. The major conclusion from this work was that increasing the proportion of nylon in the blend resulted in more burning time, implying a lower flame propagation rate. Thus, blending nylon with ASA was beneficial in terms of flammability characteristics.

References

1. Alaaeddin MH, Sapuan SM, Zuhri MYM et al (2019) Physical and mechanical properties of polyvinylidene fluoride–short sugar palm fiber nanocomposites. J Clean Prod 235:473–482. https://doi.org/10.1016/j.jclepro.2019.06.341
2. Beyer G (2007) Flame retardancy of thermoplastic polyurethane and polyvinyl chloride by organoclays. J Fire Sci 25:65–78. https://doi.org/10.1177/0734904107064602
3. Bharath KN, Basavarajappa S (2014) Flammability characteristics of chemical treated woven natural fabric reinforced phenol formaldehyde composites. Procedia Mater Sci 5:1880–1886
4. Bîrcă A, Gherasim O, Grumezescu V, Grumezescu AM (2019) Chapter 1–Introduction in thermoplastic and thermosetting polymers. In: Grumezescu V, Grumezescu AMBT-M for BE (eds). Elsevier, pp 1–28
5. Carty P, White S (1994) The importance of char-forming reactions in thermoplastic polymers. Fire Mater 18:151–166. https://doi.org/10.1002/fam.810180303
6. Carty P, White S (1996) Flammability of polymer blends. Polym Degrad Stab 54:379–381. https://doi.org/10.1016/s0141-3910(96)00066-3
7. Chapple S, Anandjiwala R (2010) Flammability of natural fiber-reinforced composites and strategies for fire retardancy: a review. J Thermoplast Compos Mater 23:871–893
8. Chaturvedi P, Kanagaraj AB, Alhammadi A et al (2021) Fabrication of PVDF–HFP-based microporous membranes by the tape casting method as a separator for flexible Li-ion batteries. Bull Mater Sci 44. https://doi.org/10.1007/s12034-021-02420-7
9. Cooke MN, Fisher JP, Dean D et al (2003) Use of Stereolithography to Manufacture Critical-Sized 3D Biodegradable Scaffolds for Bone Ingrowth. J Biomed Mater Res–Part B Appl Biomater 64:65–69. https://doi.org/10.1002/jbm.b.10485
10. Fan M, Naughton A, Bregulla J (2017) Fire performance of natural fibre composites in construction. Elsevier Ltd.
11. Gracik TD, Long GL (1992) Prediction of thermoplastic flammability by thermogravimetry. Thermochim Acta 212:163–170. https://doi.org/10.1016/0040-6031(92)80231-K
12. Haryati A, Razali N, Petrů M, et al (2021) Effect of chemically treated kenaf fibre on mechanical and thermal properties of PLA composites prepared through fused deposition modeling (FDM). Polymers (Basel) 13. https://doi.org/10.3390/polym13193299
13. Heidar Pour R, Soheilmoghaddam M, Hassan A, Bourbigot S (2015) Flammability and thermal properties of polycarbonate/acrylonitrile-butadiene-styrene nanocomposites reinforced with multilayer graphene. Polym Degrad Stab 120:88–97. https://doi.org/10.1016/j.polymdegradstab.2015.06.013

14. Kanbur Y, Tayfun U (2018) Investigating mechanical, thermal, and flammability properties of thermoplastic polyurethane/carbon nanotube composites. J Thermoplast Compos Mater 31:1661–1675. https://doi.org/10.1177/0892705717743292
15. Kannan M, Thomas S, Joseph K, Vidyapeetham AV (2015) Flame-retardant properties of nanoclay-filled thermoplastic polyurethane/polypropylene nanocomposites, pp 1–9. https://doi.org/10.1002/vnl
16. Kim NK, Dutta S, Bhattacharyya D (2018) A review of flammability of natural fibre reinforced polymeric composites. Compos Sci Technol 162:64–78
17. Korobeinichev O, Shmakov A, Paletsky A et al (2022) Mechanisms of the action of fire-retardants on reducing the flammability of certain classes of polymers and glass-reinforced plastics based on the study of their combustion. Polymers (Basel) 14. https://doi.org/10.3390/polym14214523
18. Kourtides DA, Parker JA (1978) Assessment of relative flammability and description of polymers. Polym Eng Sci 18:855–860
19. Lee CH, Salit MS, Hassan MR (2014) A review of the flammability factors of kenaf and allied fibre reinforced polymer composites. Adv Mater Sci Eng 2014. https://doi.org/10.1155/2014/514036
20. Mărieş GRE, Abrudan AM (2018) Thermoplastic polymers in product design. IOP Conf Ser Mater Sci Eng 393. https://doi.org/10.1088/1757-899X/393/1/012118
21. Muralidharan ND, Subramanian J, Rajamanickam SK et al (2024) Flame retardant characteristics of natural fibre reinforced polymer composites: a thematic review. Polym Compos 1–29. https://doi.org/10.1002/pc.28699
22. Paes LHG, Steffen TT, Becker D (2020) Comparative performance of carbon nanotube and nanoclay on thermal properties and flammability behavior of amorphous polyamide/SEBS blend. Polym Eng Sci 60:1333–1342. https://doi.org/10.1002/pen.25384
23. Pandey S, Mishra S (2023) Effect of fire-retardant fillers on the flammability and mechanical properties of wood polymer composite 10:68–73
24. Pour RH, Soheilmoghaddam M, Hassan A, Bourbigot S (2015) Flammability and thermal properties of polycarbonate/acrylonitrile-butadiene-styrene nanocomposites reinforced with multilayer graphene. Polym Degrad Stab 120:88–97
25. Prabhakar MN, Shah AUR, Song J-I (2015) A review on the flammability and flame retardant properties of natural fibers and polymer matrix based composites. Polymer (Guildf) 37:40
26. Realinho V, Haurie L, Formosa J, Velasco JI (2018) Flame retardancy effect of combined ammonium polyphosphate and aluminium diethyl phosphinate in acrylonitrile-butadiene-styrene. Polym Degrad Stab 155:208–219
27. Rigana MF, Anne TS, Balaji S et al (2021) Thermal stability and flame retardancy of epoxy composites. Epoxy Compos Fabr Charact Appl 177–215. https://doi.org/10.1002/9783527824083.ch7
28. Sessini V, Arrieta MP, Raquez JM et al (2019) Thermal and composting degradation of EVA/Thermoplastic starch blends and their nanocomposites. Polym Degrad Stab 159:184–198. https://doi.org/10.1016/j.polymdegradstab.2018.11.025
29. Shamsuri AA, Yusoff MZM, Abdan K, Jamil SNAM (2023) Flammability properties of polymers and polymer composites combined with ionic liquids. E-Polymers 23. https://doi.org/10.1515/epoly-2023-0060
30. Sharma NK, Verma CS, Chariar VM, Prasad R (2015) Eco-friendly flame-retardant treatments for cellulosic green building materials. Indoor Built Environ 24:422–432. https://doi.org/10.1177/1420326X13516655
31. Shrivastava NK, Wooi OS, Hassan A, Inuwa IM (2018) Mechanical and flammability properties of poly(lactic acid)/poly(butylene adipate-co-terephthalate) blends and nanocomposites: Effects of compatibilizer and graphene. Malaysian J Fundam Appl Sci 14:425–431. https://doi.org/10.11113/mjfas.v14n4.1233
32. Stansbury JW, Idacavage MJ (2016) 3D printing with polymers: challenges among expanding options and opportunities. Dent Mater 32:54–64. https://doi.org/10.1016/j.dental.2015.09.018

33. Tumbleston JR, Shirvanyants D, Ermoshkin N et al (2015) Continuous liquid interface of 3D objects. Science (80)347:1349–1352
34. Weng Z, Wang J, Senthil T, Wu L (2016) Mechanical and thermal properties of ABS/montmorillonite nanocomposites for fused deposition modeling 3D printing. Mater Des 102:276–283. https://doi.org/10.1016/j.matdes.2016.04.045
35. Zaripov II, Vikhareva IN, Buylova EA et al (2022) Additives to reduce the flammability of polymers. Nanotechnologies Constr 14:156–161. https://doi.org/10.15828/2075-8545-2022-14-2-156-161

Enhancing Flame Resistance of Castor Oil-Based Flexible PU Foam Through Silica Hydrogel Coating

Vinoth Kumar Selvaraj, Jeyanthi Subramanian, Vijayakumar Rajendran, S. Raja, Thangapandi Muneeswaran, S. Vignesh, and K. Anto Kumar

Abstract The study investigates the creation of flame-retardant flexible polyurethane foam (PU) made from castor oil (CO) in varying densities. This naturally flammable foam is treated using a dip coating process with concentrated hydrochloric acid (HCl) and sodium silicate to form a protective silica hydrogel layer, thereby enhancing its fire resistance. The characterization of the foam was done through thermogravimetric analysis (TGA) and Fourier transform infrared (FTIR) spectroscopy. Evaluation through the underwriter's laboratory (UL) 94 ratings and burning rate measurements demonstrated significant enhancements. Intriguingly, while non-coated foam samples received V-2 ratings, the coated counterparts exhibited markedly higher ratings of V-1 and V-0, highlighting their superior flame resistance. Specimen (S3) demonstrates greater performance, exhibiting a burning rate of 6.72 mm/min in vertical conditions and 6.16 mm/min in horizontal conditions. Notably, this sample displays a notable difference of 0.56 mm/min between its vertical and horizontal burning rates, indicating its consistent and commendable fire resistance across varying orientations. Findings revealed a direct link between foam density, burning rate, and increased flame retardancy. This innovative approach holds immense promise for an array of applications across consumer and commercial sectors, including but not limited to furniture, carpet underlay, bedding, automotive interiors, and packaging. This research lays the groundwork for creating safer and

V. K. Selvaraj · J. Subramanian (✉) · S. Vignesh · K. Anto Kumar
School of Mechanical Engineering, Vellore Institute of Technology, Chennai 600127, Tamil Nadu, India
e-mail: jeyanthi.svitc@gmail.com

V. Rajendran
Centre for Smart Energy Systems, Chennai Institute of Technology, Chennai 600069, Tamil Nadu, India

S. Raja
Center for Advanced Multidisciplinary Research and Innovation, Chennai Institute of Technology, Chennai 600069, Tamil Nadu, India

T. Muneeswaran
3D Printing, Sedaxis Advanced Materials Pvt Ltd, Chennai 600066, Tamil Nadu, India

C. Muthukumar et al. (eds.), *Flammability Performance of Biocomposites and Bionanocomposites*, Composites Science and Technology,
https://doi.org/10.1007/978-981-95-2389-4_9

more durable materials for a range of everyday products by utilizing sustainable resources and advanced flame-retardant techniques.

Keywords CO-based PU foam · Silica gel · Dip coating · Fire retardant · FTIR · TGA

1 Introduction

Flexible PU foams are nowadays broadly used for their highly porous nature and energy-absorbing properties. They possess a highly cellular structure making them a better sound-absorbing material [13–15]. The flexible polyurethane foams' properties make them suitable for various applications in various industries [11, 12, 16, 20]. Apart from these advantages, PU foam has a major concern about its flammability. Choosing the foam by understanding its flammability nature is a crucial factor. Polyurethane foams are highly flammable due to the chemicals present in them. PU foams contain mainly polyol and isocyanate which is highly flammable when exposed to fire hazards [7, 26, 27]. Its cellular structure also accounts for the flammability property. PU foam is highly flammable and can cause great damage to life and surroundings. The flammability mechanism of flexible PU foam is shown in Fig. 1.

This explosive nature of the foams can be reduced by adding the necessary additives. Incorporation of different additives in different ratios can help in reducing the flammable nature of the PU foams [19, 21, 22, 24, 28, 30]. Aleksander Hejna

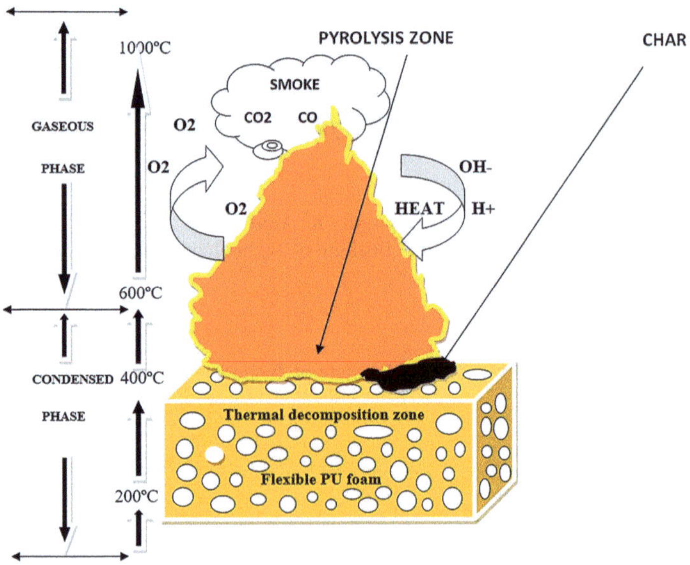

Fig. 1 Flammability process of flexible polyurethane foam

incorporated clays along with different organic compounds and noted for flammability and observed an exceptional decrease in flammability of the foam thus making it flame retardant [8]. Likewise, EHP Silva et al. added alumina trihydrate and noted the results. They observed that the PU foam with 40% alumina trihydrate obtained V-2 classification and the foam with 50% alumina trihydrate obtained HB classification as per ASTM D790-17 norms [17]. Ciecierska et al. added graphite to increase the fire-resistant property of the PU foams and witnessed a great improvement in fire resistance phenomena [5]. Yang et al. used the surface coating method to minimize the flammability of the PU foams. The surface coating method is gaining attention nowadays as it reduces flammability without altering the mechanical properties [25]. Gaan et al. added DOPO-phosphonamidate as an additive in the PU foam. To compare the performance, they used three commercially available flame retardants and compared the results. The findings showed that DOPO-phosphonamidate displayed better flame-resistant properties than commercially available flame retardants [6]. Abdel-Hakim et al. mixed metal chelate, sodium tripolyphosphate, and magnesium hydroxide with styrene butadiene rubber (SBR) composites. They observed that the horizontal burning rate of SBR composites decreased when metal chelate and magnesium hydroxide were incorporated into the foams [1]. Zygmunt-Kowalska et al. synthesized PU foam incorporating microspheres derived from coal fly ash. Their flammability tests revealed that the foam required more oxygen to burn, indicating enhanced flame resistance [32]. Many studies have been done to transform PU foam materials into flame-resistant materials. Correspondingly, Zhang et al. made three different phosphorous-containing materials phenyl phosphoryl dichloride, phenyl dichlorophosphate, and methyl phosphoryl dichloride react with 2-aminobenzothiazole. They synthesized a series of synergetic flame retardants named MPBT, PPBT, and POBT. Then, the flexible foam was synthesized with these mixtures and the flammability test was carried out. The outputs indicated that MPBT had the better flame-resistant efficiency among the three flame retardants. They noted a reduction in the peak heat release rate by 46.64%, along with an increase in the flame retardancy index to 6.88 [31]. The addition of fillers can increase the flame-resistant ability of the foams, but on the other hand, it should not cause any harm to the environment. The use of bio-based additives will be a great choice. Qi Jiang et al. coated the surface of the foam material with phytic acid, tea polyphenols, and $Fe_2\ (SO_4)_3{\cdot}xH_2O$. The outcomes proved that the foam material can pass the UL-94 V-0 rating and the peak HRR was reduced by 67.7% [10]. Xie et al. proposed a new method of preparing PU foam alumina aerogel composites with a uniform dispersion of alumina aerogel by freeze-drying method. The flammability test suggested that the heat and smoke release rate was drastically reduced as compared to normal PU foam [23]. Bellayer et al. developed a sol–gel coating for PU foam using a blend of tetraethoxysilane (TEOS), 3-aminopropyltriethoxysilane (APTES), methyltriethoxysilane (MTES), and diethyl phosphite (DEP), followed by fire testing. They found that coating protected the foam from burning thus reducing the amount of smoke released [4]. Similarly, Suhailuddin et al. used expandable graphite (EG) in combination with organo-montmorillonite (OMMT), nano-titanium dioxide (NTD), magnesium hydroxide (MGH), and MGH with OMMT. Their findings confirmed that the combination of expandable graphite and magnesium hydroxide exhibited the best flame-retardant properties, achieving a rating of 0.27 mm/s [18]. Zhang et al. incorporated aluminum diethyl phosphinate (ADP) or dicyclohexyl aluminum hypophosphate (DAH) as flame retardants in rigid PU foam and conducted flammability tests. The burning time of DAH added Rigid PU foam was 4.4 secs which was less than 12.5 secs, the burning time of ADP added Rigid PU foam [29]. Mahmood Ali Ismail et al. added melamine cyanurate flame retardant (MCFR) and organically modified montmorillonite (OMMT) clay to the PU foam. The modified PU foam was compared with unmodified PU foam for its flammability. They noted that the modified PU foam showed UL-94 ratings of V-0 and V-1 [9]. This study introduces a groundbreaking correlation between foam density and fire resistance in CO-based PU foam. It pioneers a novel dip coating technique, utilizing sodium silicate and concentrated HCl to create a protective silica hydrogel. Unlike conventional methods, this innovative approach significantly improves flame resistance while preserving the foam's flexibility and functionality. The findings highlight the potential for future material design strategies aimed at achieving superior fire safety without compromising the foam's inherent properties. Achieving superior flame retardancy ratings (V-1 and V-0) compared to traditional V-2 ratings for coated and non-coated samples respectively underscores the transformative impact of this research in various applications.

2 Experimental Approach

2.1 Materials

We acquired CO from Manali Petrochemicals Ltd in Chennai, including Empeyonate 1FL8018 and Empeyol WSF 300/1-8HLD, as well as an amine catalyst and a surfactant. Hydrochloric acid (HCl) was sourced from Srihari Chemicals Ltd, Chennai. The sodium silicate solution (water glass) was supplied by Herenba Instruments and Engineers in Ambattur, Chennai. Detailed specifications for this solution are presented in Table 1.

Table 1 Specifications of sodium silicate solution

Specific gravity at 20 °C	1.39
Assay (SiO)	25.5–28.5%
Assay (Na_2O)	Min 10%

Table 2 Chemical composition of CO-based PU foam

Polyol and CO	Isocyanate	Surfactant	Tin catalyst	Amine catalyst
6 ml	8 ml	2 μl	0.15 μl	0.60 μl

2.2 Production of CO-Based PU Foam

To create high-quality PU foam, we maintained a clean workspace. We experimented with various ingredients like castor oil, polyol, isocyanate, surfactants, and catalysts, adjusting their proportions to improve the foam's characteristics. To achieve the desired flexible foam, we mixed CO-based polyol and isocyanate in a 1:2 ratio and added sufficient surfactants, tin catalysts, and amine catalysts. Table 2 provides a detailed breakdown of the chemical composition of the CO-based PU foam.

We prepared the PU foam by carefully measuring and pouring the CO-based polyol into a beaker. Next, we introduced 2 μl of surfactant and mixed the solution thoroughly with a magnetic stirrer operating at 1300 rpm for 5 min. This process ensures a fine cell structure by controlling the size of the foam cells. We then added 0.15 μl of tin catalyst and 0.60 μl of amine catalyst, mixing the components at the same speed. These catalysts are essential for controlling the reaction rate and improving the foam's properties. For controlling density, reducing weight, and enhancing sound-absorbing qualities, we added 2 ml of distilled water (blowing agent) to the mixture. Following this, we incorporated 8 ml of isocyanate and stirred the mixture for 10 s. The solution was subsequently poured into a mold and left to expand freely at room temperature to produce standard PU foam. Subsequently, the synthesized foam was coated by dipping it in a mixture of 15 ml sodium silicate and 6 ml water and left to dry for 20 min. Concentrated hydrochloric acid was applied drop by drop to the top and bottom layers of the foam, which was then compressed in a mold. The resulting PU foam was then evaluated for its flammability characteristics.

2.3 Experimental Setup for Flammability Test

A flammability test of PU foam is conducted to examine how the foam reacts to the flame. There is a standard that needs to be followed that includes various parameters. Two commonly used tests for testing the flammability of PU foams are ASTM D635 for horizontal test and ASTM D3801 for vertical test. A flammability test is taken to measure the surface burning characteristics of PU foam. The procedure involves securing the specimen horizontally and vertically using a stand. Underneath the sample, a layer of cotton is placed on the base to assess any dripping. Using a Bunsen burner, the specimen is ignited by introducing fire at its open end for 10 s, after which the flame is removed, allowing the fire to spread along the specimen's length. A stopwatch is used to record the time it takes for each sample to burn. This process is repeated for all five samples, and the material's burning rate is calculated

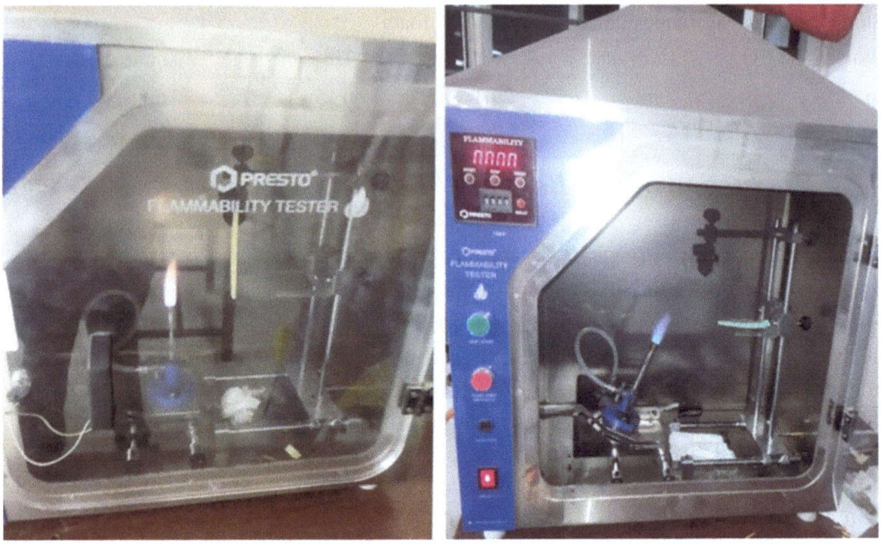

(a) (b)

Fig. 2 Experimental setup for flammability **a** Vertical burning test, **b** horizontal burning test

using the given equation [3].

$$\text{Burning Rate (mm/ min)} = 60 \times \frac{\text{length of the sample burnt(mm)}}{\text{sample burning time(s)}}.$$

UL-94 assesses the flammability of materials based on the time required for specimens to burn, employing different test methods such as vertical burn (V-0, V-1, and V-2) and horizontal burn. The flammability characteristics can vary depending on the formulations used in manufacturing the PU foams. The experimental setup for vertical and horizontal burning tests is illustrated in Fig. 2.

UL-94 encompasses several tests: 94HB, 94 V, 94HBF, 94VTM, 94HF, 94-5 V, and the radiant panel test. It assesses burning characteristics through three primary methods: surface burn (5VA & 5VB), vertical burn (V-0, V-1, V-2), and horizontal burn (H-B).

2.3.1 Vertical and Horizontal Testing

The UL-94 vertical and horizontal tests are essential standards for evaluating the flammability characteristics of materials, particularly plastics, across a range of industries. Administered by Underwriters Laboratories (UL), a prominent global entity specializing in safety consultation and certification, this test evaluates how a

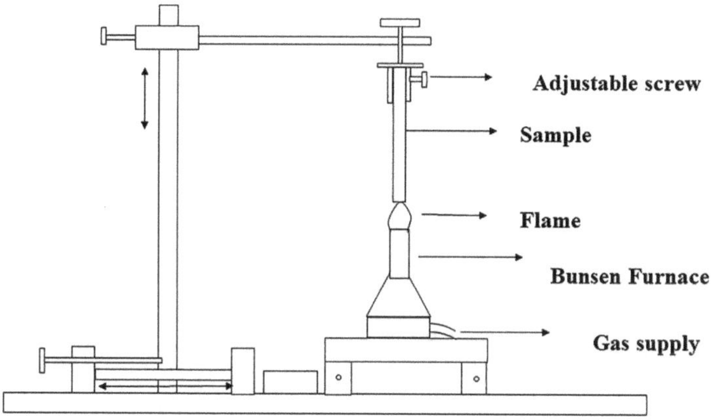

Fig. 3 UL-94 vertical test

material responds to flames when placed vertically and horizontally. In this evaluation, a sample of the material is positioned vertically and horizontally and exposed to controlled flame ignition at 90° and 45°. The assessment focuses on several factors, including the material's ignition, rate of burning, duration of combustion, and the emission of flaming particles or drips. The test adheres to specific guidelines, observing whether the flame extinguishes within a designated time after the removal of the ignition source. The UL-94 vertical and horizontal tests categorize materials into distinct ratings based on their performance, ranging from V-0 (indicating the highest flame resistance) to V-2 (representing lower flame resistance) and HB if it ceases to burn before the 100 mm mark independent of thickness. These classifications derive from criteria related to flame duration, residual burning, and the presence of droplets that could potentially ignite further flames. The specifications of the UL-94 vertical and horizontal tests are shown in Figs. 3 and 4.

3 Results and Discussion

3.1 FTIR Analysis

Fourier transform infrared (FTIR) spectroscopy is an essential technique for examining the chemical composition and structure of materials by analyzing their interaction with infrared light. In the FTIR spectra of polyurethane (PU) foam, specific peaks correspond to various molecular vibrations: Peaks in the range of 3000–2800 cm^{-1} indicate the presence of O–H groups (such as phenols or alcohols), while peaks around 3300 cm^{-1} are associated with N–H stretches from amines. The carbonyl group exhibits a peak at 1730 cm^{-1}, and the disappearance of the isocyanate (NCO) peak around 2270 and 2250 cm^{-1} suggests complete utilization of isocyanates in the

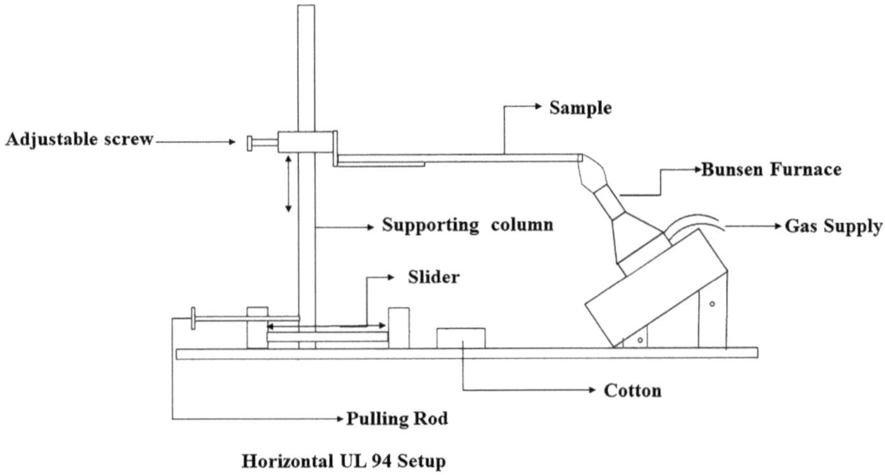

Fig. 4 UL-94 horizontal test

reaction. Additionally, a peak at 1295.38 cm^{-1} represents a C-N amide III bond, and a peak at 873.31 cm^{-1} corresponds to a bending C = O inorganic carbonate bond. The lack of a hydroxyl signal at 3450 cm^{-1} indicates full utilization of polyol. For silica gel, the FTIR spectra display bands at 3548 cm^{-1} (broad) and 1632 cm^{-1}, signifying the stretching and bending vibrations of silanol groups (Si–OH). The bands at 1094 cm^{-1} and 798 cm^{-1} correspond to the anti-symmetric and symmetric stretching modes (Si–O–Si) of [SiO_4] units. Additionally, a minor shoulder at 974 cm^{-1} is attributed to Si–OH stretching, and a band at 462 cm^{-1} is related to Si–O–Si bending vibrations [2] (Fig. 5).

3.2 TGA Analysis

Thermogravimetric Analysis (TGA) is an advanced technique utilized to examine the thermal properties of materials like polyurethane (PU) foam. It involves measuring the weight changes of a sample as it is heated at a controlled rate within a regulated environment. This technique provides valuable insights into the material's thermal stability and decomposition characteristics. The essential estimation in TGA is the adjustment of test mass as a component of temperature or time. As the sample goes through warm disintegration, debasement, or other compound changes, its weight changes, which gives significant data about its warm security, structure, and decay conduct. By subjecting the foam to controlled heating and analyzing the changes in its weight, TGA provides insights into the foam's composition and thermal stability.

Thermogravimetric analysis (TGA) was performed using a thermal gravimetric analyzer (SDT Q600, TA Instruments, USA), with a heating rate of 10 °C/min in a nitrogen atmosphere. A constant nitrogen flow of 50 mL/min was maintained, and the

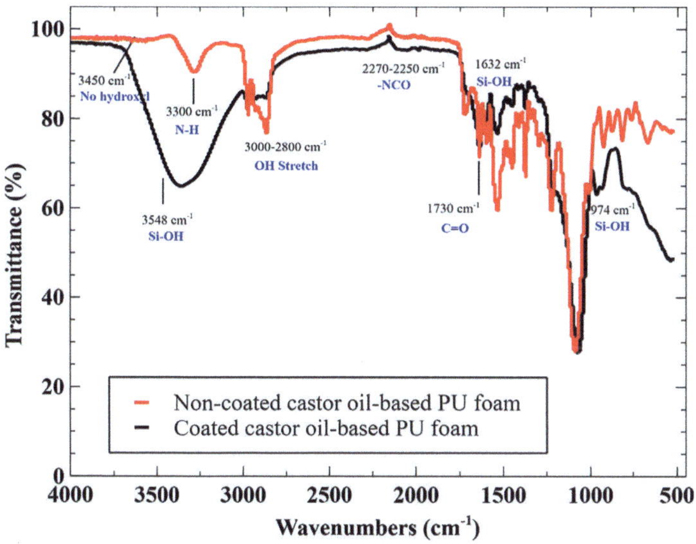

Fig. 5 FTIR graph of non-coated and coated castor oil-based PU foam

temperature was gradually increased from room temperature (25 °C) to 450 °C. Each test involved 10 mg of the sample placed in an alumina crucible. Figure 6 illustrates the TGA results for both non-coated and coated CO-based PU foam. The TGA curve for the non-coated PU foam, represented by the green line, indicates initial sublimation around 240 °C, demonstrating its non-volatility, and reaches thermal stability near 400 °C with a weight loss of approximately 75%. In contrast, the TGA curve for the coated PU foam (black line) shows that decomposition starts at a higher temperature of approximately 300 °C. This suggests that the silica gel coating has raised the melting point of the foam, resulting in a reduced weight loss of only 25% at 400 °C.

3.3 Flammability Test

Flammability assessments adhered to standardized procedures, encompassing examination of cotton ignition to appraise the composite materials' tendency to drip. The tested composites are depicted in Fig. 7, presenting burning rate data for five specimens in parallel with horizontal and vertical flame test results. The graph demonstrates remarkable similarity among the test outcomes, with marginal discrepancies observed.

Table 3 displays the UL-94 test results for castor oil-based polyurethane foam composites with five different densities. The UL-94 ratings used to evaluate the flammability of these PU foam composites include V-0, where the flame extinguishes

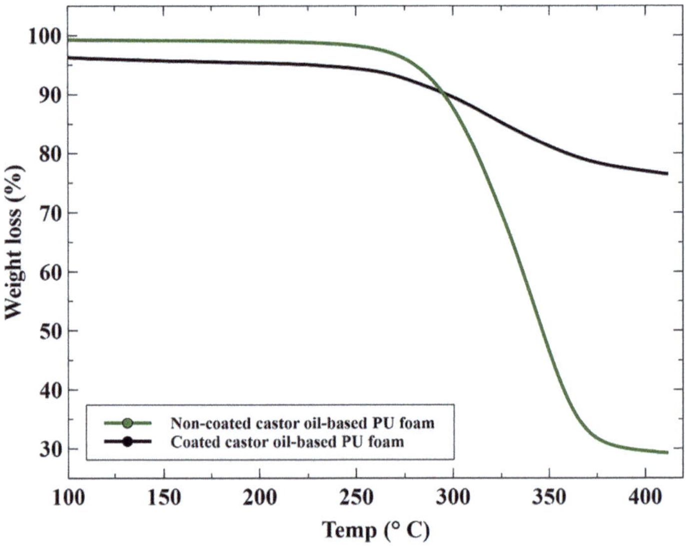

Fig. 6 TGA graph of non-coated and coated castor oil-based PU foam

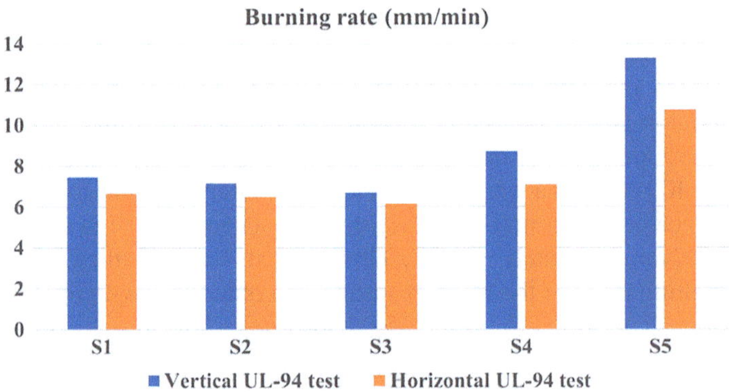

Fig. 7 Burning rate of vertical and horizontal UL-94 tests

within 10 s with no dripping; V-1, where the flame is extinguished within 30 s with no dripping; and V-2, where the flame is suppressed within 10 s but with dripping. The table provides both vertical and horizontal burning rates for CO-based PU foam with silica coating, illustrating variations between these two testing orientations. From Table 3, it is evident that the uncoated specimen has a V-2 flammability rating. However, adding silica gel to the specimen results in different ratings: Specimens S1, S2, S4, and S5 achieve a V-1 rating, while specimen S3 reaches a V-0 rating. Specimen S3 exhibits superior performance with a vertical burning rate of 6.72 mm/min and a horizontal burning rate of 6.16 mm/min, showing a minimal difference of 0.56 mm/min between the two orientations. This indicates its robust and consistent

Table 3 Flammability test for five different densities with and without coating

Specimen	Non-coated			Coated					
	Density (Kg/m^3)	UL-94 rating	Cotton ignition	Density (Kg/m^3)	UL-94 rating	Burning rate (mm/min)			Cotton ignition
						Vertical UL-94 test	Horizontal UL-94 test	Difference	
S1	14.87	V-2	Yes	125.75	V-1	7.46	6.66	0.8	No
S2	20.15	V-2	Yes	130.56	V-1	7.16	6.5	0.66	No
S3	25.61	V-2	Yes	140.61	V-0	6.72	6.16	0.56	No
S4	29.45	V-2	Yes	113.13	V-1	8.72	7.1	1.62	No
S5	35.49	V-2	Yes	101.68	V-1	13.28	10.75	2.54	No

fire resistance, suggesting that the addition of silica gel enhances the flammability performance of flexible CO-based PU foam.

4 Conclusion

In summary, this study represents a significant breakthrough in engineering flame-resistant flexible polyurethane foam derived from castor oil. This achievement stems from an innovative dip coating technique utilizing sodium silicate and concentrated hydrochloric acid, supported by meticulous analysis via thermogravimetric analysis (TGA), Fourier transform infrared (FTIR) spectroscopy, and comprehensive evaluations through Underwriter's Laboratory (UL) 94 ratings and burning rate measurements. The transition from lower V-2 to higher V-1 and V-0 ratings in treated foam, notably highlighted by the consistent and commendable fire resistance of Specimen (S3) across various orientations, signifies a substantial leap forward in flame retardancy. This establishes a crucial link between foam density, burning rate, and heightened flame resistance, offering pivotal insights for future material design strategies. This innovative approach holds immense potential across diverse consumer and commercial applications, encompassing furniture, carpet underlay, automotive interiors, bedding, and packaging. By amalgamating sustainable materials and advanced flame-retardant techniques, this research propels the development of materials that prioritize safety and resilience without compromising functionality. In conclusion, this study not only pioneers advancements in fire safety but also lays the groundwork for sustainable practices, shaping the future landscape of materials designed for enhanced safety and durability in everyday products.

Acknowledgements The authors would like to extend their gratitude to the Vellore Institute of Technology, Chennai, for providing the essential resources and facilities necessary for conducting this research.

References

1. Abdel-Hakim A, El-Basheer TM, Abdelkhalik A (2020) Mechanical, acoustical and flammability properties of SBR and SBR-PU foam layered structure. Polym Test 88:106536. https://doi.org/10.1016/j.polymertesting.2020.106536
2. Antony R, David Manickam ST, Kollu P, Chandrasekar PV, Karuppasamy K, Balakumar S (2014) Highly dispersed Cu(ii), Co(ii) and Ni(ii) catalysts covalently immobilized on imine-modified silica for cyclohexane oxidation with hydrogen peroxide. RSC Adv 4:24820–24830. https://doi.org/10.1039/c4ra01960a
3. Balan GS, Balasundaram R, Chellamuthu K, Gopan SN, Dinesh S, Vijayan V, Sathish T, Rajkumar S (2022) Flame resistance characteristics of woven jute fiber reinforced fly ash filled polymer composite. J Nanomater. https://doi.org/10.1155/2022/9704980
4. Bellayer S, Jimenez M, Prieur B, Dewailly B, Ramgobin A, Sarazin J, Revel B, Tricot G, Bourbigot S (2018) Fire retardant sol-gel coated polyurethane foam: mechanism of action. Polym Degrad Stab 147:159–167. https://doi.org/10.1016/j.polymdegradstab.2017.12.005
5. Ciecierska E, Bazarnik P, Gloc M, Kulesza M, Kowalski M, Krauze S, Lewandowska M (2015) Flammability, mechanical properties and structure of rigid polyurethane foams with different types of carbon reinforcing materials. Compos Struct. https://doi.org/10.1016/j.compstruct.2015.12.022
6. Gaan S, Liang S, Mispreuve H, Perler H, Naescher R, Neisius M (2015) Flame retardant flexible polyurethane foams from novel DOPO-phosphonamidate additives. Polym Degrad Stab 113:180–188. https://doi.org/10.1016/j.polymdegradstab.2015.01.007
7. Gao L, Zheng G, Zhou Y, Hu L, Feng G, Zhang M (2014) Synergistic effect of expandable graphite, diethyl ethylphosphonate and organically-modified layered double hydroxide on flame retardancy and fire behavior of polyisocyanurate-polyurethane foam nanocomposite. Polym Degrad Stab 101:92–101. https://doi.org/10.1016/j.polymdegradstab.2013.12.025
8. Hejna A (2021) Clays as inhibitors of polyurethane foams' flammability. Mat (Basel) 14. https://doi.org/10.3390/ma14174826
9. Ismail MHMA, Arafath BY, Majeed SSMA, Begum SS (2014) Available online at Indian journal of advances in chemical science development of flexible polyurethane nanocomposite foam with enhanced flame retardancy, pp 99–102
10. Jiang Q, Li P, Liu Y, Zhu P (2022) Green flame-retardant flexible polyurethane foam based on polyphenol-iron-phytic acid network to improve the fire safety. Compos Part B Eng 239:109958. https://doi.org/10.1016/j.compositesb.2022.109958
11. Kumar V, Jeyanthi S (2022) A comparative study of smart polyurethane foam using RSM and COMSOL multiphysics for acoustical applications: from materials to component. J Porous Mater. https://doi.org/10.1007/s10934-022-01362-7
12. Kumar V, Jeyanthi S (2022b) A study to determine electromagnetic interference- shielding effectiveness on bio-based polyurethane foam reinforced with PVDF/MgO/Ni for emerging applications, pp 1–14. https://doi.org/10.1002/app.53386
13. Rastegar N, Ershad-Langroudi A, Parsimehr H, Moradi G (2022) Sound-absorbing porous materials: a review on polyurethane-based foams. Iran Polym J (English Ed 31:83–105. https://doi.org/10.1007/s13726-021-01006-8
14. Selvaraj VK, Subramanian J (2023) Experimentation, simulation, and statistical analysis of nanofillers reinforced bio-based polyurethane foam for acoustical applications. Polym Eng Sci 1–15. https://doi.org/10.1002/pen.26273
15. Selvaraj VK, Subramanian J, Gupta M, Gayen M, Mailan Chinnapandi LB (2020) An experimental investigation on acoustical properties of organic PU foam reinforced with nanoparticles fabricated by hydrothermal reduction technique to emerging applications. J Inst Eng Ser D 101:271–284. https://doi.org/10.1007/s40033-020-00238-x
16. Selvaraj VK, Subramanian J, Jayamani E, Viswanath S, Pandurengan K, Siddharth S, Lal PJ (2023) A comparative study on biocomposites for electromagnetic interference shielding applications: a green initiative for environmental sustainability. J Mater Res Technol 26:5786–5800. https://doi.org/10.1016/j.jmrt.2023.08.290

17. Silva EHP, Souza GSC, Janes DB, Waldow G, Sales FCP, Guedes RM, Tita V, Costa RRC (2021) Flexural and flammability evaluation of a new bio-based polyurethane foam with alumina trihydrate. Proc Inst Mech Eng Part L J Mater Des Appl 235:1160–1171. https://doi.org/10.1177/1464420720986591
18. Suhailuddin SH, Aprajith K, Sanjay B, Shabeeruddin SH, Shamshath Begum S (2022) Development and characterization of flame retardant property in flexible polyurethane foam. Mater Today Proc 59:819–826. https://doi.org/10.1016/j.matpr.2022.01.115
19. Velencoso MM, Ramos MJ, Klein R, De Lucas A, Rodriguez JF (2014) Thermal degradation and fire behaviour of novel polyurethanes based on phosphate polyols. Polym Degrad Stab 101:40–51. https://doi.org/10.1016/j.polymdegradstab.2014.01.012
20. Selvaraj VK, Jeyanthi Subramanian KRP (2023) A study on flexible bio-based conductive foam for pressure sensing and electromagnetic interference applications. Mater Res Express 1–25
21. Xi W, Qian L, Chen Y, Wang J, Liu X (2015) Addition flame-retardant behaviors of expandable graphite and [bis(2-hydroxyethyl)amino]-methyl-phosphonic acid dimethyl ester in rigid polyurethane foams. Polym Degrad Stab 122:36–43. https://doi.org/10.1016/j.polymdegradstab.2015.10.013
22. Xi W, Qian L, Huang Z, Cao Y, Li L (2016) Continuous flame-retardant actions of two phosphate esters with expandable graphite in rigid polyurethane foams. Polym Degrad Stab 130:97–102. https://doi.org/10.1016/j.polymdegradstab.2016.06.003
23. Xie H, Yang W, Yuen ACY, Xie C, Xie J, Lu H, Yeoh GH (2017) Study on flame retarded flexible polyurethane foam/alumina aerogel composites with improved fire safety. Chem Eng J 311:310–317. https://doi.org/10.1016/j.cej.2016.11.110
24. Xu W, Chen R, Du Y, Wang G (2020) Design water-soluble phenolic/zeolitic imidazolate framework-67 flame retardant coating via layer-by-layer assembly technology: Enhanced flame retardancy and smoke suppression of flexible polyurethane foam. Polym Degrad Stab 176:109152. https://doi.org/10.1016/j.polymdegradstab.2020.109152
25. Yang H, Yu B, Song P, Maluk C, Wang H (2019) Surface-coating engineering for flame retardant fexible polyurethane foams : a critical review, p 176
26. Yang R, Hu W, Xu L, Song Y, Li J (2015) Synthesis, mechanical properties and fire behaviors of rigid polyurethane foam with a reactive flame retardant containing phosphazene and phosphate. Polym Degrad Stab 122:102–109. https://doi.org/10.1016/j.polymdegradstab.2015.10.007
27. Yin Z, Chu F, Yu B, Wang B, Hu Y (2022) Hierarchical Ti_3C_2Tx@BPA@PCL for flexible polyurethane foam capable of anti-compression, self-extinguishing and flame-retardant. J Colloid Interface Sci 626:208–220. https://doi.org/10.1016/j.jcis.2022.06.075
28. Zeng SL, Xing CY, Chen L, Xu L, Li BJ, Zhang S (2020) Green flame-retardant flexible polyurethane foam based on cyclodextrin. Polym Degrad Stab 178:109171. https://doi.org/10.1016/j.polymdegradstab.2020.109171
29. Zhang G, Lin X, Zhang Q, Jiang K, Chen W, Han D (2020) Anti-flammability, mechanical and thermal properties of bio-based rigid polyurethane foams with the addition of flame retardants. RSC Adv 10:32156–32161. https://doi.org/10.1039/d0ra06561g
30. Zhang M, Luo Z, Zhang J, Chen S, Zhou Y (2015) Effects of a novel phosphorus-nitrogen flame retardant on rosin-based rigid polyurethane foams. Polym Degrad Stab 120:427–434. https://doi.org/10.1016/j.polymdegradstab.2015.08.001
31. Zhang S, Chu F, Xu Z, Zhou Y, Qiu Y, Qian L, Hu Y, Wang B, Hu W (2022) The improvement of fire safety performance of flexible polyurethane foam by Highly-efficient P-N-S elemental hybrid synergistic flame retardant. J Colloid Interface Sci 606:768–783. https://doi.org/10.1016/j.jcis.2021.08.069
32. Zygmunt-Kowalska B, Zakrzewska P, Szajding A, Handke B, Kuźnia M (2023) Polyurethane foams reinforced with microspheres–assessment of the application in construction as a thermal insulation material. Thermochim Acta 726. https://doi.org/10.1016/j.tca.2023.179556

Flammability Performance of the Natural Fiber-Reinforced Hybrid Composites

Thanh Mai Nguyen Tran, Van-Ta Do, Quang Thang Do, Xiem Nguyen Thang, M. N. Prabhakar, and Jung-il Song

Abstract The chapter critically reviews the potentials of various natural fibers on the fire-resistance behavior of biocomposites and bionanocomposites. It evaluates the potential of fibers like cotton, hemp, and bamboo for flame retardation in combination with the different matrix materials and is assessed for its interaction. Material combinations and hybridization strategies of the examples given in this chapter are synergistically influential in enhancing the fire performance. It also deals with developments in natural fiber technologies with its persistent challenges on sustainability and environment-oriented issues. Future research recommendations are concerned with technological advances in composites reinforced using natural fibers. The highlighted main development covers much safer materials which are environmentally more sustainable with better flame resistance.

Keywords Biocomposites · Flame retardancy · Natural fibers · Synergistic effects · Bionanocomposites

T. M. N. Tran (✉) · J. Song
Department of Mechanical Engineering, Changwon National University, Changwon, Gyeongsangnam, South Korea
e-mail: thanhmnt@ntu.edu.vn

T. M. N. Tran · X. N. Thang
Department of Transportation Construction, Faculty of Civil Engineering, Nha Trang University, Nha Trang, Vietnam

V.-T. Do · Q. T. Do
Department of Naval Architecture and Ocean Engineering, Faculty of Transportation Engineering, Nha Trang University, Nha Trang, Vietnam

M. N. Prabhakar
The Research Institute of Mechatronics, Changwon National University, Changwon, Gyeongsangnam, South Korea

C. Muthukumar et al. (eds.), *Flammability Performance of Biocomposites and Bionanocomposites*, Composites Science and Technology,
https://doi.org/10.1007/978-981-95-2389-4_10

1 Introduction to Hybrid Composites and Natural Fiber Reinforcement in Biocomposites and Bionanocomposites

1.1 Definition and Overview of Hybrid Composites Within the Scope of Biocomposites and Bionanocomposites

In the materials science field, hybrid composites have become a new class of materials in which two or more kinds of natural fibers are reinforced in a polymer matrix to synthesize advanced composites with improved multifunctional properties. In respect to biocomposites and bionanocomposites, these hybrid composites have compositionally varied diversity and synergism arising out of interaction between constituents making up such composite materials. Basically, the differentiation between biocomposites and bionanocomposites lies in the scale and type of materials used [11, 17].

Biocomposites are prepared with natural fibers such as hemp, flax, or jute in a polymer matrix. The matrix could be the traditional nonbiodegradable polymers like polyethylene or polypropylene, or it could be bio-based polymers coming from natural resources like starch or cellulose. Natural fibers reinforce strength and lightness in biocomposites. The effectiveness of biocomposites, mainly with regard to their mechanical and thermal properties, is mostly contributed by the natural fiber and the matrix interaction [61, 167].

Bionanocomposites stretch this concept a bit more by using nanoscale materials, like nano-clays, carbon nanotubes, or graphene, within the structure of the composite. Such an addition of nanomaterials in biocomposite significantly enhances the mechanical strength, barrier properties, and thermal stability of the designed composite. This interplay among the nanoscale material-natural fiber interaction in a polymer matrix yields much better composites than their macro-scale counterparts [174, 184]. Figure 1 shows the schematic structure of a hybrid composite where various natural fibers are combined with a polymer matrix. This interaction can be noted between these components so that appropriate overall composite properties, such as improved strength and enhanced thermal stability, are thus developed.

Biocomposites and bionanocomposites reap certain advantages from natural fibers related to both renewability and biodegradability, with generally improved mechanical and thermal properties due to the optimum exploitation of bio-polymeric materials and nanomaterials. The concept addresses the emerging demands in industries like automotive, construction, and packaging and in consumer goods for green and sustainable materials [31, 98].

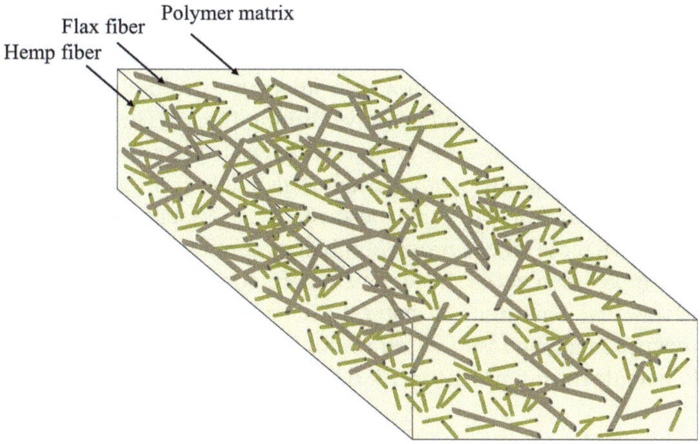

Fig. 1 Schematic diagram of hybrid composite structure

1.2 Role and Importance of Natural Fibers Specifically in Biocomposites and Bionanocomposites

Natural fibers assume significantly important roles in enhancing the mechanical properties and sustainability of both the biocomposites and bionanocomposites. They carry out functions not only as fillers but also as important intrinsic structural entities that render reinforcement properties to the composite matrix. Strengthening of a matrix by addition of fibers like hemp, flax, jute, and kenaf leads to composites with a maximum combination of strength, flexibility, and eco-compatibility.

1.2.1 Improvement of Mechanical Strength and Durability

This plays a critical role in achieving large mechanical strength improvement in biocomposites and bionanocomposites. Fibers such as hemp, flax, jute, and kenaf could also provide an imprint in composites that would quite clearly give their signature of tensile strength, stiffness, flexibility, among other properties. This means, therefore, that the stress distribution in this biopolymer or nanocomposite material enhances durability and performance accordingly. This reinforcement becomes more critical in applications where high mechanical performance is required without affecting environmental sustainability, such as for automobile components where a rise in strength cannot be allowed along with a concurrent increase in weight [4, 208].

1.2.2 Contribution to Environmental Sustainability

The resulting biocomposites and bionanocomposites explore only from the point of view of green material demand adaptive driving forces. The use of these natural fibers is that they are a renewable resource material; therefore, their carbon print is much less compared to the synthetic materials made from glass or carbon fiber-based materials. In addition to that, these fibers are biodegradable, which makes composites out of these fibers provide enhanced options with relation to the end of life of a composite material, hence even more compatible with the principles of a circular economy. This trend in natural fibers is in line with moving the world away from dependence on fossil-based resources toward greener, cleaner, and more sustainable manufacturing processes [8, 62].

1.2.3 Impact on Composite Performance

The thermal attributes of natural fibers enhance the thermal stability of the composite, most indispensable for applications that would bear wideness of temperatures, such as in construction materials or automotive parts. Besides, it can also be used for acoustic insulation. This is very suitable for the applications used when noise reduction is a must, such as in construction materials and vehicle interior applications [72, 99].

1.2.4 Applications in Various Industries

Such versatility of natural fibers makes them applicable in a number of industries. For the automotive sector, the focus is on developing lightweight vehicles that are fuel-efficient while ensuring that safety and strength parameters are not compromised; natural fiber composites can be applied in both drive interiors and the under-the-hood market [77, 157]. There are several benefits of using natural fiber composites in the construction industry: they offer thermal insulation, are good in terms of structural performance and aesthetics, and contribute to sustainable building practices [76, 183]. The use of biocomposites has gained increasing popularity in the packaging industry due to the recent trend toward being eco-friendly. Natural fibers play a substantial role as a potential substitute for traditional packaging materials due to balancing proper strength with low weight and biodegradability [96, 172].

1.3 The History and Progression of Natural Fiber-Reinforced Biocomposites and Bionanocomposites

Natural fibers have been used in composite materials since ancient times, with such fibers combined with other materials—like mud—to develop strong building materials. The modern development of natural fiber-reinforced composites began at the beginning of the twentieth century, stimulated by demands for materials that would combine the advantageous properties of both natural fibers and synthetic matrices.

1.3.1 Early Uses and Initial Developments

In the early years of the twentieth century, studies on natural fiber composites were conducted using the jute, sisal, and hemp fibers in thermoplastics. This was encouraged by their easy availability and superior mechanical properties. First, these types of natural fibers found applications in some fundamental areas, like non-structural components and insulation, with strength and the insulating ability proving quite valuable in those areas. Although environment-related benefits associated with the use of renewable resources were not primary in view, early work focused on exploiting mechanical advantages accruable from such fibers [67, 80].

1.3.2 Progress in Material Science and the Shift to Sustainability

With the course of time, through the twentieth century, synthetic materials were posing environmental problems, which deeply increased interest in studying the environment with advantages of natural fibers. This was when material science had developed new means to extend knowledge about not only the mechanical but also thermal, acoustic, and biodegradable features of natural fibers. Adding biopolymers as matrices made the potential of such composites even greater, developing materials that are not only strong and durable but also environmentally friendly, with the capability for full biodegradability. Fully sustainable composite materials had been stepped into a new approach, aligned with the increasing global interest during that time in cutting carbon footprints and environmental impacts described by Khalid et al. and Cabrera [34, 101].

1.3.3 Enhanced Fiber-Matrix Interactions

Improvement in fiber-matrix interaction has been the main direction in the research relating to natural fiber-reinforced composites. In fact, it is through this interaction that the developed composites acquire most of their characteristics. Such innovations as fiber surface treatment and creation of more compatible matrices have managed to increase significantly adhesion between fibers and matrix, hence developing composites with improved strength and performance. These developments have spawned improved stress distribution in the material, hence making for more robust and dependable composites [62, 76].

1.3.4 The Emergence of Bionanocomposites

The introduction of nanotechnology has added a different dimension to natural fiber-based composites, as bionanocomposites have been developed. These materials merge natural fibers with nanoscale fillers, which are normally nano-clays, graphene, or carbon nanotubes, in order to come up with composites that have enhanced properties. Nanoscale fillers improve the mechanical strength and barrier properties of a given composite significantly and enhance its fire retardancy for higher demanding uses. Combinations of such macro-scale fibers with nanoscale materials have given way to composites that essentially get ahead in parameters like strength, durability, and sustainability to the environment compared to traditional ones [40, 78].

1.3.5 Optimization for Specific Applications

Coupled with the advances in material science have been immense efforts toward tailoring natural fiber composites to specific industrial applications. Mostly, research laboratories of these industries, such as automotive, aerospace, and construction, have been oriented to adjust the optimum fiber types, matrices, and processing techniques so as to meet the specific demands of each particular field. This has yielded many composites that are not only strong and light but also passed rigid environmental and safety standards. In adjusting its properties to end-use applications, a wide scope has been lain for natural fiber-reinforced composites as substitutes for conventional materials in different industries [17, 211].

1.4 Overview of Flammability Challenges Specific to Biocomposites and Bionanocomposites

Biocomposites and bionanocomposites have many environmental and mechanical advantages but do have intrinsic flammability problems. Biocomposites formed by natural fibers and bio-based polymers are highly prone to risks of fire due to their organic nature itself; therefore, strategies against such risks need to be developed.

1.4.1 Intrinsic Flammability Challenges of Natural Fiber Composites

Because they are organic, biocomposites and bionanocomposites have an inherent propensity to combustibility. In fact, natural fibers, such as hemp, flax, and jute, but also the most commonly used bio-based matrices associated with them, consist of organics that can easily act like fuel in case of fire. This flammability is further worsened by the fact that these materials can undergo rapid thermal degradation, which not only exponentially weakens the structural integrity of the composite but also increases the risk of fire propagation. Whereas most synthetic fibers are artificially engineered with flame-retardant properties, such characteristics are not inherently present in natural fibers, and hence the latter are more easily ignited and catch fire [73, 79].

1.4.2 Thermal Degradation Mechanisms and Combustion

High-temperature exposure causes the natural fibers and biocomposites to decompose largely due to the breakdown of cellulose, hemicellulose, and lignin. This weakens the material because the emitted gases are volatile in nature and always support combustion. Fast thermal degradation is a serious issue with such fibers, more so in uses where the composites face elevated temperatures or flames. Understanding the mechanisms of thermal degradation and combustion processes is critical for developing effective strategies that will enhance the fire resistance of these materials [89, 149].

1.4.3 Development of Fire-Retardant Treatments

Extensive research has looked into the development of fire-retardant treatments, trying to handle a major drawback of flammability in both biocomposites and bionanocomposites. Such treatments can be either made to the natural fiber itself, the polymer matrix, or to the composite as a whole. These fire-retardant treatments have been shown to either promote charring by forming a protective barrier from heat and flames or blow out the chain reaction that facilitated combustion. Care, however, has to be exercised while using fire retardants to ensure that mechanical properties enabling the environmental benefits for composites are not negatively affected. This calls for a challenge in balancing fire resistance with the composite's overall performance and its sustainability [32, 199].

1.4.4 Exploration of New Material Combinations

Innovative material combinations certainly reduce the flammability of both biocomposites and bionanocomposites. For example, research has shown that using nanoscale fillers—especially nano-clays or carbon nanotubes—results in solid flame-retardant behavior in the composite. Such nanomaterials may enhance the thermal stability of the composite and create protective action that lessens the rate of heat and mass transportation during combustion. However, in this respect, the uniform dispersion of those nano-fillers within the composite will be inevitable in ensuring homogeneous fire-retardant properties throughout the material [10, 16].

1.4.5 Balancing Fire Retardancy with Performance and Environmental Impact

That makes the achievement of a balance among fire-retardant performance, mechanical properties, and environmental sustainability one of the most complex but extremely critical steps in the development process of biocomposites and bionanocomposites. Quite often, improved fire resistance goes with several types of trade-offs: It may mean weight increases, changes for the worse in terms of mechanical properties, or impaired biodegradability. In this respect, researchers and manufacturers have to be very careful with regard to the trade-offs involved so that materials developed pass those rigorous tests of fire safety and at the same time retain those environmental advantages which make biocomposites/bionanocomposites very compelling replacements for traditional composites [103, 200].

2 Types of Natural Fibers Used in Biocomposites and Bionanocomposites for Enhanced Fire Retardancy

The interest in the integration of natural fibers into biocomposites and bionanocomposites is not self-driven based on mechanical properties and sustainability alone; it is actually based on their real potential to improve fire retardancy. These factors are contributed specially by various kinds of natural fibers, which could be employed strategically with a view toward the improvement of fire resistance in such composites. The next section gives an overview of the most important natural fibers applied in this context and their classification with respect to their relevance for biocomposites and bionanocomposites.

2.1 *Classification of Natural Fibers: Focusing on Those Relevant to Biocomposites and Bionanocomposites*

Most natural fibers originate from plants, while some fibers from animals are used in composite materials. Their classification is important to clearly distinguish their respective roles and values in the performance of biocomposites and bionanocomposites with regard to fire retardancy.

2.1.1 Cotton (*Gossypium* spp.)

One of the widely used natural fibers all over the world, cotton is derived from the seeds of the cotton plant. Softness, absorption capabilities, and breathing/ventilation properties are the characteristics of cotton. Cotton fibers are mainly used in biocomposites in their nonwoven forms or as reinforcement in thermoplastics/thermosets. While the inherent characteristics of cotton are preferred, its fast and moderate thermal stability and propensity to ignite relative to other counterparts make it a material that will require real fire-retardant treatments when it is being used for highly fire-resistant applications [70, 189, 214].

2.1.2 Hemp (*Cannabis sativa*)

Hemp is a fiber with extraordinary tensile strength and durability, combined with resistance to wear. The fiber is derived from the stalk of *Cannabis sativa*, a plant cultivated for thousands of years for many purposes. Hemp fibers are especially valuable in the automotive and construction industries for their properties in reinforcing composite materials for both structural integrity and environmental sustainability. Besides, the hemp fibers have relatively high thermal stability, which makes them useful in high-performance applications where fire retardancy becomes a major issue [5, 71].

2.1.3 Flax (*Linum usitatissimum*)

Fibers are obtained from flax plant stems and display a high stiffness-to-weight ratio. This feature makes flax an ideal reinforcement in composite materials used for sectors like aerospace and automotive, in which weight reduction represents an added value. In addition to that, which is alongside their stiffness, flax fibers also reveal good vibration damping properties that render them very useful in noise and vibration control applications. Flax fibers can be treated with flame retardants to improve their natural thermal stability for the purpose of fire retardancy [145].

2.1.4 Kenaf (*Hibiscus cannabinus*)

Kenaf is a fast-growing plant that is native to Africa and South Asia. The fibers extracted from the stem are strong yet lightweight. Kenaf fibers become very useful in biocomposites for things such as automotive interior components, packaging, and non-structural building material applications. The fact that kenaf is easily cultivated and produces high yields makes it economically and environmentally very attractive. Although considered as having moderate thermal stability, kenaf fibers could be improved by proper fire-retardant treatments [33, 64].

2.1.5 Jute (*Corchorus* spp.)

Jute is a long, soft, shiny fiber mainly cultivated in India and Bangladesh. It is one of the cheapest natural fibers available, thus finding widespread applications in textiles, packaging, and composites. Jute fibers show good moisture absorption and respiration capacity but are rated very low in terms of fire resistance. Therefore, jute-based composites often require the addition of flame retardants to improve the performance of the composites for fire-sensitive applications. Flame retardancy for fire-sensitive applications is achieved by adding flame retardants to the jute-based composites [175].

2.1.6 Sisal (*Agave sisalana*)

Sisal fibers are extracted from the leaves of the *Agave sisalana* plant. These fibers have been found to be strong, durable, and have excellent moisture absorption properties, especially suitable for use in biocomposites parts of an automobile, in geotextiles, and as reinforcement of polymer. Sisal enjoys another added advantage of being grown in arid areas rain-fed with negligible water and chemical inputs, thus fit for sustainability. Although sisal fibers have moderate fire resistance, they can be improved by applying fire-retardant treatment [82, 141].

2.1.7 Bamboo

Bamboo fibers are extracted from different genera of the bamboo plant, a highly fast-growing and sustainable vegetal material. Bamboo fibers have been found to be strong in tension, with great flexibility; thus, they can be applied in construction materials, automotive parts, and packaging biocomposites. Additional treatments may be required to achieve the needed fire retardancy in some works, as the natural silica content in bamboo provides only some inherent fire resistance [121, 220].

2.1.8 Cellulose Nanocrystals (CNCs)

Cellulose nanocrystals are nanoscale fibers acquired from plant-based cellulose sources, for example, wood, cotton, and hemp. Nanofibers depict utmost strength and rigidity, therefore considered paramount as reinforcement agents in biocomposites. CNCs efficiently enhance the mechanical, barrier, and thermal properties of composites. In addition, they have a role in fire retardancy through facilitating char formation while minimizing thermal degradation [161, 191].

2.1.9 Microfibrillated Cellulose (MFC)

Another variant of nanoscale cellulose from plant cell walls is microfibrillated cellulose. The average MFC fibers have a high aspect ratio and surface area, making them good for their application as reinforcing agents in biocomposites. Notably, potential applications include packaging areas, from automotive parts to construction fields. One most important features is the increasing surface area that allows the fibers to bond well with the polymer matrices, imparting improved fire-retardant properties to the composite [116, 120]. Table 1 provides an overview of the mechanical properties, thermal stability, fire retardancy, and common applications of these natural fibers when used in biocomposites and bionanocomposites.

2.2 Overview of Fibers Commonly Used in Biocomposites and Bionanocomposites

In the development of biocomposites and bionanocomposites, there are many categories of natural fibers that can improve the mechanical, thermal, and environmental properties of the final products. In this paper, a review is done on some of the frequently used fibers and their unique features which make them appropriate to enhance composite materials.

2.2.1 Cotton

Cotton is very well known, ranging from traditional textiles to composite applications. In biocomposites, cotton finds use in nonwoven forms or as reinforcement in both thermoplastics and thermosets. Because of its inherent softness and high absorbency and breathability, it turns into an appealing use where comfort and flexibility are required. However, with its medium-thermal stability, cotton composites often need flame-retardant treatments to enhance the fire resistance of the products in these very demanding, high-temperature applications [93].

Table 1 Overview of natural fibers used in biocomposites and bionanocomposites

Fiber type	Mechanical properties	Thermal stability	Fire retardancy	Applications	References
Cotton (*Gossypium* spp.)	Moderate	Moderate	Weak	Textiles, non-structural composites, packaging	[68, 113]
Hemp (*Cannabis sativa*)	Strong	High	Moderate	Automotive, construction, structural composites	[5, 95, 136]
Flax (*Linum usitatissimum*)	Strong	High	Moderate	Automotive, aerospace, noise reduction	[115, 143]
Kenaf (*Hibiscus cannabinus*)	Moderate	Moderate	Weak to moderate	Automotive interiors, packaging, non-structural building	[1, 135]
Jute (*Corchorus* spp.)	Moderate	Low	Weak	Textiles, packaging, non-critical biocomposites	[56, 186]
Sisal (*Agave sisalana*)	Strong	High	Moderate	Automotive parts, geotextiles, polymer reinforcement	[87, 134, 169]
Bamboo	Strong	High	Moderate	Construction, automotive, packaging	[13, 97, 125]
Cellulose nanocrystals (CNCs)	Very strong	High	Strong	High-performance composites, fire-retardant materials	[147, 190, 191]
Microfibrillated cellulose (MFC)	Strong	High	Strong	Packaging, automotive, construction	[116, 194, 202]

2.2.2 Hemp

Such fibers are highly valued for their strength, durability, and low impact on the environment. The hemp fibers, extracted from the *Cannabis sativa* plant, find high-end application in the automotive and construction fields since they can reinforce composite materials. These show very high tensile strength and stiffness and thus are applicable for load-carrying purposes. The quite high thermal stability of hemp, therefore, contributes to the general fire-retarding ability of the composites it is used for, reducing highly extensive fire-retardant treatments [93, 139].

2.2.3 Flax

It is rigid and often used in formulating biocomposites for the automotive and aerospace industries. The fibers are highly valued since they will be able to offer the desired structural strength requirements and, at the same time, keeping the composite weight down. Another characteristic of flax fibers is their natural vibration damping properties, which ideally find application in noise reduction. Attempts to improve the fire resistance of flax-based composites result in them often being treated with flame retardants, mainly when used in environments under stringent fire safety requirements [93, 139].

2.2.4 Kenaf

Kenaf fibers are derived from the plant *Hibiscus cannabinus* and offer a unique combination of strength, flexibility, and low weight. These fibers are consumed in large quantities in biocomposites for applications related to automotive interiors, packaging, and non-structural building materials. Being a fast-growing crop with high yields, kenaf is economically and sustainably viable for producing composites. It is only moderate to show thermal stability of kenaf fibers, and a treatment with fire retardants could be carried out to optimize their performance in an application in which fire resistance is of main importance [139, 230].

2.2.5 Jute

Jute fibers are considered one of the cheapest of the natural fibers produced in South Asia. Applications using such fibers deal with textiles, packaging, and reinforcing material in polymer composites. Only due to the moisture absorption property and moderate mechanical properties will this become possible for non-critical parts in biocomposites, such as carpet backing, rope, and sacks. However, the fire resistance of jute fibers tends to be relatively low, and thus treatments with flame retardants would be in order to improve their suitability for application in situations where fire safety is a concern [137, 230].

2.2.6 Sisal

Sisal fibers derive from the plant *Agave sisalana*, characterized mainly by durability, strength, and moisture absorption capabilities. Such properties make sisal a suitable component as reinforcement material in biocomposites to be used in the manufacturing of automotive parts, geotextiles, and in the reinforcement of different polymer matrices. Sisal is therefore very suitable for growth in arid climates and is also sustainable in that it requires very little water and does not need any chemical fertilizers or pesticides. Some treatment that increases the fire performance of sisal includes the use of flame retardants, mostly in highly demanding fire safety applications [93].

2.2.7 Bamboo

Extracted from a fast-growing plant that is treasured for its renewability and large tensile strength, bamboo fibers find increasing applications in biocomposites for construction materials, automotive parts, and packaging. The natural silica content in bamboo inherently provides some fire-retarding capability, but for applications where high fire safety standards prevail, this will require additional treatments to achieve the required degree of fire-retarding capability [93, 139].

2.2.8 Cellulose Nanocrystals (CNCs)

Cellulose nanocrystals are nanoscale fibers made from plant cellulose, known to be one of the most abundant polymers and to offer very high strength and stiffness. These characteristics render the material best suited and valuable in application areas that require high mechanical strength as well as fire retardancy toward the enhancement of mechanical and barrier properties in biocomposites. The retardation of thermal degradation and char formation are also enhanced by CNCs [93, 139].

2.2.9 Microfibrillated Cellulose (MFC)

Microfibrillated cellulose is nanoscale fibers from plant cell walls. MFC represents high aspect ratio and large surface area and is therefore classified as a material that can reinforce biocomposites. Such applications relate to packaging, automotive, and construction materials, where the improvement of the bonding between fibers and polymer matrices is a bottleneck in applications. MFC contributes to improving the fire performance of composites by improving flame-retardant dispersion and reducing the general flammability of the material [139, 230].

2.3 Properties of Each Fiber in the Context of Biocomposites and Bionanocomposites

Natural fibers for biocomposites and bionanocomposites will have to be chosen based on thermal stability, mechanical strength, and fire resistance. Each fiber contributes specifically to the overall performance of the composite material. These properties are discussed in detail to show their importance in the development of biocomposites and bionanocomposites.

2.3.1 Thermal Stability

- Hemp and Flax Fibers: Hemp and flax fibers have higher ignition temperatures and thermal stabilities due to their cellular structure and chemical composition. That consequently makes them useful in high-temperature or fire-resistance areas, for the simple reason that such fibers are difficult to ignite as compared to others, followed by a lower burning rate [16, 166].
- Kenaf and Jute Fibers: Most of the fibers have shown very bad heat stability; however, when fire-retardant chemicals are used, their performance improves drastically. These fire-retardant chemicals enhance either char-layer development, thus tending to act like insulation against heat and flames, or alter the steps of thermal decomposition in such a manner as to delay ignition and lower flammability [19, 102].
- Cotton: Although cotton fibers are of medium-thermal stability, which seems to increase the flammability tendency somewhat, many high-temperature resistances can be earned by incorporating fire-retardant treatments. This should make it more applicable in areas where fire safety is of utmost concern [222].
- Sisal and Bamboo Fibers: Both sisal and bamboo fibers show good thermal stability, with the bamboo fiber specially advantaged due to the presence of its natural siliceous content, which translates into better resistance against thermal degradation. As such, they can both be quite useful in applications that involve exposure to a wide range of temperature variations [25, 222].

2.3.2 Mechanical Strength

- Hemp Fibers: With the presence of high tensile strength, hemp fibers are long and strong in cellulose fibers, so as to be very useful for applications that require a high load-bearing capacity. Their mechanical properties, therefore, make hemp fibers a kind of preferred reinforcement for composites, particularly within the automotive, construction, and other industrial sectors [25].
- Flax Fibers: These are stiff fibers, due to the high cellulose content and particular microstructure. This stiffness is important in fields where rigidity is required, such as in automotive structural parts or aircraft applications. Moreover, flax fiber contribution to vibration damping boosts the added value of their application in those areas [162].
- Kenaf and Jute Fibers: These find a good balance between strength and flexibility. It renders good semi-structural as well as non-load-bearing applications of the composite material. These fibers are easily moldable into complex shapes. Though tensile strength may not be equivalent to hemp or flax, these serve as good reinforcement that finds an application for producing composites meant for duty applications which are less demanding in nature [20].
- Cotton: Although cotton fibers are not as strong as some other plant fibers like hemp or flax, they have sufficient mechanical strength to substantiate non-structural applications. Moreover, in terms of strength, they can be greatly

improved when combined with other fiber reinforcement or polymer matrices, thus creating a number of possibilities for their use in several kinds of composites [222].
- Sisal and Bamboo Fibers: Sisal fibers indicate high mechanical strength that is very durable. Contrary, bamboo fibers display exceptionally high strength and stiffness. The indication of these properties serves to prove that the two materials are appropriate in carrying structural applications and mostly in industries where mechanical performance has a vital role [25].

2.3.3 Fire Retardancy

- Composition and Fire Retardancy: The inherent composition of fibers, such as hemp, flax, kenaf, and jute, significantly impacts their fire-resistant properties. As an instance, the lignin in jute and kenaf fibers forms char during combustion, which acts as a barrier to flame spread by slowing it down [162].
- Cotton, Sisal, and Bamboo: Cotton and sisal fibers may undergo fire-retardant treatments that further enhance their natural fire resistance. There is some level of built-in fire retardancy in bamboo due to its intrinsic siliceous content, but this can be further improved through treatments to meet specified fire safety requirements for a given product [20].
- Enhancing Fire Retardancy: Such fibers can further be modified to enhance their retarding ability. These include the techniques of applying fire-retardant chemicals, incorporating fire-resistant minerals, and blending it with polymers with fire-resistant properties in order for use in industries such as construction, automobiles, and aeronautic industries that bear strict fire safety standards [18].
- Cellulose Nanocrystals (CNCs) and Microfibrillated Cellulose (MFC): Both can make significant contributions toward fire retardancy in composites. CNCs favor the formation of char, while MFC improve the distribution of flame retardants within the composite. Such nanocellulose-based materials have shown high thermal stability and mechanical reinforcement, further enhancing the fire-resistance properties of the biocomposites and bionanocomposites [24, 39, 181].

The unique advantages which each natural fiber contributes to thermal stability, mechanical strength, and fire retardancy suffice for a variety of applications. One can develop biocomposites and bionanocomposites that ensure the developed products, with deep understanding of these properties, retain their mechanical and thermal stress-bearing capacity while they meet stringent fire safety standards as well.

2.4 Comparative Analysis of Fire-Retardant Capabilities Specific to Biocomposites and Bionanocomposites

Any development of biocomposites or bionanocomposites shall be based on fire-retardant properties in relation to different natural fibers, matrix materials, additives, and nanomaterials to have safer and more resilient material development. This comparative analysis is done with some research to identify fire-retardant properties for hemp, flax, kenaf, jute, cotton, sisal, bamboo, cellulose nanocrystals, and microfibrillated cellulose in terms of enhancements of the said properties using combinations of materials.

2.4.1 Fire Retardancy in Different Natural Fibers

- Hemp and Flax: Represent fibers that exhibit highly inherent fire-retardant properties due to their high-temperature ignition. Thus, it finds prominent uses in the high-temperature applications of the automotive and aerospace industries [164].
- Kenaf and Jute: Although Kenaf and Jute have the least inherent fire resistance than other natural fibers, several recent studies demonstrated that their potential for fire retardancy can be significantly improved either by chemical treatment or incorporating fire-resistant material into the composite matrix. This can make them increasingly suitable for construction-related applications where strict fire safety regulations are a factor [128, 185].
- Cotton: Its ignition point is relatively lower compared to the other fibers; however, its fire resistance can still be effectively improved with the use of fire-retardant chemicals. These treatments are indispensable to create new uses of cotton in many consumer and industrial applications where improved fire safety is mandatory [81, 118].
- Sisal and Bamboo: While treatment processes are increasingly making sisal fibers more fire-resistant, the former will be helped by the inherent siliceous content of the latter which provides a certain level of natural reticence to fires. Their characteristics thus fit for use within construction and the automotive sectors [185].
- CNCs and MFC: Due to high char formation tendencies, cellulose nanocrystals and microfibrillated cellulose have demonstrated high thermal degradation rates, which therefore mainly impart fire retardancy into the composites. Developing fire-retardant systems based on these renewable materials could become a very promising way to allow for biocomposites with better fire retardancy and reduced ecological footprint [128, 178, 193].

2.4.2 Role of Matrix Material and Additives

One of the main parameters by which biocomposites and bionanocomposites are defined with regard to fire-resistance properties is the choice of matrix material. There can be large differences in thermal and combustion properties between biopolymers and synthetic polymers, such that the total fire resistance of the composite could be affected accordingly. Incorporating additives that are fire-retardant remains one of the most common approaches to enhancing fire resistance. Such additives act either by inducing char formation, creating physical barriers, or chemical interference with the combustion process. Therefore, it becomes very important in the construction sector, especially for residential and public buildings, due to the fire safety regulations that are particularly stringent [111, 128].

2.4.3 Improvement with Nanomaterials in Bionanocomposites

Probably one of the most promising research areas in this connection is the use of nanomaterials that can be embedded into a composite matrix, thereby improving its fire retardancy. Such nanomaterials can achieve this by forming some kind of protective barrier around the surface of the composite material, accordingly limiting its exposure to oxygen and heat and thus slowing down the process of combustion. The effectiveness of the latter depends on a controlled nanomaterial distribution and concentration in the matrix to ensure optimal fire resistance [66, 203].

2.4.4 Synergistic Effects in Fire Retardancy

There is evidence of synergistic effects when natural fibers are used in association with either nano-enhanced matrices or fire-retardant additives. The synergy shown by the combination between such kinds of fibers, the specific matrix, and additive or nanomaterial contributions will be the origin of the unexpected characteristics that motivate the research in fire resistance of composites. Such a composite material does not ignite easily, and in case of ignition, it has reduced flame spread, smoke generation, and emission of toxic gases during combustion—properties that keep pace with the growing global demand for sustainable yet safe materials [16, 193, 203]. Figure 2 shows fire retardancy mechanisms in biocomposites and bionanocomposites. It is observed that the interaction among natural fibers, matrix materials, and nanomaterials lead to an improvement in fire retardancy.

The potentials of different natural fibers with respect to their fire-retardant properties, together with matrix materials, and the eventual improvement by the incorporation of additives and nanomaterials, must be fully comprehended. In doing so, designers can strategically develop biocomposites and bionanocomposites to meet a set of predetermined criteria—defining safety and application standards that will help to accommodate growing global demand for both sustainable and safe materials.

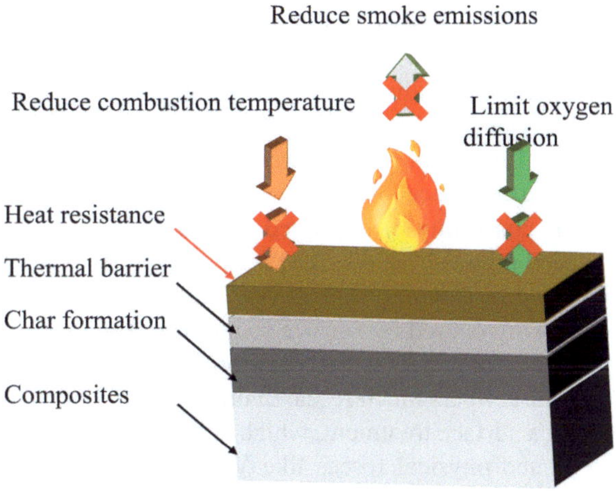

Fig. 2 Mechanisms of fire retardancy in biocomposites and bionanocomposites

3 Interaction Between Natural Fibers and Matrix Materials in Biocomposites and Bionanocomposites

3.1 Fundamentals of Matrix-Fiber Interactions in Biocomposites and Bionanocomposites

3.1.1 Importance of Interfacial Adhesion

- Critical Role in Performance: Interfacial adhesion between the involved natural fibers and the matrix plays a critical role in defining the resulting mechanical strength and overall efficiency of biocomposites and bionanocomposites. Effective interfacial adhesion would guarantee efficient transfer of the stress and strain from the matrix across the interface to the fibers for the cohesion needed to maintain integrity under mechanical loads. Recent studies demonstrate that the bond is instrumental in enhancing mechanical properties; therefore, any weak bond results in unsatisfactory durability and stress performance of these composites, as observed in several works such as Lee et al., Andrew, and Dhakal [14, 117].
- Determining Factor for Composite Integrity: Broad transmission of loads from the matrix to the reinforcing fibers requires strong interfacial adhesion; hence, it becomes a central determinant in the maintenance of composite integrity and its lifespan against a raft of mechanical stresses. The development pertaining to bast fiber-reinforced biocomposites has very recently come up with the fact that consolidated fibrous structures with the polymer matrix have surprising improvements in mechanical properties evincing strong fiber-matrix adhesion in the same [201].

3.1.2 Mechanisms of Fiber-Matrix Adhesion

- Chemical Bonding: Because the functional groups of fibers and matrix are compatible, it is possible to form chemical bonds at a molecular level. The development of new functional groups through recent treatments of natural fibers has given rise to strong covalent bonds or hydrogen bonds with the matrix, hence increasing the possibility of improved chemical adhesion that enhances overall composite performance [14, 63, 201].
- Physical Bonding: The physical adhesion mechanisms include mechanical interlocking caused by surface roughness of the natural fibers that interlock with the matrix material. It has been reported that in recent studies on modification of natural fibers much research on fiber surface roughness has been done and called for a surface treatment, which enhances better mechanical interlocking and other related physical forces like Van der Waals and electrostatic interactions hence improving fiber-matrix adhesion [12].

3.1.3 Influence of Matrix Material

- Biopolymer Matrices in Biocomposites: Most of the time, a biopolymer acts as a matrix material in biocomposites, and parameters of compatibility and bondability with natural fibers may differ from synthetic polymers. As most natural fibers are hydrophilic in nature and biopolymers range from rather hydrophobic to rather hydrophilic, quality adhesion is considerably affected by these phenomena. The increasing demand for biocomposites due to environmental awareness and sustainability issues puts forward the need for enhancement in fiber–matrix adhesion of such materials [14].
- Nano-fillers in Bionanocomposites: Nano-fillers may be incorporated into bionanocomposites and are likely to drastically change the matrix's properties. Some recent research has demonstrated that the addition of nano-fillers, such as nano-clays, graphene, and carbon nanotubes, can improve mechanical properties and increase matrix–natural fiber interactions. The nano-fillers provide additional bonding sites and, depending on their concentration, could also shift the viscosity of the matrix and its curing behavior to obtain stronger, more resistant composites [184].

3.1.4 Processing Techniques and Their Effects

- Effect of Manufacturing Processes: The production methods used to develop biocomposites and bionanocomposites in general are very crucial during the quality determination of the fiber-matrix interface. For example, processes such as extrusion, injection molding, and 3D printing have certain parameters which affect to a great extent the bonding degree between the fibers and the matrix. Recent developments in the use of biofiber-based polymer composites by different industries and research on the comparison of various recycling processes ought to prepare a pathway for the optimization of the manufacturing technique to attain desired properties in produced materials and set sustainability goals [60].

- Surface Treatments and Compatibility: The surface treatments of the natural fiber, such as alkali treatment, silanization, or acetylation, are very important to allow it better bond with the matrix. Some of these treatments may clean impurities, change the surface roughness, introduce functional groups, improve bonding, and so forth. Studies that involve modification of natural fibers are still reflecting that such treatments are necessary for maximum improvement in interfacial adhesion for enhanced overall performance in both biocomposites and bionanocomposites [12].

3.2 Chemical and Physical Bonding Mechanisms at Fiber-Matrix Interfaces in These Composites

3.2.1 Chemical Bonding Mechanisms

- Types of Chemical Bonds: At the interface fiber-matrix, it can occur by covalent, ionic, or secondary mechanisms, like hydrogen bonding. The latter types of bonds, namely, covalent and ionic bonds, are very strong and predominantly contribute to the composite strength. The advances in recent times on the interphase engineering of polymer composites were based on developing hierarchical fibers with nanostructures and handling fiber surfaces by polymer deposition that enhances these kinds of chemical bonds to improve composite performance [92].
- Surface Treatments for Advanced Bonding: Chemical treatments on the natural fibers are conducted in order to gain better compatibility with the matrix and assisted in providing better chemical bonds. As an example, silane treatments can impart functional groups on the surface of the fiber that can react with a matrix to form strong covalent bonds. Other types of treatments include the use of coupling agents or compatibilizers, which help build a bridge between the hydrophilic nature of natural fibers and mostly hydrophobic matrix materials. Several researches today focus on these treatments to enhance the bonding properties by improving the chemical interaction between fiber and matrix, hence improving bonding and general composite performance [187].
- Impact on Composite Properties: The enhanced chemical bonding at the fiber-matrix interface increases the effectiveness of the load transferred between the matrix and the fibers. This is what sustains the composite's integrity and mechanical performance. Besides, strong chemical bonds reduce moisture absorption by natural fibers, hence increasing the durability of the composite material. The latest research into plant fiber interfaces and the mechanisms of bonding has pointed out the importance of the fiber-matrix interface sometimes referred to as the reaction or diffusion zone in defining the mechanical properties for the composite and

thereby defining the necessity for proper relations in regard to chemical bonding [229].

3.2.2 Physical Bonding Mechanisms

- Hydrogen Bonding: Hydrogen bonding is another secondary bonding that can take place within a fiber-matrix interface, especially if there are polar groups available on the fiber surface and the matrix. Though weak as compared to covalent bonds, hydrogen links make their contribution toward cohesion and hence to the overall stability of the composite. Microscopic investigation of the interface in carbon fiber-reinforced polymer composites suggests understanding mechanisms governing degradation, including hydrogen bonding, is a key requirement for accomplishing any long-term improvement [124].
- Mechanical Interlocking: This technique is highly influenced by the texture and roughness of the surface of natural fibers. Rougher fiber surfaces offer more crevices and points for the matrix material to latch onto, hence improving the physical fiber-matrix bond. This mechanism is very important when there is little chemical compatibility between the fiber and the matrix, so it provides an alternative means of holding the fiber within the composite [48, 140].
- Van Der Waals Forces: These forces, though relatively weak, can still facilitate the adhesion between fiber and matrix when they are very close. Although very weak individually, collectively these forces could slightly add to the adhesion in a composite [114, 140].

3.2.3 Effect of Fiber Surface Properties

- Surface Roughness: The natural roughness of the fiber, as in the case of jute or hemp, is high and thus provides good mechanical interlocking. Surface treatments will modify this and optimize it for better bonding. Nanotechnology and computational modeling are advances that have made significant improvement in the understanding of deterioration mechanisms and durability of FRP composites, underscoring the role of surface roughness in mechanical interlocking and its effects on long-term performance [123].
- Surface Chemistry: The surface chemistry of the fiber is important in the degree to which it can form chemical bonds with the matrix. Therefore, fiber surface treatment processes like alkalization remove non-cellulosic materials from the surface, making more reactive groups available to increase potentially specific chemical bonding. One of the basic factors controlling the short- and long-term performance of FRP composites is that at the fiber-matrix interface, since the effectiveness of shear lag stresses transferred from the matrix into the fibers depends on the quality of this interface. Therefore, understanding and optimizing surface chemistry for better bonding assume great importance [124].

3.3 *Impact of Fiber-Matrix Interactions on Thermal Degradation and Fire Behavior in Biocomposites and Bionanocomposites*

The interactions at the fiber-matrix interface have a serious effect on the thermal degradation and fire dynamics of biocomposites and bionanocomposites. Interactions govern the resistance to heat, retardation of flame front, and suppression of other harmful emission formation in combustion.

3.3.1 Interfacial Adhesion's Significance on Thermal Properties

Improved Thermal Stability: Another way in which the composites demonstrate improved thermal stability is through the development of strong bonding between the fibers and the matrix. Scientific research has revealed that in cases where fibers are perfectly bonded and distributed uniformly within the matrix, they act to efficiently disrupt heat flow and retard flame propagation, therefore enhancing resistance to thermal degradation. The intensity of this interfacial bonding is a very critical factor in deciding the decomposition behavior of the composite material. Moreover, a strong interface can ensure uniform degradation, which hence can reinforce the fire resistance by not allowing non-uniform breakdown of material under the high temperature [128].

3.3.2 Role in Flame Propagation

- Barriers to Flame Spread: The natural fibers in the biocomposites act as a physical barrier that helps in the compartmentalization of the flames in a very limited area, thus retarding their spread throughout the material. The incorporation of these fibers in the matrix in an effective manner results in an increase in the ability of the composites to reduce the spread of fire. Current studies that have been conducted on flammability for natural fiber-reinforced composites underscore the need to optimize fiber-matrix interaction in an enhancement program of flame retardancy. This goes to show that the prevention of the spread of fire needs an interface that is aptly engineered, and hence, it limits fire spread drastically [128, 185].
- Char Formation and Insulation: Some natural fibers are prone to produce a protective char layer in a heated condition. This char offers an insulating barrier protecting the inner layers of the composite from direct heat and flame impingement. Successful charring and insulation significantly based on the quality of the fiber-matrix interaction. Recently, studies have been augmented to optimal development of these interactions, promoting effective char formation in combustion for maximum fire performance of the composite [49].

3.3.3 Enhancement with Nanomaterials in Bionanocomposites

- Formation of Protective Char Layer: Several research studies have indicated the considerable increase in fire retardancy upon using nanomaterials, particularly nano-clays, or carbon nanotubes as additives bionanocomposites. They help develop a more stable and protective char layer with that, thus offering an additional barrier to heat and flames. Most of the developments in this field have aimed for the strategic enhancement of thermal stability and flame resistance in composites through strategic integration of nanomaterials with natural fibers, and the resultants demonstrate much superior defense to fire [74, 128].
- Better Dispersion and Interaction: The efficiency of nanomaterials in the improvement of the thermal stability and diffusion of fire considerably depends on the dispersion and interaction of the material in the composite. Well-dispersed nanomaterials lead to a structure much more homogeneous throughout a composite, which results in very good thermal and fire-resistant property through the material. The flame-resistant property results from a lack of oxygen supply into the combustion zone, thereby hindering true material burning and flame spread during a fire. Such thermal and the flame-resistant properties are constant throughout the material, making this homogeneity very important to ensure that a composite can, in fact, resist high temperatures or spread a drop flame under flaming conditions [49].

3.3.4 Affecting Smoke and Toxic Gas Emissions

- Smoke Reduction: It's the fiber-matrix interface configuration that might help reduce or repress smoke and the emission of toxic gases in the event of combustion. Effective dispersion and bonding at the interface lead to an enhanced quality of the combustion process, where it may be possible to eliminate or greatly reduce the emission of smoke and harmful gases. This becomes critically important when such applications are employed, where safety is of prime importance and directed to causing the least amount of smoke. Summary researchers and other advances have indicated recent advances where people have noted a need for further undertaken research in the better understanding as also happening with the view of subsequently reducing from smoke and other emitted gases [49].
- Material Compositional Dependency: Composition of fibers, the matrix, and nanomaterials or fired retardants incorporated is key in defining the behavior of these composite regarding emissions of smoke and gas. Current research is fine-tuning these constituents in order to obtain the right balance between performance, safety, and environmental impact. Proper selection and optimization of such materials can lead to the development of composites that meet fire safety standards and at the same time minimize the environmental impact from their use [74, 185]. Table 2 summarizes the various effects of fiber-matrix interactions on thermal degradation, fire behavior, and emissions in biocomposites and bionanocomposites.

Table 2 Impact of fiber-matrix interactions on thermal degradation, fire behavior, and emissions in biocomposites and bionanocomposites

Aspect	Effect of fiber-matrix interaction	Significance	References
Thermal stability	The strong bonding between the fibers and the matrix disrupts heat flow and retards flame propagation	It improves thermal degradation resistance and provides uniform breakdown	[128]
Flame propagation	The effective fiber incorporation creates a physical barrier to compartmentalize flames	It helps in reducing the flare-up of fire across the material and thus improves fire safety	[128, 185]
Char formation and insulation	Such quality fiber-matrix interaction provides a way for char to form in an effective insulating manner	It protects the inner layers from heat and improves fire performance	[49]
Nanomaterial enhancement	Nano-clays are among the nanomaterials that increase the stability of the char and develop extra barriers against heat	It significantly improves fire retardancy and thermal stability	[74, 185]
Smoke and toxic gas emissions	Proper dispersion and bonding reduce the amount of smoke and toxic emissions when burned	Crucial for safety, particularly when low-emission work is realized	[49]
Material composition	Selection of fibers and matrix, as well as their combination with nanomaterials, can modify smoke, gas emissions, and general fire behavior	Balances performance, safety, and environmental impact	[74, 185]

3.4 Analytical Methods for Examining Fiber-Matrix Interactions Specific to Biocomposites and Bionanocomposites

Advanced analytical techniques are required to assess the fiber-matrix interactions in biocomposites and bionanocomposites and should be utilized to analyze the microstructure, chemical bonding, and thermal properties of these materials at a detailed level. Some of the highly important methods to understand and optimize these interactions are as follows.

3.4.1 Scanning Electron Microscopy (SEM)

- Morphological Analysis: An SEM is a very important tool. It can be used with high resolution and, therefore, be of aid in visualizing the microstructure of both biocomposites and bionanocomposites. Recent developments have focused on its potential to allow detailed observations of the fiber-matrix interface, describing how fibers are integrated with the matrix and measuring the effectiveness of surface treatments on mechanical properties. Facilities of SEM include the capturing images at micro and nanoscales; hence, an instrument is necessary since one would wish to study the interface implications among other furnishing that occur after fiber modifications before finally studying its implications on the overall performance of composites [201].

- Voids and Dispersion Detection: SEM is also effective for void, crack, and unbound region detection at the fiber-matrix interface, which provides key indicators in terms of quality and integrity. The tool proves quite helpful regarding the dispersion of the fibers/nano-fillers within the matrix to obtain uniform material property distribution. Voids or inconsistent dispersion in the matrix can considerably affect the mechanical and thermal properties of the composite, which renders SEM as one crucial quality control tool for ensuring performance optimization of materials [206].
- Surface Modifications: SEM is rather effective in assessing the effect, in terms of the surface treatments, on the fibers—like etching or coating of the fibers—in order to enhance their interfacial adhesion and compatibility with the matrix. In improving fields of biocomposites, the role of SEM in the evaluation of such surface treatments and the changes in fiber-matrix bonding has become very important. It provides insight into how these changes eventually bring an improvement in performance for said composites [158].

3.4.2 Fourier Transform Infrared Spectroscopy (FTIR)

- Chemical Bonding Analysis: FTIR spectroscopy has been mostly used to identify chemical bonds and functional groups at the fiber–matrix interface. This technique comes in very useful to interpret the nature and types of chemical interactions resurging as adhesion which underpins biocomposites and bionanocomposites. Characterization studies research the effective surface treatments and strong interfacial adhesion of a specified bond between inter-fiber and matrix to result in good mechanical performance of a composite and its sustainability against the environment [14].
- Chemical Change Monitoring: FTIR also plays a prime role in monitoring the chemical changes in fiber and matrix, such as those due to surface treatments or the addition of compatibilizers. Of a similar interest are cases of chemical changes leading to the generation of new bonds or the removal of an unwanted group that is important in gaining satisfactory performance out of advanced biocomposites. By verifications of such changes, the use of FTIR will confirm that chemical modifications have occurred and as a result enhanced material change [14].

3.4.3 Thermogravimetric Analysis (TGA)

- Thermal Stability Measurement: TGA is one of the most efficient techniques used in the determination of thermal stability for biocomposites and bionanocomposites through its recording of weight loss against an increased temperature. This determination gives valuable data about the decomposition temperature and thermal stability of composites, most valued for its high-temperature applications. The data from TGA help explain how interactions between fiber and interface could sum up to the composite's thermal behavior and were further utilized in the development of materials with improved thermal resistance [206].
- Degradation Behavior: TGA also provides knowledge about degradation behavior associated with the composite regarding the way of its breakdown behavior under thermal stress. The resistance to degradation of the composite provides evidence about how efficient fiber-matrix interaction proves and significantly affects the behavior of being fire-retardant materials. Degradation patterns and temperatures usually help to tune fiber-matrix bonding for improved thermal stability and fire resistance of the composite [206].

3.4.4 Differential Scanning Calorimetry (DSC)

- DSC Thermal Transition Analysis: Dealt with thermal transition analysis of the composite which includes glass transition temperature (T_g), melting points, and crystallization patterns. The quality of fiber-matrix adhesion could contribute significantly toward this—an important parameter for performance improvement of the developed composite under thermal stress. Knowledge of these thermo-transitions is essential in predicting material behavior with varying thermal environments so as to enhance its suitability to a particular application [86].
- Impact on Fire Behavior: The thermal transitions can elucidate just how the composite may behave under fire conditions. For example, normally high T_g values correspond to better thermal stability, so these materials, as a rule, show higher qualities of resistance against deformation and degradation in high-temperature conditions. Since recently, bionanocomposites are more and more getting involved with industrial application, one may say the detailed study of such thermal properties is indispensable for ensuring safety and desired behavior [86].

4 Case Studies of Hybrid Composites with Various Natural Fibers in Biocomposites and Bionanocomposites

4.1 Cotton and Hemp

The synergistic effects of various material compositions toward increasing flame retardancy in cotton fibers have been testified to by a few fundamental investigations carried out in the emerging area of biocomposites and bionanocomposites. Of these, one piece of research by Mateusz Barczewski et al. deals with the incorporation of materials like expanded vermiculite and ammonium polyphosphate into epoxy/flax-cotton composites. Together, they greatly reduced the heat release rates through improved thermal stability; hence, a prominent improvement in their resultant composite is noted for its fire-resistance properties [28].

Similarly, Suheyla Kocaman et al. investigated the use of cotton waste blended with flame-retardant additives in phenolic novolac epoxy composites. Their findings have revealed that char formation will be more significant, hence improving the flame resistance capacity of the composite [108]. Another critical study by Jie Xu et al. exemplifies an application of carbonaceous nanomaterials and, in particular, single-walled carbon nanohorns in combination with ammonium polyphosphate on cotton fabrics—this technology can realize high improvements in fire resistance at very low additive levels [218].

These developments are expanded upon by Carlos Rafael Silva de Oliveira et al., who proposed a new kaolinite-TiO_2 nano-hybrid composite applied to cotton via a solvothermal process, improving the fire resistance of the fabric dramatically [44]. Other than that, Na Li et al. have put forward the development of a strange Schiff base synthesized SiDP nanoparticle that imparts flame retardancy and antibacterial properties to cotton fabrics in their work [119]. Finally, Dangge Gao et al. investigate a composite of poly(acrylic acid), modified attapulgite, and zinc oxide, which can improve the flame-retardant and UV protection performances of cotton fabrics [54].

These works collectively highlight significant improvement in flammability performance through synergistic effects of different composite materials in cotton fiber-based biocomposites and bionanocomposites. This study opens more ways to develop safer, more sustainable, and eco-friendly fire-resistant materials, developing the idea that new material combinations and hybridization techniques can alter the functional properties of cotton-based composites.

Parallel to this, the same kind of growth was observed in hemp-based composites. In the domain of hemp's biocomposites and bionanocomposites, a piece of research by Sakil Mahmud et al. is concentrated on a high-temperature shape memory polymer reinforced by both glass and natural hemp fibers. This study primarily found significant bending strength and flame retardancy coupled with excellent shape memory characteristics of composites compared to those with only hemp fibers. Irrespective, the study can still promote the potential application of different blends involving natural and synthetic fibers in significantly improving mechanical properties together with fire resistance in hemp-based composites [130].

These improvements in hemp composites reflect work in cotton-based composites in a quest for advanced materials with improved flammability and enhanced sustainability in biocomposites and bionanocomposites. It further justifies hybrid materials' role in building up composites that not only meet safety needs but also meet the expectations of environmental sustainability.

4.2 Flax

Flax fiber composites have gained much interest in the domain of biocomposites and bionanocomposites due to their potential to couple mechanical strength with enhanced flame retardancy. Actually, a number of studies have been devoted to investigating such properties, which resulted in quite important developments in material performance.

Shivakumar Rachoti et al. studied hybrid composites with flax and sisal fibers and achieved noteworthy mechanical property improvements, but at a cost: the fire retardancy was clearly reduced. This highlights the dilemma of trying to improve two foremost properties which often seem mutually incompatible [170]. In contrast, Wiwat Pornwannachai et al. investigated flax-reinforced polypropylene/polylactic acid composites and found that an organophosphonate flame-retardant agent provided an appropriate balance between fire resistance and mechanical performance [165].

Further developments were shown by Imran Ali et al., who introduced graphene nanoplatelets into flax-polypropylene composites, achieving dramatic improvements in mechanical strength and fire performance. This work has demonstrated that the incorporation of nanomaterials certainly is one of the potent ways toward optimization of the properties of composites [9]. Similarly, N. K. Kim and D. Bhattacharyya investigated flax fiber composite fire reaction and post-fire impact properties and reported that the same was very dependent on the surface composition and application of flame-retardant layers [105].

Kamaraj et al. [90] proved the positive effect of graphene on flax fabric-reinforced epoxy composites concerning the flame retardancy aspect but also in terms of water absorption, which often plays a key role in maintaining the structural integrity of a composite under various environmental conditions. Zhuo Li et al. focused their research on Dimethyl Methylphosphonate and the nano-sized Montmorillonite synergistic effects on flax fiber composites. Their work highlighted improved fire retardancy and mechanical performance, in particular, under hydrothermal aging destined to degrade composite material performance rather frequently [122].

This aspect was furthered by Charlotte Campana et al., who studied the environmental factor influences on flax composites, aiming to ascertain how hygrothermal aging impacts flax fiber/epoxy composites that were subjected to flame-retardant treatments. Their findings underline the delicate balance needed to maintain fire safety while gaining structure under environmental stressors [35].

In another work, Thanh Mai Nguyen Tran et al. studied the influence of the addition of biowastes onto treated flax fiber-based polypropylene composites. The flame retardancy and mechanical properties were visibly enhanced, resulting in more eco-friendly composite materials [133]. The same authors, in another work by Mai Nguyen Tran et al., examined the effect of hybrid eco-friendly reinforcements on mechanical strength and improvement in fire-resistance properties of polypropylene composites [132]. By development this research, there authors used design of experiment (DOE) Box-Behnken and optimization method NSGA-II for multi-target optimum to finding optimization equations, highly enhanced mechanical, thermal and flame-retardant properties [131].

Daniele Battegazzore et al. put forwarded a new route based on layer-by-layer nanostructured interphase for enhancing the strength and flame resistance of biocomposites as reported in this research. The approach opened a new avenue to modify flax-based composite interface that underpinned superior material performance [29]. Moreover, C. Dreyer and D. Luca Motoc assessed the reaction-to-fire parameters of hybrid composites based on flax and synthetic fibers and concluded that materials in such a combination had reduced heat release and improved fire retardancy [52].

Finally, M. N. Prabhakar et al. proposed a hybrid approach for the improvement of fire retardancy with thermal stability in flax-based biocomposites for outdoor engineering applications. Hybrid composites with flax fabrics subjected to surface pre-treatment by various flame retardants in a vinyl ester resin matrix show enhanced flame-retardant and improved thermal properties [166].

Such studies, in sum, have conclusively opened up newer possibilities of future work in flax fiber composites by proposing different innovative techniques to improve flame retardancy and mechanical performance while maintaining environmental sustainability. These developments further expand the horizon of applications of flax-based biocomposites, breaking conventional boundaries and potentially opening vistas for high-performance fire-resistant materials for various industrial sectors.

4.3 Kenaf

The research on biocomposites and bionanocomposites containing kenaf fibers delivers much information regarding the former's potential applications in enhancing mechanical strength, thermal stability, and flame retardancy. A collection of articles has investigated the varied aspects associated with kenaf-based composites that have proven the fiber's versatility and efficiency in improving composite material properties.

M. J. Suriani et al. studied kenaf fiber/PET yarn-reinforced epoxy composites with significantly improved tensile strength and fire resistance. The research has shown that kenaf fibers clearly improve mechanical properties and flammability resistance of epoxy-based composites, thus broadening their field of application [200].

Tamil Moli Loganathan et al. prepared the hybrids of composites consisting of Cyrtostachys renda and kenaf fibers with phenolic resin modified by MWCNT. Experimental results indicated improved thermal stability and fire-retardant characteristics, marking success in incorporating MWCNTs for improvement in the thermal properties of kenaf-based composites [126].

In the recent effort of M. N. M. Azlin et al., the focus was on woven kenaf/polyester-reinforced PLA hybrid laminates, and detailed research was conducted on thermal stability, dynamic mechanical properties, and flammability characteristics. Hybrid laminates were used in this work for concurrent enhancement in mechanical performance with fire resistance, particularly where any degradation in thermal stability of the material could have serious consequences [23].

The influence of unbleached and bleached nanocellulose on thermal and flammability properties in polypropylene-reinforced kenaf core hybrid polymer bionanocomposites was investigated by Fatimah Athiyah Sabaruddin et al. Their work has shown that thermal behavior and fire resistance depend very strongly on the content of lignin, therefore underlining the necessity of fiber treatment for the optimization of composite properties [180].

In another work, the ability of using kenaf flour in polypropylene/graphene nanoplatelet composites for their electrical, thermal, flammability properties has been constructed by Christopher Igwe Idumah et al. The incorporation of graphene nanoplatelets has improved the thermal and electrical conductivity of the composite substantially and, hence, enhanced the fire retardancy, proving that kenaf flour is very useful in high-performance bionanocomposites [75].

Finally, Nur Adilah Abu Hassan et al. performed valuable research on the improvement of mechanical and combustion properties and acoustic insulation in biopolymer composite foams by reinforcing kenaf fiber. This work elaborates the potential of kenaf fibers in improving the discontinuous properties of composite materials about specifications for which applications are intended to combine mechanical strength with their fire resistance and sound insulation [2].

These studies collectively add to the rapidly accumulating knowledge of kenaf as a major resource in composite material development. In comparison with other natural fibers, kenaf fibers represent a balanced solution between sustainability and performance, which makes their potential very promising for the manufacture of advanced, environment-friendly composites related to different industrial applications.

4.4 Jute

In the last decade, appreciable interest in jute-based biocomposites and bionanocomposites has been developed because of the numerous studies on its improvements in fitting different industrial applications. These improvements have been targeted in relation to flame retardancy, mechanical strength, and chemical resistance, making jute very valuable in composite material development.

Shahood Uz Zaman et al. investigated the development of jute/unsaturated polyester chemical- and fire-resistant composites. In that procedure, it has been observed that through the hybrid fluorocarbon treatment and incorporation of zinc oxide nanoparticles, the compatibility of jute fibers with the resin matrix has been raised to a manifold level, and overall performance of composites enhanced. Therefore, it can be quite tended from this research that combining nanoparticles with traditional fiber treatment can realize better composite materials [207].

Jalalah et al. worked on the incorporation of eco-friendly fire retardants like micro-integrated Triphenyl Phosphate and Aluminum Trihydroxide into epoxy resins for jute composites. Results had indicated significant improvements in both fire-resistance and mechanical properties, thus establishing the incorporation of eco-friendly additives for performance improvements in jute-based composites [83].

Hürrem Canıtezoffers et al. studied the effects of silicon dioxide and zinc borate on the properties of jute/epoxy composites. The outcome of the study demonstrated that with such potential, the flame retardancy, thermal insulation, and mechanical strength offered these composites with the enhancement of safety and durability during extreme applications [36].

M. Rahul Kumar and G. Anbuchezhiyan conducted a comparative study of flexural strength and fire resistance in new types of composites, combining the eggshell powder filler and jute fiber in a glass fiber-reinforced polymer (GFRP). Their results evidence large improvement of mechanical performance and fire resistance, meaning that natural additives like eggshell powder could greatly contribute toward the improved performance of jute-based composites [173].

Yanli Dou et al. developed an intumescent flame-retardant layer, using β-cyclodextrin, for jute/polypropylene composites. This kind of new technology improved flame retardancy and the mechanical properties very much for the composites, endowing the jute-based family of materials with a totally new approach to improved safety and performance [50].

In another follow-up study, a new phosphorus/nitrogen-containing flame-retardant film was prepared for jute fiber composites. This showed considerable progress not only in flame retardation but also in mechanical properties, thus emphasizing the necessity for the emergence of advanced flame-retardant treatment for natural fiber composites [51].

Muhammad Muzammil Azad et al. reported the use of nano-silica particles in composites make from jute and thermoplastic starch. Chemical and structural analyses of nano-silica demonstrated simultaneous improvements in flame retardancy, thermal resistance, and mechanical properties, thereby indicating the potential for significant improvement of the overall properties of jute-based biocomposites using these nanomaterials [22].

All these studies collectively show advancement in the field of jute-based composites regarding many novel approaches to make these composites flame-retardant, mechanically strong, and chemically resistant. The research done proves that jute is a versatile, effective, and green material for various industrial applications and also its potential to serve as a forerunner to developing advanced biocomposites and bionanocomposites.

4.5 Sisal

In recent years, much attention has been given to the very intensive development of sisal-based biocomposites and bionanocomposites; several studies have been carried out intended for enhancing properties relative to various industrial applications. The study underscores creative ways in which flame retardancy, mechanical strength, and interfacial characteristics of sisal fiber-based composites could be enhanced to be better adapted to uses such as sustainability, performance, and safety–critical applications.

Zhenhua Wang et al. addressed sisal fiber-reinforced polypropylene composites with an intumescent flame-retardant system, which made the fire-resistance and mechanical properties of the said composites greatly enhanced for highly demanding uses in the automobile and construction industries where flame retardancy is a major requirement. This research reiterates the importance of adding flame retardants to natural fiber composites so that they may meet rigorous safety standards [216].

In their study, Zhanying Sun and Mingming Wu analyzed the effect of sol–gel treatment on sisal fiber-reinforced polypropylene composites. The research was conducted on improving mechanical properties of sisal fibers through surface treatments to enhance their interaction with the PP matrix. This research signifies how some surface modification techniques can be applied to improve the performance of NFCs by ensuring improved compatibility and stronger interfacial bonding [197].

In another similar work, Prakash Krishnaiah et al. studied the polypropylene composites with surface-treated sisal fibers and halloysite nanotubes. The current work deals with the combination of alkali and ultrasound treatments with silane grafting on nanotubes to improve mechanical, dynamic mechanical, and thermal properties of such composites. Again, this multifaceted approach elucidates the potential for combining advanced fiber treatments with nanomaterials to create high-performance biocomposites [112].

Feng Kang et al. constructed a flame-retardant three-dimensional network within sisal/polypropylene composites in a unique way. The modification of the sisal fibers by ammonium polyphosphate, polyethyleneimine, and nano-silica increased the composite's fire retardancy to a great extent without affecting its mechanical properties or harming the environment. This new approach will not only ensure better safety in use but also fit into tightened requirements for eco-friendliness [91].

These studies have greatly advanced the state of the art in sisal-based composites and pointed out innovative ways to improve flame retardancy, mechanical strength, and interfacial properties. These advances will make great strides toward the possible application of sisal fibers in various industries related to automotive, packaging, and construction, where the balance among sustainability, mechanical performance, and safety is critical. These studies answer with unique solutions to the challenging use of sisal fibers in composite materials toward an effective application in environmentally friendly and high-performance products.

4.6 Bamboo

Recent works in bamboo-based biocomposites and bionanocomposites have aimed directly at better flame retardancy coupled with mechanical attributes at the forefront of the development of sustainable composites. These investigations greatly stress novel techniques and treatments to upgrade bamboo fibers for performance in a wide range of applications.

Qianqian Niu et al. have developed phosphorus-silicon aerogel-treated bamboo fiber reinforcement PLA composites. The desired improvements in the flame retardancy and the mechanical properties made the composite fit for the high-performance sector. The research has proved the potential of blending cutting-edge treatments with aerogel and natural fibers for obtaining outstanding material properties [152].

K. Venkata Chalapathi et al. developed improved bamboo fabric/vinyl ester composites through surface treatment and hybridization of flame retardants. In their study, very significant improvement was shown in the mechanical strength and fire resistance, hence confirming that surface modification and hybridization play vital roles in optimizing the properties of bamboo-based composites [38].

By contrast, Lu Fang et al. developed the bamboo fiber-reinforced polypropylene composite in which melamine pyrophosphate with aluminum hypophosphate was incorporated. The newly added flame retardants were highly efficient in improving the flame retardancy of the composites while having a medium effect on the performance of mechanical properties. Thereby, this work holds one key feature: tension between fire-resistance and mechanical properties in composite materials [53].

Silica aerogel was grafted onto bamboo fiber by Qianqian Niu et al.; therefore, this modification enhanced the flame retardancy and mechanical properties of PLA composites. The modification using silica aerogel expresses its remarkable role in enhancing the overall performance of natural fiber-based composites [153].

Ziling Teng et al. deposited a cerium-doped titanate nanotube coating on bamboo fibers, leading to significant enhancements in their flame retardancy and thermal stability. This work provides an excellent example of how advanced nanotechnology can be applied to improve the thermal properties of bamboo-based materials [204].

Wenwen Guo et al. prepared a high-strength and flame-retardant composite from natural bamboo fibers, thus realizing improved performance in both strength and flame resistance. This identifies the potential of processing techniques in enhancing the inherent properties of bamboo fibers [59].

Liang Zhang et al., in their study, prepared PLA composites reinforced by bamboo charcoal grafted with phosphate and urea. This modification has remarkably improved the flame retardancy without an appreciable loss in mechanical strength of the composite, which gives hope for bamboo charcoal as one of the contender reinforcing fillers in biocomposites [224].

Qianqian Niu et al., among others, have taken the enhancement of flame retardance of bamboo fiber-reinforced PLA composites even further by applying interfacial silicon-nitrogen aerogel. This technique needed to maintain the composite's mechanical properties and only offer elevated fire resistance to be a robust solution able to compete in high-performance applications [151].

The work by Shanshan Wang et al., whereby bamboo charcoal combined with aluminum hypophosphite was edited to enhance the flame retardancy of the PLA composites, resulted in significant heat release reduction and flame resistance. In this way, the present work has highly illustrated the importance of additive selection for optimizing the fire performance of bamboo-based composites [215].

Junbao Yu et al. also added layers of glass fiber in the bamboo scrimber composites, which substantially upgraded their mechanical characteristics, incremental aging resistance, and resistance to fire after burning. Mechanical enhancement, aging resistance, and fire resistance have greatly been improved, and the overall durability and safety make these composites much more appropriated than before for long and heavy demands [223].

Siew Sand Chee et al. developed hybrid nanocomposites filled with bamboo, kenaf, nano-clay, and epoxy. The results presented a sample showing enhanced thermal stability and flame retardancy, hence suggesting a synergistic benefit when putting various fibers and nanomaterials together [42].

According to Chenmin Zheng et al., these phenomena, whether they improve the flame retardancy of bamboo fiber materials or not, were effectively investigated using a layer-by-layer technique with $H_2Ti_2O_5 \cdot H_2O$ nanotubes. This method developed effective flame-retardant coatings for a substantial enhancement in the safety and durability of bamboo-based materials [227].

Yaowen Zhang et al. further improved flame-retardant bamboo fabric grafted with a novel compound, ATTPMA, to maintain its mechanical performance while significantly enhancing its flame resistance. This provides an obvious indication of the potential of grafting techniques established by chemical methods to improve flame retardancy without compromising mechanical properties [225].

Lastly, Chaohua Peng et al. synthesized transparent, robust organic-silica hybrid coatings on bamboo substrates to bring out bamboo's ablaze-retardant and anti-fungal belongings exclusive of compromising its naturalness. The trend is gaining importance because this category of materials is needed everywhere, from aerospace to common household items [159].

Taken together, all these current studies have supported the field by providing innovative solutions for improving flame retardancy, mechanical strength, and overall utility of these bamboo-based composites and nanocomposites. The increase in the performance of bamboo-based materials could play a very vital role toward their increased scope of application in the industrial sector, which has become the need of the hour in this world running toward a sustainable and environmentally green material trend.

4.7 Cellulose Nanocrystals

One of the major advancements in making biocomposites and bionanocomposites for improving flame-retardant properties is cellulose nanocrystals. Recent studies highlighted that the synergistic effects from combing CNCs with a number of materials can effectively increase the thermal stability, mechanical strength, and flammability characteristics.

Previously, by Dilpreet S. Bajwa et al., an efficient enhancement in terms of thermal, mechanical, and flammability properties of the composite was established, resulting from the incorporation of nano-zinc oxide with CNCs into poly(lactic acid) (PLA). Indeed, the study is very relevant for application in fields where a high level of fire resistance is required, and at the same time, high mechanical performance should be obtained [26].

A novel intumescent flame retardant (CNC@MEL-PA), in the modification with melamine and phytic acid, was developed by Yunyun Zhang et al., which improved the thermal stability and smoke-suppressing efficiency of the composites, making them more applicable for fire-sensitive environments [226].

Zhang et al. reported further research in the tailoring of such bamboo pulp paper with a cellulose nanofibril-based flame retardant. The work demonstrated that this treatment not only significantly increased flame resistance but also improved greatly the mechanical strength of the papers, proving the high versatility of CNC-based methodologies in boosting the properties of natural fiber-based materials [217].

Wang et al. developed bioinspired nacre-mimetic lamellar zing barriers in nanocellulose-based composites, providing phenomenal improvements not only in flame retardancy but also in mechanical stability, hence evidencing the strength of the motivation offered by natural structures in material design [212].

Dhamodharan et al. explored the applicability of cellulose nanocrystals-graphene oxide hybrids as nano-fillers into the matrix of solvent-free polymer nanocomposites. They concluded that significant improvements in the mechanical and thermal properties resulted from synergistic interactions between CNC and GO, which make these hybrid materials quite promising for applications demanding high performance [45].

Shuai Jiang et al. further developed cellulose composite foams with added biopolymer sodium alginate and boric acid, resulting in the achievement of self-extinguishing properties with enhanced thermal insulation. In this regard, this research study further elaborates on CNC as the precursor for development into an advanced fire-retardant material [84].

Carlos Rafael Silva de Oliveira et al. reported an eco-responsible flame-retardant treatment of cotton fabric using a kaolinite-TiO_2 nano-hybrid composite. Such a method can be expected to render an efficient resistance against flames to a fabric, alongside ensuring no mechanical property of the fabric is compromised, thus confirming the environmental and functional performances that treatments noticeable to be based on CNC would develop [44].

Hu et al. studied the incorporation of both high thermal conductivity and flame resistance into nanocellulose composites using functionalized boron nitride nanosheets and found that such composites show enhanced thermal conductivity and flame resistance and hence are suitable for use in both the aforementioned fields [69].

Medinaet et al. developed novel eco-friendly foam materials based on poly(vinyl alcohol), clay, and cellulose nanofibrils, showing good flame retardancy and mechanical strength, which underlines extremely promising prospects for the application of CNCs in sustainable material development [138].

Chunxiang Ding et al. improved the flame retardancy and mechanical strength of wood-plastic composites using an anionic polyelectrolyte hybrid. This method was quite effective in developing both these factors, so as to obtain more robust and safe composites for use [47].

Ruey Shan Chen et al. have done similar work by much improving the thermal stability and flame retardancy of the polypropylene matrix composites upon incorporating graphite nanoplatelets and magnesium hydroxide. That is to say, synergistic effects can be provoked by combinations of CNC with other nanoparticles to achieve higher material properties [43].

Li et al. introduced a new hybrid nanoparticle for cellulose, SiDP, which demonstrated improved flame-retardant and antibacterial activities. This dual functionality make it especially for application in textiles and other consumer products [119].

Kim et al. developed eco-friendly nanocellulose-embedded polymer composite foam, along with retarded flame propagation and simultaneously lower emissions level of toxic gases. So, this development is very important for application in the areas of more concern about fire safety and environmental impacts [104].

Wang et al. prepared biomimetic metamaterial cellulose nanofiber aerogels with excellent mechanical, flame-retardant, and thermal insulations, fabricated from natural structures. The present work is, thus, a prototype showing that CNCs have the capacity for developing advanced multi-property materials [213].

Wenwen Guo et al. prepared composite foams of lightweight cellulose nanofibers and phosphorus-hybridized graphene nanosheets. The flame-retardant performance of such composite foams made from these materials was as excellent as ensilication. This innovation further expands the potential applications of CNCs in high-performance materials [58].

Gorur et al. presented a route to the fast preparation of cellulose nanofibril hybrids with gas barrier and flame-retardant feature, showing energy storage capacity. Further, our study is another piece of evidence confirming the vast potentials CNCs can generate in achieving multifunctional composites [57].

Xuanzheng Zhou et al. achieved markedly improved fire resistance for wood-plastic composites by using a nanostructured hybrid coating, leading to unique enhancements in time to ignition and overall fire safety. This example illustrates the combined effect of introducing interfaces into a matrix for superior CNC composite performance [228].

Qianqian Peng et al. proposed a biomimetic hierarchical-ordered cellulose nanoclaw hybrid aerogel with high strength and thermal insulation. Such novel designs would enlighten the capabilities of CNCs in producing advanced materials with structural and thermal performances [160].

Lastly, Yan et al. developed "robust-soft" anisotropic aerogels of nano-fibrillated cellulose with improved mechanical, flame-retardant, and thermal insulating properties. This work uses NCC to the very edge of possible, as it represents real multiple functionalities in one material [221].

In general, the findings from this study corroborated one another and collectively supported the potentiation of cellulose nanocrystals and hybrids in varying forms, including aerogels, nanocomposites, coatings, and foams, to most often impart flame retardancy. The combination of CNCs with materials like zinc oxide, graphene oxide, and melamine, among other nanoparticles and biopolymers, leads to the formation of composites with synergistic effects, implicating overall better thermal stability, mechanical strength, and lower flammability. These advances formed a section not only pertinent to the field of material science but also bring about sustainability for industries on safe turf for their materials.

4.8 Microfibrillated Cellulose

Microfibrillated cellulose is currently one of the most significant constituents in the preparation of biocomposites and bionanocomposites. History has shown that MFC is able to contribute to the building of highly sustainable, flame-retardant, mechanically strong materials. Many pioneering works have exhibited the potential of MFC as a reinforcing material for enhancing the diverse properties of composite materials so that they could be of use in a wide range of applications.

Jiaming Sun et al. prepared tubular cellulose aerogels from kapok fibers that showed outstanding compressive strength, heat preservation, and flame retardancy. The unique tubular configuration formed in these aerogels, in conjunction with the use of ammonium phosphate ($(NH_4)_2HPO_4$), afforded superior properties to those of similar materials in applications related to exterior wall insulation and vehicle interiors, where mechanical strength and fire resistance are of vital importance [196].

He et al. improved the mechanical properties of PLA by using epoxidized microfibrillated cellulose as both reinforcement filler and interfacial compatibilizer in the field of PLA composites. This composite showed outstanding performance with regard to tensile strength and elongation, therefore improving toughness to make PLA tougher for more demanding applications [65].

Lou et al. focused on the development of fully bio-based FRs for epoxy resin composites through surface functionalization by chitosan and sodium phytate onto microcrystalline cellulose. This newly developed method caused a peak heat release rate, total heat release reduction, and smoke production reduction if combined, besides mechanical property improvement in the composites' specifications. These results indicate that the bio-based materials have huge potential for developing environmentally friendly fire-resistant composites [127].

A green composite of cellulose long filaments combined with vanillin-derived epoxy resin was prepared by Samia Adil et al. It showed a large improvement in flexural strength and modulus, coupled with outstanding flame-retardant characteristics and hydrophobicity, making it ideal for high-performance structural applications that require both mechanical durability and fire safety [3].

Nam Kyeun Kim et al. applied bio-based FRs of extracellular polymeric substances from wastewater sludge in PP composites. This route not only further improved the fire performance of their PP composites but also increased their tensile moduli, hence making great strides toward sustainable flame retardancy in polymer materials [106].

Oruç Köklükaya et al. deposited multilayer thin films onto low-density wood fiber networks through a layer-by-layer assembly technique, achieving improved flame retardancy. It is reported that the coated fiber networks showed self-extinguishing behavior and reduced total heat release in studies by Köklükaya et al., which shows that such materials have potential for creating sustainable, fire-resistant materials to be used in many industrial applications [109].

Oskars Platnieks et al. have radicalized the development of sustainable polymer composites by combining microcrystalline cellulose and nano-fibrillated cellulose fillers into a bio-based poly(butylene succinate) matrix. In some ways, this new class of composites will have improved mechanical and thermal features that may signify better biodegradability and hence meet the increasing public demand for eco-friendly materials in a number of industries [163].

Finally, Vanja Kokol et al. investigated the screen-printing method for treating flame-resistant fabrics using MFC. In this technique, moisture management, strength, and abrasion resistance were improved while maintaining their flame resistance properties. It improved the thermophysiological comfort and made those kinds of fabrics appropriate for protective clothing where both safety aspects and comfort are required [110].

These studies, when taken in conjunction, underline the exact breadth of scope within which microfibrillated cellulose holds potential for highly performant yet sustainable material development. This research represents a leap ahead in the development of biocomposites and bionanocomposites and significantly contributes toward eco-friendly solutions in their functionality for varied applications.

4.9 Synergistic Enhancement of Flammability Performance in Biocomposites and Bionanocomposites: A Comprehensive Synthesis

The case studies presented throughout this chapter point to one central theme in the development of biocomposites and bionanocomposites: the importance of synergistic effects in improving flammability performance. Even though several types of natural fibers have been explored, including cotton, hemp, flax, kenaf, jute, sisal, bamboo, cellulose nanocrystals, and microfibrillated cellulose, synergistic interactions have still remained the key factor in dramatically enhancing flame retardancy and thermal stability.

Central to these developments is the fact that different materials—from nano-fillers and flame-retardant additives to hybrid mixtures of natural and synthetic fibers—are combined in a strategically effective way. For example, the addition of expanded vermiculite, ammonium polyphosphate, carbon-based nanomaterials, and nano-hybrid composites in cotton composites, and silica aerogels, phosphorus-silicon aerogels, and nanostructured interphases in bamboo composites, reflect strategic balancing between the choice of materials and functional improvement.

Moreover, the use of CNCs and MFC either by themselves or in combination with other materials marks an emerging trend that exploits both inherent properties to obtain flame retardancy and synergistically enhance the whole mechanical and thermal properties of the composites. There are several advantages for such synergistic material combinations where this strategy has to be adopted.

Those studies on the whole bear out the message that effective flammability performance improvement requires neither natural fiber type nor additive type, but are involved in their interactive interplay, thus leading to a synergistic effect that produces materials with very high levels of fire resistance and retains or even improves mechanical, thermal, and structural integrity. Such holistic improvement is urgently required to make them fit for industrially demanding applications that need high performance, safety, and sustainability.

In other words, there is an intrinsic identification of the development of biocomposites and bionanocomposites toward improved flammability performance with a focus on synergistic effects derived from new combinations of materials. This trend will not only progress the cause of materials science but also meet global ambitions in sustainable development by paving paths for the realization of safer and more environmentally friendly materials across an array of applications.

5 Assessing the Synergistic Effects on Flammability Performance in Biocomposites and Bionanocomposites

5.1 Theoretical Framework for Understanding Synergistic Effects in These Materials

According to most researchers, synergy in biocomposites and bionanocomposites is a phenomenon in which the effects of mixing different constituents separately give better fire-retardant effects than the sum of them. This is created by the coordinated interaction of natural fibers, the polymer matrix, and nano-fillers, whereby each of its elements contributes a unique effect in increasing the fire resistance of the composite material.

- Role of Natural Fibers: The main fibers of this category are cotton, hemp, flax, and bamboo; these contribute to fire resistance in composites by virtue of the inherent formation of char under combustion conditions. The fibers form a carbonaceous char and act as an insulating thermal barrier in the event of combustion, which slows down the heat release rate of the underlying substrate. This char layer also restricts the flow of oxygen and the release of flammable volatiles to the combustion zone, resulting in an effective reduction in the overall flammability of the composite [16, 30].
- Contribution of the Matrix: It is also realized that the polymer matrix, particularly if bio-based, itself plays a significant role in imparting the fire-retardant properties of the composite. Under a high-temperature environment, the matrix degrades without propagating the degradation reactions and produces compounds that either dilute the flammable gases or help to form a protective char. The formed char layer thus insulates the internal structure of the composite from direct flame contact, rendering the composite highly flame-retardant [188, 209].
- Impact of Nano-fillers: Their addition—nano-clays, carbon nanotubes, graphene nanoplatelets, etc.—to bionanocomposites increases their thermal stability and resistance to fire. This means that the nano-fillers form a tortuous path for gases and heat, thus forming a solid barrier that can protect a composite from direct penetration by heat and a flame. Besides that, nano-fillers can play a catalytic role during the process of char formation and, most probably, provide reinforcement to the structural integrity of the developed char layer, thus providing fire and heat resistance to the materials [16, 107, 111].
- Incorporation of Cellulose Nanocrystals (CNCs) and Microfibrillated Cellulose (MFC): The introduction of CNC and MFC opens a new route toward fire-retardant synergism. The high aspect ratio and inherent thermal stability of CNCs could contribute to the development of a dense char layer, an effective thermal barrier. Additionally, it might influence the pyrolysis pathway of the matrix material into one that produces a residue richer in carbon, which is less ignitable. Being fibrous in nature and having relatively high specific surface area, MFC contributes to the mechanical integrity of the char, phenomena that further hinders the process of heat and mass transfer during combustion. Besides, this hydrophilic nature of MFC encourages uniform dispersion of other flame-retardant additives inside the composite, hence giving an empirical superior flame-retardant performance when compared against non-hydrophilic performance [88, 155, 184]. Figure 3 shows a representation of synergism stemming from the combination of natural fibers, polymer matrix, nano-fillers, CNCs, and MFC in biocomposites and bionanocomposites, explaining how these components interact to impart better flammability performance in the material.

Thus, these synergistic effects for bio or bionanocomposites occur through the diversity of the distinct qualities, starting from natural fibers up to matrix-mediated protective roles, barrier functions of nano-fillers, thermal stability, and structural support. This comprehensive approach not only increases the performance of flame

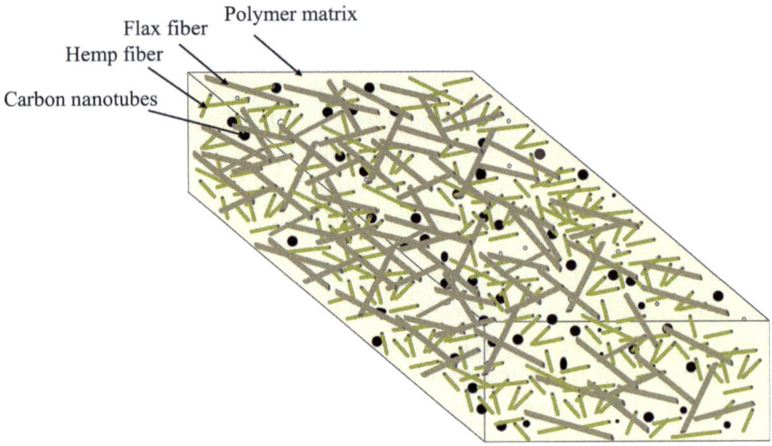

Fig. 3 Conceptual diagram of synergistic effects in flammability performance of biocomposites and bionanocomposites

retardants for the material but maintains or enhances other essential properties such as mechanical strength, temperature stability, and environmental sustainability.

5.2 Experimental Methods and Techniques for Flammability Evaluation Specific to Biocomposites and Bionanocomposites

Experimental methods and techniques to determine flammability in biocomposites and bionanocomposites are very diverse, and each of them could give quite unique information concerning fire behavior. These techniques will become indispensable in establishing whether new synergistic flame-retardant strategies work and proving these materials' safety for performance under various applications [107, 144, 219].

- Cone Calorimetry: This is one of the most vital techniques to determine the flammability characteristics of any material, as it provides, in a direct way, all the detailed information regarding some key parameters such as HRR, smoke production, total heat release, and ignition time. The knowledge pertaining to how a material may contribute to the rise and spread of a fire can thus be viewed in much detail in cone calorimetry through the simulation of real scenarios of fires. Consequently, it has proved to be very useful while assessing the effectiveness of flame-retardant treatments in biocomposites.
- Thermogravimetric Analysis (TGA): TGA is one of the principal tools utilized in evaluating thermal stability and decomposition by the continuous measurement of weight changes in a material during heating. This delivers very important data related to the intrinsic flame-retardant properties, directly connected with the

decomposition temperature and the amount of residue that is formed. TGA helps in understanding how a material breaks down under heat stress to give insights into fire performance.

- Limiting Oxygen Index (LOI) and UL-94 Tests: While the LOI test quantifies the minimum oxygen concentration to support the combustion of a material, it offers a fast and reliable flammability assessment. On the other hand, the UL-94 test looks at burning rate, dripping behavior, and afterglow characteristics in specific orientations. Such testing through these methods makes for practical assessments of fire performance in different use-case scenarios, particularly within consumer goods and electronics.
- Microscale Combustion Calorimetry (MCC): MCC is a unique technique that characterizes the combustible behavior of small samples of materials by quantifying their heat release capacity. This approach is very useful during the early development stages of a material since it provides fundamental information on fire hazard and is thus extremely useful in screening new formulations.
- Pyrolysis-Combustion Flow Calorimetry (PCFC): PCFC is a coupling of pyrolysis with combustion calorimetry; it allows one to assess pyrolysis products and their combustion characteristics in one single measurement. On the other hand, this technique gives more detailed information on the flame-retardant mechanisms inside the material and is, therefore, useful in learning how different additives and fillers affect the general fire performance of biocomposites and bionanocomposites.
- Oxygen Bomb Calorimetry: The measurement by oxygen bomb calorimetry provides the total heat of combustion, an integral parameter related to processes connected with the general energy material content. Hence, it is particularly relevant for evaluating the effectiveness of flame-retardant additives working on the principle of reducing combustible material content available at the time of outbreak of a fire.
- Smoke Density Test: Smoke production can be a major safety concern in case of fires. Thus, smoke density measurement can become very vital. Smoke density tests determine the volume and opacity of smoke generated with the burning process and turn out highly critical information for materials intended to be used in enclosed areas like buildings and vehicle interiors.
- Infrared Thermography: This technique is considered non-intrusive and involves an infrared camera that monitors the temperature distribution and heat flow at the surface of a material while it is undergoing combustion. It becomes very useful for identifying hotspots and evaluating the thermal-protective performance of flame-retardant treatments by providing visual insight into how the composite handles heat internally.
- Radiant Panel Test: This is the test where a trial sample is put in contact with a radiant heat source and evaluated for surface spread of flames. Therefore, this method would be particularly useful for evaluating materials that are exclusively used for floor covering and wall paneling where surface flame spread becomes a significant safety parameter.

- Heat Release Rate (HRR) Curve Analysis: The HRR curves developed from cone calorimetry provide an insight into the efficiency in combustion and flame-reducing efficacy of the material. This analysis randomizes understanding toward peak heat release rate and time to peak HRR—both very critical factors in fire safety engineering.

Each of these methods gives a different perspective about the flammability performance of biocomposites and bionanocomposites and provides an improved judgment, as part of an ensemble, of how these materials behave under exposure to fire for the eventual development of safer, more flame-retardant biocomposite materials for various industrial applications.

5.3 Data Analysis Demonstrating Synergistic Effects in Fire Retardancy Within Biocomposites and Bionanocomposites

In the domain of biocomposites and bionanocomposites, synergism in fire retardancy was found to exist through analysis of data from different methods of experiments. These synergistic effects can be noted through key indicators that give insights into how the interplay between different materials affects the overall fire-resistant properties of such composites, as shown below [107, 144, 219]:

- Analysis of Cone Calorimetry Data: Cone calorimetry allows the evaluation of the most detailed metrics of fire performance, such as peak heat release rate, total released heat, and time to ignition. Obvious evidence of synergism can be realized where a composite exhibit reduced peak HRR compared to its constituents or an equivalent non-synergistic combination. This will mean that there is an extra enhancement on the fire-retardant effect due to combined influences of natural fibers and nano-fillers. In addition, increase in time to ignition and decrease in the total heat release also support the synergistic effects, since these parameters are indicative of enhanced resistance to combustion.
- Thermogravimetric Analysis (TGA) Measurements: Valuable information derived from TGA provides data for evaluating the thermal stability of the said material. In case the degradation temperature of a composite material is observed to be higher in comparison with the individual constituents in the composite, that would imply better thermal stability, and such behavior can be ascribed to synergistic effects of the combined materials. A higher char residue at the end of a TGA run would indicate better charring ability, besides being important in fire retardancy since the char layer protects against further decomposition.
- Analysis of LOI and UL-94 Results: This should clearly indicate improved flame retardancy showing a value of LOI that is higher in the composite than its individual components with a synergistic effect. In the UL-94 test, achieving a rating of V-0 or any higher rating and dripping behavior reduction manifest further

improved fire-retardant performance due to combined action from the composite's constituents.

- Smoke Density Test Conclusions: The smoke density test may further go on to indicate synergistic effects with regard to smoke production and its opacity reduction. A composite that gives off less dense and less opaque smoke as compared to that from its components may indicate effective synergistic action in the suppression of smoke, which is very important in improving safety in environments where fires are prone to occur.
- Heat Release Rate (HRR) Curve Analysis: The HRR curve obtained from cone calorimetry would lead to deep insights into the combustion behavior of the composite material. If the peak of the HRR curve is retarded or if its increase rate is flatter, it will also indicate a synergistic effect in retarding progressive combustion. This behavior is indicative of improved fire resistance and a more controlled combustion process.
- Comparative Analysis with Microscale Combustion Calorimetry (MCC): Clearly, MCC data shows the reduced heat release capacity of composite material compared to the sum of its constituents, thus revealing synergistic behavior of flame retardancy. A reduction in the heat release capacity would suggest that when combined, such materials quench combustion more effectively than if used independently.
- Pyrolysis-Combustion Flow Calorimetry (PCFC): The PCFC better explains the gases involved in pyrolysis and their subsequent combustion process. From there, synergistic effects can be adduced as changes in the kinds and amount of combustible gases emanating from the composite, in comparisons to those from its respective components. Any reduction in flammable gas or change of composition toward by-product involvement that is less combustible means an increase in flame retardancy resulting from synergistic interactions.

These different analyses of data bring into focus the very existence and importance of synergistic effects on enhancing fire retardancy properties in both biocomposites and bionanocomposites. Researchers and material scientists, by understanding the nature of synergistic interactions, are better positioned to formulate composites with improved fire-resistant properties, thus meeting higher safety and performance requirements for use in various applications.

5.4 Discussion on Mechanisms Contributing to Enhanced Flame-Retardant Performance in These Composites

The better flame-retardant behavior of biocomposites and bionanocomposites is the result of a large number of different underlying mechanisms, where each of the mechanisms plays a significant role in the synergistic interaction among the different constituents of the materials. Knowledge and the understanding of such mechanisms are of tremendous importance for the optimization of composite design parameters intended for advanced fire safety applications.

- Combustion Mechanism and Natural Fiber: One of the main mechanisms by which these composites present their flame retardancy properties is by char development by natural fibers. Natural fibers pyrolyze into a carbonaceous char under the action of the heat. The formed char acts as a kind of barrier insulation to retard the attack of the heat and flame on the material burning. The char layer is also responsible for hindering oxygen access and the release of flammable volatiles, aiding in important molecules responsible for the propagation of a fire [6, 111, 178].
- Barrier Formation by Nano-fillers: The chemical addition of nano-fillers, such as nano-clays, carbon nanotubes, or graphene nanoplatelets, would be effective in enhancing their flame-retardant ability. These nano-fillers form a tortuous path along which gases expand; they increase the difficulty for volatile compounds to escape and heat and oxygen to enter the material. This barrier effect does not only have a resistance to the outflow of combustible gases, but it also could delay the heat transfer inside the unburned material, thus decelerating combustion in an overall manner [46, 94, 179].
- Enhancement of Thermal Stability: Nano-fillers improve the thermal stability of composites. They improve the decomposition temperature of the material so that the composite can now be exposed to higher temperature without the initiation of degradation. Such improved thermal stability is very significant in enhancing the resistance against ignition and reduces burning rates of the material, which is very essential in fire-retardant applications [37, 94, 182].
- Dilution of Flammable Gases in Bio-based Matrices: Some bio-based matrices in composites decompose upon heating in non-flammable or less flammable gases. These gases mix with the flammable gases in the combustion zone, which effectively dilutes their concentration. This actually reduces their overall flammability of the gas mixture, slowing down or even totally inhibiting the combustion process [111, 178, 193].
- Catalytic Effects and Additives: Some of the flame-retardant additives or constituents of the composites can possess catalytic effects on the charring process. Such additives shall improve the formation of char, conditioned at being more expanded and stable, or shall technically influence the pyrolysis pathway to degrade products being of less flammable nature. From this, catalytic effects significantly contribute to the intensification of the flame-retardant properties of composites [77, 85, 146].
- Synergistic Interactions: The synergistic effects exhibited in these composites are not simply additive but are the results of strategic interactions between the various constituents. For example, when natural fibers are combined with nano-fillers, they can together create a stronger and more effective char layer than all the individual components on their own. Simultaneously, there is the interaction of bio-based matrices with additives for the retardation of flame, which may involve the dilution of flammable gases in a much effective way, especially by the creation of protective layers. These may be key synergistic interactions important

in pushing flammability performances of these latter classes of advanced materials up to significantly higher levels [21, 77, 150].

Understanding these mechanisms and being able to skillfully hit them makes it possible for researchers to optimize the design of this material for superior fire retardancy in biocomposites and bionanocomposites. The synergy that ensues in these materials is thus not an additive but rather careful interaction between the components, leading to higher performance in other flammability performances, making it possible to be more safe and upright in function for various applications.

6 Future Prospects and Challenges in Natural Fiber-Reinforced Biocomposites and Bionanocomposites

6.1 New Trends and Innovations in Natural Fiber Composites

Second-generation natural fiber-based composites have started developing very fast due to new invention methods, wherein sustainability has emerged as the prime objective. Recent trends reflect the dynamic evolution of this field effectively bracketed through the comprehensive case studies mentioned in Sect. 4:

- High-Performance Natural Fibers Development: Substantial effort is increasingly directed to improving the mechanic properties of these natural fibers so that they are comparable to, or even outperform, their synthetic counterparts. The efforts entail changes at the molecular or structural level in fibers through techniques of chemical treatments and surface modifications to hybrid fiber blends. These developments are toward advanced fiber strength, fire resistance, and compatibility with different matrices, hence making natural fiber composites more strong and versatile in functionality [7, 17, 148].
- Advances in Bio-based Resins: Development of bio-based resin systems would provide enhanced compatibility and performance with natural fibers. This would be an area of improvement where resins enhance adhesion to natural fibers, the flame-retardant properties, and mechanical properties of the produced composite material. These innovations appear to be of some importance, to expand the application spectrum of natural fiber composites, into areas that emphasize environmental sustainability [15, 31, 111].
- Integration of Smart Technologies and Sensing Capabilities: Another budding trend gaining a lot of prominence is the integration of smart technologies within natural fiber composites. This can be attained by giving them sensing capabilities through the integration of sensors or smart materials into the composite matrix, which monitors critical parameters such as structural health, environmental conditions, and performance characteristics, in real time. It is envisioned that such smart composites would apply in many sectors, such as automotive, aerospace, and civil

engineering, where continuous monitoring is essential for maintenance and safety purposes [41, 171, 177].

- Enhancing Flame Retardancy Through Material Hybridization: A trend that arises in several case studies, as reported in Sect. 4 of this chapter, is a move toward the creation of some hybrid composites comprising natural fibers and flame-retardant material types, such as nano-fillers and resins, with special flame-retardant properties. Therefore, this new approach could apply synergistically to improve flame retardancy without loss of mechanical strength or environmentally friendly sustainability. It has been a very important conjunction with nano-fillers, such as nano-clays, carbon nanotubes, and graphene nanoplatelets, in combination with natural fibers.
- Focus on Life Cycle Assessment and Circular Economy: Environmental awareness about the role of materials during their life cycle has driven researchers in the LCA of natural fiber composites. This is aimed at better understanding and reducing the environmental footprint of such materials from productions to the final disposal stage. Other efforts are on the increase in enhancing recyclability or biodegradability of these composites and that their production and application must be in line with the goals of a circular economy [156, 195, 210].
- Application-Specific Tailoring: This has been another prominent trend in making the natural fiber composites application-specific. It implies application tailoring of composition, structure, and properties for various industries like automotive, construction, and electronic applications. It is about the optimization of characteristics or attributes of performance, for example, strength-to-weight ratio, durability, and thermal stability, with regard to certain desired end-use needs [142, 154, 192].

These trends reflect deep striving to move the boundaries of natural fiber composites further and put them as rather competitive with synthetic materials, keeping up a strong focus on sustainability. Also, innovations in the line of fiber modification, resin development, integration of smart technology, and material hybridization are bringing out new, much more advanced, and environmentally responsible composite materials across various industries.

6.2 *Challenges Specific to Scaling Up Production and Ensuring Consistent Quality in Biocomposites and Bionanocomposites*

The scaling up of production of biocomposites and bionanocomposites brings with it several serious problems in ensuring uniformity in quality at larger volumes. There are some challenges that need to be sorted out for actual commercialization and mother warming these advanced materials, as discussed in Bangar and Whiteside, Garavand et al., and Trotta et al. [27, 55, 205]:

- Variability in Quality of Natural Fibers: The intrinsic variability within the different categories of natural fibers, arising from differences in agricultural practice, geographical location, and climatic conditions, presents a core issue toward producing large quantities. This can mean variations in critical fiber properties, most significantly strength, moisture content, and chemical composition—factors that are crucial for determining the quality and performance of composites. To mitigate this defect, rigorous quality control at all levels, coupled with the standardization of sourcing raw material, is required for used natural fibers to be uniform in their properties.
- Process Optimization for Mass Production: Increased production scales currently raise the need for optimized processing techniques that would be able to handle bigger volumes without structural and mechanical damages to the fibers. This comprises developing improved fabrication methods which recognize the unique properties of natural fibers and nano-fillers to ensure their uniform dispersion and proper alignment within the matrix. Moreover, scaling often leads to continuity from batch processes that require very proper calibration in order to maintain consistency with regard to the composite properties across large-scale production.
- Balancing Cost-Effectiveness: While scaling up, cost effectiveness becomes more necessary, especially in the context of novel materials or state-of-the-art technologies. The addition of nano-fillers, bio-based resins, or smart technologies naturally raises high production costs. Some options for cost balancing would be to develop cheaper methods for the synthesis of nano-fillers, finding local natural fibers to avoid transportation costs, and process optimization in such a way that it reduces waste generation and electric power consumption.
- Ensuring Uniformity in Composite Properties: One major challenge remains to obtain consistency in properties for biocomposites and bionanocomposites with large production volumes. This requires a good standardization of the production procedure with associated quality control measures that are tight. Parameters relating to fiber treatment, matrix formulation, and composite curing processes all have to be very tightly controlled—and this would ensure batch-to-batch variation of the composite material within specified limits.
- Integrated Scaling of Advanced Technologies: Advanced technologies in terms of smart sensing elements or new flame-retardant treatments applied to large production volumes bring along a variety of other challenges. This comprises technical challenges during the integration of sensor components without damage to the composite structure and homogenous distribution with continuous functionality up until the end of life of the flame-retardant additives.
- Supply Chain Management and Availability of Raw Materials: Scalability in production requires efficient management of its supply chain. This encompasses the steady and sustainable supply of raw materials, like natural fibers and bio-based resins, and the management of logistics so that there are no bottlenecks in production. High-quality raw materials, especially specialized fibers or eco-friendly resin systems, may be difficult to procure in large enough quantities.
- Compliance with Environmental and Safety Regulations: It is then more difficult to comply with environmental and safety legislations when production increases.

Some of these include regulations that apply to emission, waste disposal, and workers' safety in the use of nanomaterials and chemical additives. This implies that proper attention should be given to seeing that all routes of production and/or final products must meet the minimum requirements or standards in force for environmental and safety impacts.
- Market Acceptance and Consumer Perception: Lastly, the issues associated with market acceptance and consumer perception are also attached to scaling up in production. Therefore, customers have to be made aware of the benefits of using biocomposites and bionanocomposites, and the seven misconceptions prevailing about their performance against traditional materials need to be addressed in order to create a bigger market.

Looking forward, the challenges need to be overcome satisfactorily for large-scale production of biocomposites and bionanocomposites with consistency in quality and at a below threshold cost for market acceptances. It is only after conquering all the problems that these materials will become semi-classical in various industries and can contribute to a more sustainable and environmentally softer future.

6.3 *Potential Environmental Impact and Sustainability Considerations in This Field*

Considering that the concept of sustainability serves as an essential guiding principle in the rapidly developing domain of natural fiber-based composites, environmental impact is the major challenge for such materials if their long-term viability is considered with regard to global sustainability goals. The key considerations are [14, 168, 176]:

- Life Cycle Assessment (LCA): The natural fiber composite can only identify the real ecological impact when a life cycle assessment is made from cradle to grave. LCAs present information on energy use, emissions, and solid wastes that are created in productive procedures for its whole lifecycle. This analysis puts forward insights of areas where environmental impacts can be mitigated and speaks to decisions about material selection, manufacturing processes, and end-of-life management strategies.
- Advancements in Biodegradability and Recycling: The improvement in the biodegradability of natural fiber composites is one of the lead directions. Materials are developed that, while at the end of life, will decompose without leaving a lot to burthen the environment again. In parallel, techniques about the recycling of composites after use are worked on. This can include the development of effective mechanical, chemical, or biological recycling processes that recover useful fibers and matrix materials from spent composite materials and allow for their re-use to make new products.

- Reducing Carbon Footprint: This will be one of the principal sustainability goals for the manufacturing and consumption of natural fiber composites. This can find methods wherein energy use during manufacturing processes would be minimal, sourcing materials locally to reduce emissions associated with transportation, and having renewable sources of unpolluted energy. The use of natural fibers themselves is already a factor in reducing carbon footprint because it is based on renewable resources that, while growing, have the potential to capture carbon dioxide.
- Utilization of Renewable and Bio-based Materials: Much of this industry now explores renewable and bio-based material resources, which involves the application of bio-resins and natural fibers. Seen next to synthetic materials, these have less environmental impact. These materials lessen dependence on fossil resources, reduce the generation of wastes through biodegradability, and reinforce other environmental goals as well.
- Environmental Impact of Additives and Treatments: While developing flame-retardant treatments and other performance-enhancing additives, it is very important to know their environmental impact. Research is inclined toward finding and using eco-friendly additives that do not emit hazardous substances during production, at the point of use, or during disposal. This signifies that the performance benefit associated with these additives does not come at a cost to environmental health.
- The Calculation of the Ecological Footprint: Parallel activities to evaluate ecological footprint with aspects like quantum of water, land, and biodiversity, besides carbon emission, are becoming common for natural fiber composites. In this way, the fiber procurement related to agriculture and forest practices will be made sustainable so that the natural fiber sourcing does not result in any ecological imbalances or degradation of natural eco-systems.
- Promoting Circular Economy Principles: Natural fiber composite production and the whole life cycle are being aligned with circular economy principles. Designing easy disassembly and recyclability of composites, encouraging the use of recycled material in composite production, and business models that enable return and re-utilization drive such strategies to help reduce wastes and maximize lifecycle values for natural fiber composites.

In taking into account these sustainability concerns, the development of natural fiber composites will be in an environmentally responsible direction so that such materials can go on and play their essential contribution toward global sustainability goals. This approach not only adds to the environmental profile of natural fiber composites but also puts them at the top in industries desiring to reduce their ecological impact.

6.4 Future Research Directions and Potential for Technological Advancements in Natural Fiber-Reinforced Biocomposites and Bionanocomposites

The future of natural fiber-reinforced biocomposites and bionanocomposites seems very promising, as several research directions and technological innovations provide impetus for further scientific development. Many of these focus areas, identified by the recent case studies, have revealed the prospect of improving performance, sustainability, and applicability for those materials [100, 129, 198]:

- Innovative Hybrid Material Systems: One of the most promising lines of inquiry would concern the development of new hybrid composite systems integrating natural fibers with other eco-friendly and high-performance materials. This would include the integration of natural fibers with advanced polymers, bio-based resins, and nano-enhanced materials for developing appropriate mechanical strength, durability, and environmental sustainability. In this regard, future research would then be more focused on synergism from these hybrid systems in tailoring attributes to diverse applications spanning various industrial sectors.
- Advancing Nanotechnology Integration: Nanotechnology remains key to the betterment of the properties in natural fiber composites. Future studies will be carried out with a view toward exploiting nanomaterials such as nano-clays, graphene, and carbon nanotubes for higher fire resistance, improved thermal stability, and enhanced mechanical strength. One major problem at this junction will be developing proper techniques that could incorporate these nanomaterials into the matrix of natural fiber composites with homogeneous distribution and strong bonding at the fiber-matrix interface.
- Industrial Application-Based Customization: Developing the attributes of natural fiber composites to suit the exacting requirements of industries like aerospace, automotive, and construction—this has been evolving further into significance in various research studies. This comprises the tailoring of the interface between the fiber and the matrix, perfecting the fiber orientation and arrangement, and finally, fine-tuning the composite structure for obtaining required strength, weight, and performance for a given application.
- Overcoming Technological and Processing Barriers: One of the key challenges to be addressed, as identified in this roadmap action plan, is related to existing technological challenges in processing, fiber treatment, and interfaces or compatibility between different composite constituents. It will aim to develop advanced fabrication techniques and methods for improving fiber pre-treatment while enhancing the interfacing between natural fibers and matrices. Such innovations could include advancements in 3D printing of natural fiber composites, automated manufacturing processes, and environmentally friendly fiber treatment techniques.
- Focus on Flame Retardancy and Functionalization: Very few will be focused on flame retardancy and functionalization based on the results from recent research.

Future research shall hence be carried out in the continued effort of improving upon the natural fiber composite's flame retardancy by developing some innovative flame-retardant additives and suitable surface treatments, especially functional nanomaterials that can enhance fire retardancy without any effects on the mechanical properties or the environment.

- Smart and Multifunctional Composites: A very exciting future research avenue may involve smart and multifunctional natural fiber composites. These could include embedded sensors, actuators, or other composite matrix-embodied smart materials that admit in real time for structural health reality monitoring or environmental responsiveness among other functional attributes attached to them. All these innovations can offer composites with good mechanical performance, endowing more features to the designed material in terms of safety, observation, and adaptation.
- Sustainability and Environmental Impact Studies: This would involve further reviews in light of sustainability and environmental impact that relate to natural fiber composites. This will entail detailed life cycle assessments, strategies for end-of-life disposal or recycling, and explorations of possibilities exploiting renewable sources of energy in the line of production. Every one of these efforts is rather very germane to ensuring full-fledged progress in natural fiber composites aligning with global goals on sustainability.
- Scalability and Industrial Application: Finally, research in scaling up the production process with stable quality and regard for the environment is pursued. It involves developing large-scale manufacturing processes, rigorous methodologies of quality control, and market analysis that will aid in the broad industrial applications of these materials. Maintaining this scalability without compromise in performance and sustainability will be important to gaining further adoption of natural fiber composites.

It is some of the current research directions and technological advances in natural fiber-reinforced biocomposites and bionanocomposites, which will just go on to change with better performance, sustainability, and applicability pertaining to a wide range of industries.

Acknowledgements The Basic Science Research Program through the National Research Foundation of Korea, funded by the Ministry of Science Education 2018R1A6A1A03024509, 2023R1A2C1006234, and 2022R1I1A1A01069282 supported this work.

References

1. Abbas AGN, Aziz FNAA, Abdan K et al (2022) Kenaf fibre reinforced cementitious composites. Fibers 10:3. https://doi.org/10.3390/FIB10010003
2. Abu Hassan NA, Ahmad S, Chen RS, Natarajan VD (2022) Synergistically enhanced mechanical, combustion and acoustic properties of biopolymer composite foams reinforcement by

kenaf fibre. Compos Part A Appl Sci Manuf 155:106826. https://doi.org/10.1016/J.COMPOSITESA.2022.106826

3. Adil S, Kumar B, Pham DH, Kim J (2024) Biobased, cellulose long filament-reinforced vanillin-derived epoxy composite for high-performance and flame-retardant applications. Int J Biol Macromol 256:128411. https://doi.org/10.1016/J.IJBIOMAC.2023.128411
4. Agarwal J, Mohanty S, Nayak SK (2021) Influence of cellulose nanocrystal/sisal fiber on the mechanical, thermal, and morphological performance of polypropylene hybrid composites. Polym Bull 78:1609–1635. https://doi.org/10.1007/S00289-020-03178-4/METRICS
5. Ahmed ATMF, Islam MZ, Mahmud MS et al (2022) Hemp as a potential raw material toward a sustainable world: a review. Heliyon 8:e08753. https://doi.org/10.1016/J.HELIYON.2022.E08753
6. Akter M, Uddin MH, Tania IS (2022) Biocomposites based on natural fibers and polymers: a review on properties and potential applications. J Reinforced Plast Compos 41:705–742. https://doi.org/10.1177/07316844211070609
7. Alam MA, Sapuan SM, Ya HH et al (2021) Application of biocomposites in automotive components: a review. Biocomposite Synth Compos Automot Appl 1–17. https://doi.org/10.1016/B978-0-12-820559-4.00001-8
8. Alam MR, Alimuzzaman S, Shahid MA et al (2023) Collagen/Nigella sativa/chitosan inscribed electrospun hybrid bio-nanocomposites for skin tissue engineering. J Biomater Sci Polym Ed 34:1517–1538. https://doi.org/10.1080/09205063.2023.2170139
9. Ali I, Kim NK, Bhattacharyya D (2021) Effects of graphene nanoplatelets on mechanical and fire performance of flax polypropylene composites with intumescent flame retardant. Molecules 26:4094. https://doi.org/10.3390/MOLECULES26134094
10. Ali N, Hoque ME (2022) Bionanocomposites in the automotive and aerospace applications, pp 237–253. https://doi.org/10.1007/978-981-16-8578-1_13
11. Alias AH, Norizan MN, Sabaruddin FA et al (2021) Hybridization of MMT/lignocellulosic fiber reinforced polymer nanocomposites for structural applications: a review. Coatings 11:1355. https://doi.org/10.3390/COATINGS11111355
12. Amiandamhen SO, Meincken M, Tyhoda L (2020) Natural fibre modification and its influence on fibre-matrix interfacial properties in biocomposite materials. Fibers Polym 21:677–689. https://doi.org/10.1007/S12221-020-9362-5/METRICS
13. Amjad AI (2024) Bamboo fibre: a sustainable solution for textile manufacturing. Adv Bamboo Sci 7:100088. https://doi.org/10.1016/J.BAMBOO.2024.100088
14. Andrew JJ, Dhakal HN (2022) Sustainable biobased composites for advanced applications: recent trends and future opportunities—a critical review. Compos Part C Open Access 7:100220. https://doi.org/10.1016/J.JCOMC.2021.100220
15. Anwer AH, Ahtesham A, Shoeb M et al (2023) State-of-the-art advances in nanocomposite and bio-nanocomposite polymeric materials: a comprehensive review. Adv Colloid Interface Sci 318:102955. https://doi.org/10.1016/J.CIS.2023.102955
16. Ao X, Vázquez-López A, Mocerino D et al (2024) Flame retardancy and fire mechanical properties for natural fiber/polymer composite: a review. Compos Part B Eng 268:111069. https://doi.org/10.1016/J.COMPOSITESB.2023.111069
17. Aravind D, Senthilkumar K, Chandrasekar M et al (2022) A compendious review on fatigue properties of the bio-nanocomposites. Polym Compos 43:6803–6816. https://doi.org/10.1002/PC.26859
18. Asim M, Paridah MT, Chandrasekar M et al (2020) Thermal stability of natural fibers and their polymer composites. Iran Polym J 29(7):625–648. https://doi.org/10.1007/S13726-020-00824-6
19. Asyraf MRM, Rafidah M, Azrina A, Razman MR (2021) Dynamic mechanical behaviour of kenaf cellulosic fibre biocomposites: a comprehensive review on chemical treatments. Cellulose 28:2675–2695. https://doi.org/10.1007/S10570-021-03710-3/METRICS
20. Attia NF (2022) Sustainable and efficient flame retardant materials for achieving high fire safety for polystyrene composites. J Therm Anal Calorim 147:5733–5742. https://doi.org/10.1007/S10973-021-10948-3/METRICS

21. Awang NA, Mohamed MA, Salleh WNW (2023) Polyvinyl alcohol/cellulose-based biocomposites and bionanocomposites. Polyvinyl alcohol-based biocomposites and bionanocomposites, pp 81–130. https://doi.org/10.1002/9781119593218.CH4
22. Azad MM, Ejaz M, Shah A ur R et al (2022) A bio-based approach to simultaneously improve flame retardancy, thermal stability and mechanical properties of nano-silica filled jute/thermoplastic starch composite. Mater Chem Phys 289:126485. https://doi.org/10.1016/J.MATCHEMPHYS.2022.126485
23. Azlin MNM, Sapuan SM, Zuhri MYM et al (2022) Thermal stability, dynamic mechanical analysis and flammability properties of woven kenaf/polyester-reinforced polylactic acid hybrid laminated composites. Polymers 14:2690. https://doi.org/10.3390/POLYM14132690
24. Azman Mohammad Taib MN, Hamidon TS, Garba ZN et al (2022) Recent progress in cellulose-based composites towards flame retardancy applications. Polymer (Guildf) 244:124677. https://doi.org/10.1016/J.POLYMER.2022.124677
25. Bachtiar EV, Kurkowiak K, Yan L et al (2019) Thermal stability, fire performance, and mechanical properties of natural fibre fabric-reinforced polymer composites with different fire retardants. Polymers 11:699. https://doi.org/10.3390/POLYM11040699
26. Bajwa DS, Shojaeiarani J, Liaw JD, Bajwa SG (2021) Role of hybrid nano-zinc oxide and cellulose nanocrystals on the mechanical, thermal, and flammability properties of poly (lactic acid) polymer. J Compos Sci 5:43. https://doi.org/10.3390/JCS5020043
27. Bangar SP, Whiteside WS (2021) Nano-cellulose reinforced starch bio composite films—a review on green composites. Int J Biol Macromol 185:849–860. https://doi.org/10.1016/J.IJBIOMAC.2021.07.017
28. Barczewski M, Sałasińska K, Raś W et al (2023) The effect of hybridization of fire retarded epoxy/flax-cotton fiber laminates by expanded vermiculite: structure-property relationship study. Adv Ind Eng Polym Res 6:181–194. https://doi.org/10.1016/J.AIEPR.2023.01.005
29. Battegazzore D, Frache A, Carosio F (2020) Layer-by-layer nanostructured interphase produces mechanically strong and flame retardant bio-composites. Compos Part B Eng 200:108310. https://doi.org/10.1016/J.COMPOSITESB.2020.108310
30. Bazan P, Salasińska K, Kuciel S (2021) Flame retardant polypropylene reinforced with natural additives. Ind Crops Prod 164:113356. https://doi.org/10.1016/J.INDCROP.2021.113356
31. Bhanderi K, Joshi J, Suthar V et al (2023) Eco-friendly polymer nanocomposites based on bio-based fillers: preparation, characterizations and potential applications. Biodegradable and biocompatible polymer nanocomposites. Processing, characterization, and applications, pp 173–203. https://doi.org/10.1016/B978-0-323-91696-7.00015-5
32. Bhat R, Saxena P (2023) Natural fiber–reinforced polymer nanocomposites. Nanomaterials for sustainable tribology, pp 115–141. https://doi.org/10.1201/9781003306276-8
33. Birhanie ZM, Xiao A, Yang D et al (2021) Polysaccharides, total phenolic, and flavonoid content from different kenaf (*Hibiscus cannabinus* L.) genotypes and their antioxidants and antibacterial properties. Plants 10. https://doi.org/10.3390/PLANTS10091900
34. Cabrera FC (2021) Eco-friendly polymer composites: a review of suitable methods for waste management. Polym Compos 42:2653–2677. https://doi.org/10.1002/PC.26033
35. Campana C, Léger R, Sonnier R et al (2022) Effect of hygrothermal ageing on the mechanical and fire properties of a flame retardant flax fiber/epoxy composite. Polymers 14:3962. https://doi.org/10.3390/POLYM14193962
36. Canıtez H, Yüksel Yılmaz AN, Çelik Bedeloğlu A (2023) The effect of silicon dioxide and zinc borate on the flame retardancy, thermal and mechanical properties of jute/epoxy hybrid composite. J Nat Fibers 20. https://doi.org/10.1080/15440478.2022.2131683
37. Chakraborty A, Ghalsasi P, Radha P (2023) Insight into nano-fillers and their reinforcement onto polylactic acid. J Inorg Organomet Polym Mater 33:1119–1133. https://doi.org/10.1007/S10904-023-02605-Z/METRICS
38. Chalapathi KV, Il SJ, Prabhakar MN (2022) Impact of surface treatments and hybrid flame retardants on flammability, and thermal performance of bamboo fabric composites. J Nat Fibers 19:2129–2139. https://doi.org/10.1080/15440478.2020.1798849

39. Chandrasekaran S, Cruz-Izquierdo A, Castaing R et al (2023) Facile preparation of flame-retardant cellulose composite with biodegradable and water resistant properties for electronic device applications. Sci Rep 13(1):1–12. https://doi.org/10.1038/s41598-023-30078-0
40. Chaturvedi S, Kataria A, Chaudhary V et al (2023) Bionanocomposites reinforced with cellulose fibers and agro-industrial wastes. Cellulose fibre reinforced composites. Interface engineering, processing and performance, pp 317–342. https://doi.org/10.1016/B978-0-323-90125-3.00017-3
41. Chausali N, Saxena J, Prasad R (2022) Recent trends in nanotechnology applications of bio-based packaging. J Agric Food Res 7:100257. https://doi.org/10.1016/J.JAFR.2021.100257
42. Chee SS, Jawaid M, Alothman OY, Yahaya R (2020) Thermo-oxidative stability and flammability properties of bamboo/kenaf/nanoclay/epoxy hybrid nanocomposites. RSC Adv 10:21686–21697. https://doi.org/10.1039/D0RA02126A
43. Chen RS, Mohd Amran NA, Ahmad S (2019) Reinforcement effect of nanocomposites with single/hybrid graphene nanoplatelets and magnesium hydroxide: thermal stability, flame retardancy and mechanical performance. J Therm Anal Calorim 137:79–92. https://doi.org/10.1007/S10973-018-7935-Y/METRICS
44. de Oliveira CRS, Batistella MA, Guelli Ulson de Souza SM de A, Ulson de Souza AA (2021) Functionalization of cellulosic fibers with a kaolinite-TiO_2 nano-hybrid composite via a solvothermal process for flame retardant applications. Carbohydr Polym 266:118108. https://doi.org/10.1016/J.CARBPOL.2021.118108
45. Dhamodharan D, Dhinakaran V, Ghoderao PN et al (2022) Synergistic effect of cellulose nanocrystals-graphene oxide as an effective nanofiller for enhancing properties of solventless polymer nanocomposites. Compos Part B Eng 238:109918. https://doi.org/10.1016/J.COMPOSITESB.2022.109918
46. Dharini V, Periyar Selvam S, Jayaramudu J, Sadiku Emmanuel R (2022) Functional properties of clay nanofillers used in the biopolymer-based composite films for active food packaging applications—review. Appl Clay Sci 226:106555. https://doi.org/10.1016/J.CLAY.2022.106555
47. Ding C, Pan M, Chen H et al (2020) An anionic polyelectrolyte hybrid for wood-polyethylene composites with high strength and fire safety via self-assembly. Constr Build Mater 248:118661. https://doi.org/10.1016/J.CONBUILDMAT.2020.118661
48. Doineau E, Cathala B, Benezet JC et al (2021) Development of bio-inspired hierarchical fibres to tailor the fibre/matrix interphase in (bio)composites. Polymers (Basel) 13:1–34. https://doi.org/10.3390/POLYM13050804
49. Dorez G, Taguet A, Ferry L, Lopez-Cuesta JM (2013) Thermal and fire behavior of natural fibers/PBS biocomposites. Polym Degrad Stab 98:87–95. https://doi.org/10.1016/J.POLYMDEGRADSTAB.2012.10.026
50. Dou Y, Li X, Zhang T, Xu H (2021) An intumescent flame-retardant layer with β-cyclodextrin as charring agent and its flame retardancy in jute/polypropylene composites. Polym Bull 78:4281–4296. https://doi.org/10.1007/S00289-020-03315-Z/METRICS
51. Dou Y, Zhong Z, Huang J et al (2023) A new phosphorous/nitrogen-containing flame-retardant film with high adhesion for jute fiber composites. Polymers (Basel) 15:1920. https://doi.org/10.3390/POLYM15081920/S1
52. Dreyer C, Motoc DL (2020) Reaction-to-fire parameters of CE/epoxy blend-based CF/FF and BF/FF hybrid composites. J Therm Anal Calorim 142:661–669. https://doi.org/10.1007/S10973-020-09359-7/METRICS
53. Fang L, Lu X, Zeng J et al (2020) Investigation of the flame-retardant and mechanical properties of bamboo fiber-reinforced polypropylene composites with melamine pyrophosphate and aluminum hypophosphite addition. Materials 13:479. https://doi.org/10.3390/MA13020479
54. Gao D, Zhao P, Lyu B et al (2020) Composite based on poly(acrylic acid)/modified attapulgite/zinc oxide as a flame retardant of cotton fabrics. Cellulose 27:2873–2886. https://doi.org/10.1007/S10570-019-02948-2/METRICS
55. Garavand F, Nooshkam M, Khodaei D et al (2023) Recent advances in qualitative and quantitative characterization of nanocellulose-reinforced nanocomposites: a review. Adv Colloid Interface Sci 318:102961. https://doi.org/10.1016/J.CIS.2023.102961

56. Gopinath A, Senthil Kumar M, Elayaperumal A (2014) Experimental investigations on mechanical properties of jute fiber reinforced composites with polyester and epoxy resin matrices. Procedia Eng 97:2052–2063. https://doi.org/10.1016/J.PROENG.2014.12.448
57. Gorur YC, Francon HS, Sethi J et al (2022) Rapidly prepared nanocellulose hybrids as gas barrier, flame retardant, and energy storage materials. ACS Appl Nano Mater 5:9188–9200. https://doi.org/10.1021/ACSANM.2C01530/ASSET/IMAGES/MEDIUM/AN2C01530_M004.GIF
58. Guo W, Hu Y, Wang X et al (2019) Exceptional flame-retardant cellulosic foams modified with phosphorus-hybridized graphene nanosheets. Cellulose 26:1247–1260. https://doi.org/10.1007/S10570-018-2127-2/METRICS
59. Guo W, Kalali EN, Wang X et al (2019) Processing bulk natural bamboo into a strong and flame-retardant composite material. Ind Crops Prod 138:111478. https://doi.org/10.1016/J.INDCROP.2019.111478
60. Gupta M, Jain A, Kamineni JN, Burela RG (2022) Advances and applications of biofiber-based polymer composites. Advances in bio-based fiber. Moving towards a green society, pp 575–602. https://doi.org/10.1016/B978-0-12-824543-9.00002-5
61. Gupta MK, Ramesh M, Thomas S (2021) Effect of hybridization on properties of natural and synthetic fiber-reinforced polymer composites (2001–2020): a review. Polym Compos 42:4981–5010. https://doi.org/10.1002/PC.26244
62. Haris NIN, Hassan MZ, Ilyas RA et al (2022) Dynamic mechanical properties of natural fiber reinforced hybrid polymer composites: a review. J Mater Res Technol 19:167–182. https://doi.org/10.1016/J.JMRT.2022.04.155
63. Hasan KF, Horváth PG, Bak M, Alpár T (2021) A state-of-the-art review on coir fiber-reinforced biocomposites. RSC Adv 11:10548–10571. https://doi.org/10.1039/D1RA00231G
64. Hazrol MD, Sapuan SM, Zuhri MYM et al (2021) Recent development in kenaf (*Hibiscus cannabinus*)-based biocomposites and their potential industrial applications: a review. Des Sustain Green Mater Process 329–368. https://doi.org/10.1016/B978-0-12-819482-9.00007-1
65. He L, Song F, Li DF et al (2020) Strong and tough polylactic acid based composites enabled by simultaneous reinforcement and interfacial compatibilization of microfibrillated cellulose. ACS Sustain Chem Eng 8:1573–1582. https://doi.org/10.1021/ACSSUSCHEMENG.9B06308/ASSET/IMAGES/LARGE/SC9B06308_0003.JPEG
66. He W, Song P, Yu B et al (2020) Flame retardant polymeric nanocomposites through the combination of nanomaterials and conventional flame retardants. Prog Mater Sci 114:100687. https://doi.org/10.1016/J.PMATSCI.2020.100687
67. Holbery J, Houston D (2006) Natural-fiber-reinforced polymer composites in automotive applications. JOM 58:80–86. https://doi.org/10.1007/S11837-006-0234-2/METRICS
68. Horrocks AR, Anand SC (2016) Handbook of technical textiles, vol 2. Technical textile applications, p 453
69. Hu D, Liu H, Ding Y, Ma W (2021) Synergetic integration of thermal conductivity and flame resistance in nacre-like nanocellulose composites. Carbohydr Polym 264:118058. https://doi.org/10.1016/J.CARBPOL.2021.118058
70. Huang G, Huang JQ, Chen XY, Zhu YX (2021) Recent advances and future perspectives in cotton research. Annu Rev Plant Biol 72:437–462. https://doi.org/10.1146/ANNUREV-ARPLANT-080720-113241
71. Hussain T, Jeena G, Pitakbut T et al (2021) Cannabis sativa research trends, challenges, and new-age perspectives. iScience 24:103391. https://doi.org/10.1016/J.ISCI.2021.103391
72. Ibrahim MI, Alamry KA, Hussein MA et al (2023) An efficient antimicrobial performance of hybrid functionalized MWCNTs/GNPts bionanocomposites based on poly(3-hydroxybutyrate-co-3-hydroxyvalerate). ChemistrySelect 8:e202205011. https://doi.org/10.1002/SLCT.202205011
73. Idumah CI (2022) Emerging advancements in flame retardancy of polypropylene nanocomposites. J Thermoplast Compos Mater 35:2665–2704. https://doi.org/10.1177/0892705720930782/ASSET/IMAGES/LARGE/10.1177_0892705720930782-FIG16.JPEG

74. Idumah CI, Hassan A (2016) Emerging trends in flame retardancy of biofibers, biopolymers, biocomposites, and bionanocomposites. Rev Chem Eng 32:115–148. https://doi.org/10.1515/REVCE-2015-0017/ASSET/GRAPHIC/J_REVCE-2015-0017_CV_002.JPG
75. Idumah CI, Hassan A, Ogbu JE et al (2018) Electrical, thermal and flammability properties of conductive filler kenaf–reinforced polymer nanocomposites. J Thermoplast Compos Mater 33:516–540. https://doi.org/10.1177/0892705718807957
76. Idumah CI, Obele CM, Ezeani EO (2021) Understanding interfacial dispersions in ecobenign polymer nano-biocomposites. Polym Technol Mater 60:233–252. https://doi.org/10.1080/25740881.2020.1811312
77. Igwe Idumah C, Nwabanne JT, Tanjung FA (2021) Novel trends in poly (lactic) acid hybrid bionanocomposites. Clean Mater 2:100022. https://doi.org/10.1016/J.CLEMA.2021.100022
78. Ilyas RA, Sapuan SM, Nurazzi NM et al (2021) Macro to nanoscale natural fiber composites for automotive components: research, development, and application. Biocomposite and synthetic composites for automotive applications, pp 51–105. https://doi.org/10.1016/B978-0-12-820559-4.00003-1
79. Ilyas RA, Zuhri MYM, Norrahim MNF et al (2022) Natural fiber-reinforced polycaprolactone green and hybrid biocomposites for various advanced applications. Polymers 14:182. https://doi.org/10.3390/POLYM14010182
80. Islam MS, Kovalcik A, Hasan M, Thakur VK (2015) Natural fiber reinforced polymer composites. Int J Polym Sci 2015. https://doi.org/10.1155/2015/813568
81. Islam MS, van de Ven TGM (2021) Cotton-based flame-retardant textiles: a review. BioResources 16:4354–4381. https://doi.org/10.15376/BIORES.16.2.ISLAM
82. Jadoun S, Riaz U (2022) Polymer-based green composites and their applications. Green sustainable process for chemical and environmental engineering and science: green composites: preparation, properties, and allied applications, pp 123–145. https://doi.org/10.1016/B978-0-323-99643-3.00007-3
83. Jalalah M, Zulfiqar S, Afzal A et al (2023) Microencapsulation based fire retardant eco-friendly jute composite. Polym Polym Compos 31. https://doi.org/10.1177/09673911221147702/ASSET/IMAGES/LARGE/10.1177_09673911221147702-FIG11.JPEG
84. Jiang S, Zhang M, Li M et al (2021) Cellulose-based composite thermal-insulating foams toward eco-friendly, flexible and flame-retardant. Carbohydr Polym 273:118544. https://doi.org/10.1016/J.CARBPOL.2021.118544
85. Jose A, Radhakrishnan EK (2023) Polyvinyl alcohol/polylactic acid-based biocomposites and bionanocomposites. Polyvinyl alcohol-based biocomposites and bionanocomposites, pp 151–178. https://doi.org/10.1002/9781119593218.CH6
86. Joseph B, Krishnan S, Sagarika VK et al (2020) Bionanocomposites as industrial materials, current and future perspectives: a review. Emergent Mater 3:711–725. https://doi.org/10.1007/S42247-020-00133-X/METRICS
87. Joseph K, Dias R, Filho T et al (1999) A review on sisal fiber reinforced polymer composites. Rev Bras Eng Agrícola Ambient 3:367–379. https://doi.org/10.1590/1807-1929/AGRIAMBI.V3N3P367-379
88. Kabir II, Wang C, Yuen ACY, Yeoh GH (2022) Cellulose-based flame retardants for polymeric materials. Bio-based Flame-Retardant Technol Polym Mater 97–131. https://doi.org/10.1016/B978-0-323-90771-2.00002-X
89. Kahali P, Montazer M, Kamali Dolatabadi M (2022) Sustainable copper oxide/Tragacanth gum bionanocomposites with multi-purpose catalytic activities on textile. J Appl Polym Sci 139:e52781. https://doi.org/10.1002/APP.52781
90. Kamaraj M, Dodson EA, Datta S (2020) Effect of graphene on the properties of flax fabric reinforced epoxy composites. Adv Compos Mater 29:443–458. https://doi.org/10.1080/09243046.2019.1709679
91. Kang F, Han H, Wang H et al (2024) Construction of a flame retardant three-dimensional network structure in sisal/polypropylene composites. Ind Crops Prod 209:117973. https://doi.org/10.1016/J.INDCROP.2023.117973

92. Karger-Kocsis J, Mahmood H, Pegoretti A (2015) Recent advances in fiber/matrix interphase engineering for polymer composites. Prog Mater Sci 73:1–43. https://doi.org/10.1016/J.PMATSCI.2015.02.003
93. Karimah A, Ridho MR, Munawar SS et al (2021) A review on natural fibers for development of eco-friendly bio-composite: characteristics, and utilizations. J Mater Res Technol 13:2442–2458. https://doi.org/10.1016/J.JMRT.2021.06.014
94. Karki S, Gohain MB, Yadav D, Ingole PG (2021) Nanocomposite and bio-nanocomposite polymeric materials/membranes development in energy and medical sector: A review. Int J Biol Macromol 193:2121–2139. https://doi.org/10.1016/J.IJBIOMAC.2021.11.044
95. Kaur G, Kander R (2023) The sustainability of industrial hemp: a literature review of its economic, environmental, and social sustainability. Sustainability 15:6457. https://doi.org/10.3390/SU15086457
96. Kaur G, Sharma S, Mir SA, Dar BN (2021) Nanobiocomposite films: a "greener alternate" for food packaging. Food Bioprocess Technol 14:1013–1027. https://doi.org/10.1007/S11947-021-02634-X/METRICS
97. Kaur N, Saxena S, Gaur H, Goyal P (2017) A review on bamboo fiber composites and its applications. In: 2017 international conference on infocom technologies and unmanned systems (trends and future directions) (ICTUS), pp 843–849. https://doi.org/10.1109/ICTUS.2017.8286123
98. Kausar A (2021) Progress in green nanocomposites for high-performance applications. Mater Res Innov 25:53–65. https://doi.org/10.1080/14328917.2020.1728489
99. Kazemi H, Mighri F, Park KW et al (2022) Natural rubber biocomposites reinforced with cellulose nanocrystals/lignin hybrid fillers. Polym Compos 43:5442–5453. https://doi.org/10.1002/PC.26846
100. Kenned JJ, Sankaranarayanasamy K, Kumar CS (2021) Chemical, biological, and nanoclay treatments for natural plant fiber-reinforced polymer composites: a review. Polym Polym Compos 29:1011–1038. https://doi.org/10.1177/0967391120942419/ASSET/IMAGES/LARGE/10.1177_0967391120942419-FIG14.JPEG
101. Khalid MY, Imran R, Arif ZU et al (2021) Developments in chemical treatments, manufacturing techniques and potential applications of natural-fibers-based biodegradable composites. Coatings 11:293. https://doi.org/10.3390/COATINGS11030293
102. Khan A, Sapuan SM, Siddiqui VU et al (2023) A review of recent developments in kenaf fiber/polylactic acid composites research. Int J Biol Macromol 253:127119. https://doi.org/10.1016/J.IJBIOMAC.2023.127119
103. Khan A, Sapuan SM, Yusuf J et al (2023) An examination of cutting-edge developments in Bamboo-PLA composite research: a comprehensive review. Renew Sustain Energy Rev 188:113832. https://doi.org/10.1016/J.RSER.2023.113832
104. Kim H, Park J, Minn KS et al (2020) Eco-friendly nanocellulose embedded polymer composite foam for flame retardancy improvement. Macromol Res 28:165–171. https://doi.org/10.1007/S13233-020-8020-5/METRICS
105. Kim NK, Bhattacharyya D (2023) Fire reaction and post-fire impact properties of flax fibre reinforced composites containing intumescent flame retardants. J Reinf Plast Compos. https://doi.org/10.1177/07316844231158112/ASSET/IMAGES/LARGE/10.1177_07316844231158112-FIG9.JPEG
106. Kim NK, Bhattacharyya D, van Loosdrecht M, Lin Y (2023) Enhancement of fire resistance and mechanical performance of polypropylene composites containing cellulose fibres and extracellular biopolymers from wastewater sludge. Polym Test 127:108185. https://doi.org/10.1016/J.POLYMERTESTING.2023.108185
107. Kim Y, Lee S, Yoon H (2021) Fire-safe polymer composites: flame-retardant effect of nanofillers. Polymers 13:540. https://doi.org/10.3390/POLYM13040540
108. Kocaman S, Soydal U, Ahmetli G (2021) Influence of cotton waste and flame-retardant additives on the mechanical, thermal, and flammability properties of phenolic novolac epoxy composites. Cellulose 28:7765–7780. https://doi.org/10.1007/S10570-021-04037-9/METRICS

109. Köklükaya O, Carosio F, Durán VL, Wågberg L (2020) Layer-by-layer modified low density cellulose fiber networks: a sustainable and fireproof alternative to petroleum based foams. Carbohydr Polym 230:115616. https://doi.org/10.1016/J.CARBPOL.2019.115616
110. Kokol V, Vivod V, Peršin Z et al (2021) Screen-printing of microfibrillated cellulose for an improved moisture management, strength and abrasion resistant properties of flame-resistant fabrics. Cellulose 28:6663–6678. https://doi.org/10.1007/S10570-021-03915-6/FIGURES/8
111. Kovačević Z, Flinčec Grgac S, Bischof S (2021) Progress in biodegradable flame retardant nano-biocomposites. Polymers (Basel) 13:1–30. https://doi.org/10.3390/POLYM13050741
112. Krishnaiah P, Manickam S, Ratnam CT et al (2020) Surface-treated short sisal fibers and halloysite nanotubes for synergistically enhanced performance of polypropylene hybrid composites. J Thermoplast Compos Mater 35:2089–2104. https://doi.org/10.1177/0892705720946063
113. Kumar P, Ram CS, Srivastava JP et al (2021) Synthesis of cotton fiber and its structure. Natural and synthetic fiber reinforced composites: synthesis, properties and applications, pp 17–36. https://doi.org/10.1002/9783527832996.CH2
114. Kun D, Kárpáti Z, Fekete E, Móczó J (2021) The role of interfacial adhesion in polymer composites engineered from lignocellulosic agricultural waste. Polymers (Basel) 13. https://doi.org/10.3390/POLYM13183099
115. Kymäläinen HR, Sjöberg AM (2008) Flax and hemp fibres as raw materials for thermal insulations. Build Environ 43:1261–1269. https://doi.org/10.1016/J.BUILDENV.2007.03.006
116. Lavoine N, Desloges I, Dufresne A, Bras J (2012) Microfibrillated cellulose—its barrier properties and applications in cellulosic materials: a review. Carbohydr Polym 90:735–764. https://doi.org/10.1016/J.CARBPOL.2012.05.026
117. Lee CH, Khalina A, Lee SH (2021) Importance of interfacial adhesion condition on characterization of plant-fiber-reinforced polymer composites: a review. Polymers (Basel) 13:1–22. https://doi.org/10.3390/POLYM13030438
118. Li D, Wang ZH, Zhu YS et al (2022) Synergistically improved flame retardancy of the cotton fabric finished by silica-coupling agent-zinc borate hybrid sol. J Ind Text 51:8297S-8322S. https://doi.org/10.1177/15280837211028800/ASSET/IMAGES/LARGE/10.1177_15280837211028800-FIG10.JPEG
119. Li N, Kang G, Liu H et al (2022) Hybrid nanoparticles based on novel Schiff Base for durable flame retardant and antibacterial properties. Compos Part B Eng 238:109905. https://doi.org/10.1016/J.COMPOSITESB.2022.109905
120. Li T, Chen C, Brozena AH et al (2021) Developing fibrillated cellulose as a sustainable technological material. Nature 590(7844):47–56. https://doi.org/10.1038/s41586-020-03167-7
121. Li Z, Chen C, Xie H et al (2021) Sustainable high-strength macrofibres extracted from natural bamboo. Nat Sustain 53(5):235–244. https://doi.org/10.1038/s41893-021-00831-2
122. Li Z, Li Y, Shen Y et al (2022) Synergic effects of dimethyl methylphosphonate (DMMP) and nano-sized montmorillonite (MMT) on the flammability and mechanical properties of flax fiber reinforced phenolic composites under hydrothermal aging. Compos Sci Technol 230:109487. https://doi.org/10.1016/J.COMPSCITECH.2022.109487
123. Lin K, Wang Z (2023) Multiscale mechanics and molecular dynamics simulations of the durability of fiber-reinforced polymer composites. Commun Mater 4(1):1–16. https://doi.org/10.1038/s43246-023-00391-2
124. Liu L, Du M, Liu F (2023) Recent advances in interface microscopic characterization of carbon fiber-reinforced polymer composites. Front Mater 10:1124338. https://doi.org/10.3389/FMATS.2023.1124338/BIBTEX
125. Liu Y, Li H, Chen Q et al (2020) Effect of bamboo flour on flame retardancy and smoke suppression of polypropylene/ammonium polyphosphate composites. Front Mater 7:574924. https://doi.org/10.3389/FMATS.2020.574924/BIBTEX
126. Loganathan TM, Sultan MTH, Ahsan Q et al (2022) Thermal degradation, visco-elastic and fire-retardant behavior of hybrid *Cyrtostachys renda*/kenaf fiber-reinforced MWCNT-modified phenolic composites. J Therm Anal Calorim 147:14079–14096. https://doi.org/10.1007/S10973-022-11557-4/FIGURES/9

127. Lou G, Ma Z, Dai J et al (2021) Fully biobased surface-functionalized microcrystalline cellulose via green self-assembly toward fire-retardant, strong, and tough epoxy biocomposites. ACS Sustain Chem Eng 9:13595–13605. https://doi.org/10.1021/ACSSUSCHEMENG.1C04718/ASSET/IMAGES/LARGE/SC1C04718_0008.JPEG
128. Madyaratri EW, Ridho MR, Aristri MA et al (2022) Recent advances in the development of fire-resistant biocomposites—a review. Polymers (Basel) 14. https://doi.org/10.3390/POLYM14030362
129. Mahmud S, Hasan KMF, Jahid MA et al (2021) Comprehensive review on plant fiber-reinforced polymeric biocomposites. J Mater Sci 56:7231–7264. https://doi.org/10.1007/S10853-021-05774-9/METRICS
130. Mahmud S, Konlan J, Deicaza J, Li G (2023) Hybrid hemp/glass fiber reinforced high-temperature shape memory photopolymer with mechanical and flame-retardant analysis. Sci Rep 13(1):1–15. https://doi.org/10.1038/s41598-023-44710-6
131. Mai Nguyen Tran T, Dang XP, Prabhakar MN, Song JI (2024) Multi-objective optimization of hybrid polypropylene composites for enhanced mechanical, thermal, and flame-retardant properties. Polym Compos. https://doi.org/10.1002/PC.28746
132. Mai Nguyen Tran T, Prabhakar MN, Lee DW, Song JI (2023) Effect of hybrid eco-friendly reinforcement and their size on mechanical and flame retardant properties of polypropylene composites for technical applications. Polym Compos. https://doi.org/10.1002/PC.27930
133. Mai Nguyen Tran T, Prabhakar MN, Song JI (2021) Influence of biowaste additive and treated short woven flax fibers on the flame retardancy and mechanical properties of PP composites. Ind Crops Prod 174:114176. https://doi.org/10.1016/J.INDCROP.2021.114176
134. Makinde-Isola BA, Taiwo AS, Oladele IO et al (2023) Development of sustainable and biodegradable materials: a review on banana and sisal fibre based polymer composites. J Thermoplast Compos Mater 37:1519–1539. https://doi.org/10.1177/08927057231186324
135. Malik K, Ahmad F, Gunister E (2021) A review on the kenaf fiber reinforced thermoset composites. Appl Compos Mater 28(2):491–528. https://doi.org/10.1007/S10443-021-09871-5
136. Manaia JP, Manaia AT, Rodriges L (2019) Industrial hemp fibers: an overview. Fibers 7:106. https://doi.org/10.3390/FIB7120106
137. Manu T, Nazmi AR, Shahri B et al (2022) Biocomposites: a review of materials and perception. Mater Today Commun 31:103308. https://doi.org/10.1016/J.MTCOMM.2022.103308
138. Medina L, Carosio F, Berglund LA (2019) Recyclable nanocomposite foams of poly(vinyl alcohol), clay and cellulose nanofibrils—mechanical properties and flame retardancy. Compos Sci Technol 182:107762. https://doi.org/10.1016/J.COMPSCITECH.2019.107762
139. Mochane MJ, Magagula SI, Sefadi JS, Mokhena TC (2021) A review on green composites based on natural fiber-reinforced polybutylene succinate (PBS). Polymers (Basel) 13. https://doi.org/10.3390/POLYM13081200
140. Mohammed M, Rasidi MSM, Mohammed AM et al (2022) Interfacial bonding mechanisms of natural fibre-matrix composites: an overview. BioResources 17:7031–7090. https://doi.org/10.15376/BIORES.17.4.MOHAMMED
141. Molina A, Kothari A, Odundo A, Prakash M (2023) *Agave sisalana*: towards distributed manufacturing of absorbent media for menstrual pads in semi-arid regions. Commun Eng 2:1–14. https://doi.org/10.1038/s44172-023-00130-y
142. Mondal A, Nayak AK, Chakraborty P et al (2023) Natural polymeric nanobiocomposites for anti-cancer drug delivery therapeutics: a recent update. Pharmaceutics 15:2064. https://doi.org/10.3390/PHARMACEUTICS15082064
143. More AP (2021) (2021) Flax fiber–based polymer composites: a review. Adv Compos Hybrid Mater 51(5):1–20. https://doi.org/10.1007/S42114-021-00246-9
144. Mouritz A, Gibson A (2006) Fire tests for composites. Fire properties of polymer composite materials, pp 325–358. https://doi.org/10.1007/978-1-4020-5356-6_11
145. Mueed A, Shibli S, Korma SA et al (2022) Flaxseed bioactive compounds: chemical composition, functional properties, food applications and health benefits-related gut microbes. Foods 11.https://doi.org/10.3390/FOODS11203307

146. Nagalakshmaiah M, Afrin S, Malladi RP et al (2019) Biocomposites: present trends and challenges for the future. Green composites for automotive applications, pp 197–215. https://doi.org/10.1016/B978-0-08-102177-4.00009-4
147. Nagarajan KJ, Ramanujam NR, Sanjay MR et al (2021) A comprehensive review on cellulose nanocrystals and cellulose nanofibers: pretreatment, preparation, and characterization. Polym Compos 42:1588–1630. https://doi.org/10.1002/PC.25929
148. Naghdi R (2021) Advanced natural fibre-based fully biodegradable and renewable composites and nanocomposites: a comprehensive review. Int Wood Prod J 12:178–193. https://doi.org/10.1080/20426445.2021.1945180
149. Nazrin A, Sapuan SM, Zuhri MYM et al (2022) Flammability and physical stability of sugar palm crystalline nanocellulose reinforced thermoplastic sugar palm starch/poly(lactic acid) blend bionanocomposites. Nanotechnol Rev 11:86–95. https://doi.org/10.1515/NTREV-2022-0007/MACHINEREADABLECITATION/RIS
150. Nimbagal V, Banapurmath NR, Sajjan AM et al (2021) Studies on hybrid bio-nanocomposites for structural applications. J Mater Eng Perform 30:6461–6480. https://doi.org/10.1007/S11665-021-05843-9/METRICS
151. Niu Q, Yue X, Cao W et al (2022) Interfacial silicon-nitrogen aerogel raise flame retardancy of bamboo fiber reinforced polylactic acid composites. Int J Biol Macromol 222:2697–2708. https://doi.org/10.1016/J.IJBIOMAC.2022.10.051
152. Niu Q, Yue X, Guo Z et al (2022) Flame retardant bamboo fiber reinforced polylactic acid composites regulated by interfacial phosphorus-silicon aerogel. Polymer (Guildf) 252:124961. https://doi.org/10.1016/J.POLYMER.2022.124961
153. Niu Q, Yue X, Guo Z et al (2022) Strengthening and flame retarding effect of bamboo fiber modified by silica aerogel on polylactic acid composites. Constr Build Mater 340:127696. https://doi.org/10.1016/J.CONBUILDMAT.2022.127696
154. Olvera D, Monaghan MG (2021) Electroactive material-based biosensors for detection and drug delivery. Adv Drug Deliv Rev 170:396–424. https://doi.org/10.1016/J.ADDR.2020.09.011
155. Omran AAB, Mohammed AABA, Sapuan SM et al (2021) Micro- and nanocellulose in polymer composite materials: a review. Polymers 13:231. https://doi.org/10.3390/POLYM13020231
156. Ortega F, Versino F, López OV, García MA (2021) Biobased composites from agro-industrial wastes and by-products. Emergent Mater 5(3):873–921. https://doi.org/10.1007/S42247-021-00319-X
157. Othman FEC, Hassan SA (2023) Hybrid biocomposites—combining synthetic nanofillers and natural fibers in composite structures. Synthetic and natural nanofillers in polymer composites. Properties and applications, pp 259–272. https://doi.org/10.1016/B978-0-443-19053-7.00011-1
158. Pabón Rojas JJ, Ramón Valencia BA, Bolívar Osorio FJ, Ramirez D (2021) The role of fiber-matrix compatibility in vacuum processed natural fiber/epoxy biocomposites. Cellulose 28:7845–7857. https://doi.org/10.1007/S10570-021-04055-7/METRICS
159. Peng C, Zhong J, Ma X et al (2022) Transparent, hard-wearing and bio-based organic/silica hybrid coating for bamboo with enhanced flame retardant and antifungal properties. Prog Org Coat 167:106830. https://doi.org/10.1016/J.PORGCOAT.2022.106830
160. Peng Q, Lu Y, Li Z et al (2022) Biomimetic, hierarchical-ordered cellulose nanoclaw hybrid aerogel with high strength and thermal insulation. Carbohydr Polym 297:119990. https://doi.org/10.1016/J.CARBPOL.2022.119990
161. Peng S, Luo Q, Zhou G, Xu X (2021) Recent advances on cellulose nanocrystals and their derivatives. Polymers (Basel) 13. https://doi.org/10.3390/POLYM13193247
162. Pérez-Chávez R, Sánchez-Aguilar J, Calderas F et al (2022) A statistical approach to the development of flame retardant and mechanically strong natural fibers biocomposites. Polym Degrad Stab 201:109991. https://doi.org/10.1016/J.POLYMDEGRADSTAB.2022.109991
163. Platnieks O, Gaidukovs S, Barkane A et al (2020) Bio-based poly(butylene succinate)/microcrystalline cellulose/nanofibrillated cellulose-based sustainable polymer composites:

thermo-mechanical and biodegradation studies. Polymers 12:1472. https://doi.org/10.3390/POLYM12071472

164. Pornwannachai W, Ebdon JR, Kandola BK (2018) Fire-resistant natural fibre-reinforced composites from flame retarded textiles. Polym Degrad Stab 154:115–123. https://doi.org/10.1016/J.POLYMDEGRADSTAB.2018.05.019
165. Pornwannachai W, Ebdon JR, Kandola BK (2019) Fire-resistant flax-reinforced polypropylene/polylactic acid composites with optimized fire and mechanical performances. J Thermoplast Compos Mater 33:898–914. https://doi.org/10.1177/0892705718815538
166. Prabhakar MN, Naga Kumar C, Dong Woo L, Jung-IL S (2022) Hybrid approach to improve the flame-retardant and thermal properties of sustainable biocomposites used in outdoor engineering applications. Compos Part A Appl Sci Manuf 152:106674. https://doi.org/10.1016/J.COMPOSITESA.2021.106674
167. Prabhu L, Krishnaraj V, Sathish S et al (2021) A review on natural fiber reinforced hybrid composites: chemical treatments, manufacturing methods and potential applications. Mater Today Proc 45:8080–8085. https://doi.org/10.1016/J.MATPR.2021.01.280
168. Prakash B, Soman R, Arun Sasi S, Mohan M (2022) Impact of bionanocomposites on the environment. Bionanocomposites for food packaging applications, pp 435–452. https://doi.org/10.1016/B978-0-323-88528-7.00023-X
169. Qu W, Niu B, Lv C, Liu J (2024) A review of sisal fiber-reinforced geopolymers: preparation, microstructure, and mechanical properties. Molecules 29:2401. https://doi.org/10.3390/MOLECULES29102401
170. Rachoti S, Akash GKG (2022) Mechanical and fire retardant behaviour of Flax/Sisal fiber hybrid composites. Mater Today Proc 64:32–36. https://doi.org/10.1016/J.MATPR.2022.03.474
171. Raghuvanshi S, Khan H, Saroha V et al (2023) Recent advances in biomacromolecule-based nanocomposite films for intelligent food packaging—a review. Int J Biol Macromol 253:127420. https://doi.org/10.1016/J.IJBIOMAC.2023.127420
172. Rahmadiawan D, Abral H, Yesa WH et al (2022) White ginger nanocellulose as effective reinforcement and antimicrobial polyvinyl alcohol/ZnO hybrid biocomposite films additive for food packaging applications. J Compos Sci 6:316. https://doi.org/10.3390/JCS6100316
173. Rahul Kumar M, Anbuchezhiyan G (2022) Comparative study of flexural strength and fire resistance of novel eggshell powder blended GFRP and jute fibre blended GFRP with GFRP. Mater Today Proc 69:1055–1060. https://doi.org/10.1016/J.MATPR.2022.08.166
174. Rajeshkumar G, Seshadri SA, Ramakrishnan S et al (2021) A comprehensive review on natural fiber/nano-clay reinforced hybrid polymeric composites: materials and technologies. Polym Compos 42:3687–3701. https://doi.org/10.1002/PC.26110
175. Ramesh M, Deepa C (2024) Processing and properties of jute (*Corchorus olitorius* L.) fibres and their sustainable composite materials: a review. J Mater Chem A. https://doi.org/10.1039/D3TA05481K
176. Ramesh M, Deepa C, Kumar LR et al (2022) Life-cycle and environmental impact assessments on processing of plant fibres and its bio-composites: a critical review. J Ind Text 51:5518S-5542S. https://doi.org/10.1177/1528083720924730/ASSET/IMAGES/LARGE/10.1177_1528083720924730-FIG5.JPEG
177. Rana MS, Khatun MA, Biswas BK et al (2023) Alginate-based nanocomposites for smart technology of food packaging. Polysaccharides, pp 212–233. https://doi.org/10.1201/9781003265054-12/ALGINATE-BASED-NANOCOMPOSITES-SMART-TECHNOLOGY-FOOD-PACKAGING-MD-SOHEL-RANA-AFROZA-KHATUN-BIPLOB-KUMAR-BISWAS-SHAHANAZ-PARVIN-MD-WASIKUR-RAHMAN-AZIZUR-KHAN-SK-MD-ALI-ZAKER-SHAWON
178. Rashid M, Chetehouna K, Cablé A, Gascoin N (2020) Analysing flammability characteristics of green biocomposites: an overview. Fire Technol 57(1):31–67. https://doi.org/10.1007/S10694-020-01001-0
179. Saba N, Jawaid M, Asim M (2019) Nanocomposites with nanofibers and fillers from renewable resources. Green composites for automotive applications, pp 145–170. https://doi.org/10.1016/B978-0-08-102177-4.00007-0

180. Sabaruddin FA, Tahir PM, Sapuan SM et al (2020) The effects of unbleached and bleached nanocellulose on the thermal and flammability of polypropylene-reinforced kenaf core hybrid polymer bionanocomposites. Polymers 13:116. https://doi.org/10.3390/POLYM13010116
181. Sai T, Ran S, Guo Z et al (2022) Recent advances in fire-retardant carbon-based polymeric nanocomposites through fighting free radicals. SusMat 2:411–434. https://doi.org/10.1002/SUS2.73
182. Sanni SE, Sadiku ER, Okoro EE (2022) Thermal stabilities of bionanocomposites at elevated temperatures, pp 51–68. https://doi.org/10.1007/978-981-16-8578-1_3
183. Satdive A, Tayde S, Tonde S et al (2022) Bionanocomposites in the construction and building applications, pp 293–310. https://doi.org/10.1007/978-981-16-8578-1_16
184. Seydibeyoğlu MÖ, Dogru A, Wang J et al (2023) Review on hybrid reinforced polymer matrix composites with nanocellulose, nanomaterials, and other fibers. Polymers 15:984. https://doi.org/10.3390/POLYM15040984
185. Shah AUR, Prabhakar MN, Il SJ (2017) Current advances in the fire retardancy of natural fiber and bio-based composites—a review. Int J Precis Eng Manuf Technol 4:247–262. https://doi.org/10.1007/S40684-017-0030-1
186. Shahinur S, Hasan M, Ahsan Q, Haider J (2020) Effect of chemical treatment on thermal properties of jute fiber used in polymer composites. J Compos Sci 4(2):132. https://doi.org/10.3390/JCS4030132
187. Sharma H, Kumar A, Rana S et al (2023) Critical review on advancements on the fiber-reinforced composites: ROLE of fiber/matrix modification on the performance of the fibrous composites. J Mater Res Technol 26:2975–3002. https://doi.org/10.1016/J.JMRT.2023.08.036
188. Shen J, Liang J, Lin X et al (2021) The flame-retardant mechanisms and preparation of polymer composites and their potential application in construction engineering. Polymers 14:82. https://doi.org/10.3390/POLYM14010082
189. Sheri V, Kumar M, Jaconis S, Zhang B (2023) Antioxidant defense in cotton under environmental stresses: unraveling the crucial role of a universal defense regulator for enhanced cotton sustainability. Plant Physiol Biochem 204:108141. https://doi.org/10.1016/J.PLAPHY.2023.108141
190. Shojaeiarani J, Bajwa D, Shirzadifar A (2019) A review on cellulose nanocrystals as promising biocompounds for the synthesis of nanocomposite hydrogels. Carbohydr Polym 216:247–259. https://doi.org/10.1016/J.CARBPOL.2019.04.033
191. Shojaeiarani J, Bajwa DS, Chanda S (2021) Cellulose nanocrystal based composites: a review. Compos Part C Open Access 5:100164. https://doi.org/10.1016/J.JCOMC.2021.100164
192. Shoueir KR, El-Desouky N, Rashad MM et al (2021) Chitosan based-nanoparticles and nanocapsules: overview, physicochemical features, applications of a nanofibrous scaffold, and bioprinting. Int J Biol Macromol 167:1176–1197. https://doi.org/10.1016/J.IJBIOMAC.2020.11.072
193. Sienkiewicz A, Czub P (2020) Flame retardancy of biobased composites—research development. Materials (Basel) 13:1–30. https://doi.org/10.3390/MA13225253
194. Siró I, Plackett D (2010) Microfibrillated cellulose and new nanocomposite materials: a review. Cellulose 17(3):459–494. https://doi.org/10.1007/S10570-010-9405-Y
195. Sukumaran NP, Gopi S (2021) Overview of biopolymers: resources, demands, sustainability, and life cycle assessment modeling and simulation. Biopolymers and their industrial applications, pp 1–19. https://doi.org/10.1016/B978-0-12-819240-5.00001-8
196. Sun J, Wu Z, An B et al (2021) Thermal-insulating, flame-retardant and mechanically resistant aerogel based on bio-inspired tubular cellulose. Compos Part B Eng 220:108997. https://doi.org/10.1016/J.COMPOSITESB.2021.108997
197. Sun Z, Mingming W (2019) Effects of sol-gel modification on the interfacial and mechanical properties of sisal fiber reinforced polypropylene composites. Ind Crops Prod 137:89–97. https://doi.org/10.1016/J.INDCROP.2019.05.021
198. Suriani MJ, Ilyas RA, Zuhri MYM et al (2021) Critical review of natural fiber reinforced hybrid composites: processing, properties, applications and cost. Polymers 13:3514. https://doi.org/10.3390/POLYM13203514

199. Suriani MJ, Sapuan SM, Ruzaidi CM et al (2021) Flammability, morphological and mechanical properties of sugar palm fiber/polyester yarn-reinforced epoxy hybrid biocomposites with magnesium hydroxide flame retardant filler. Text Res J 91:2600–2611. https://doi.org/10.1177/00405175211008615/ASSET/IMAGES/LARGE/10.1177_00405175211008615-FIG7.JPEG
200. Suriani MJ, Zainudin HA, Ilyas RA et al (2021) Kenaf fiber/pet yarn reinforced epoxy hybrid polymer composites: morphological, tensile, and flammability properties. Polymers 13:1532. https://doi.org/10.3390/POLYM13091532
201. Syduzzaman M, Sultana Rumi S, Faiza Fahmi F et al (2023) Mapping the recent advancements in bast fiber reinforced biocomposites: a review on fiber modifications, mechanical properties, and their applications. Results Mater 20:100448. https://doi.org/10.1016/J.RINMA.2023.100448
202. Taheri H, Mastali M, Falah M et al (2022) Microfibrillated cellulose as a new approach to develop lightweight cementitious composites: rheological, mechanical, and microstructure perspectives. Constr Build Mater 342:128008. https://doi.org/10.1016/J.CONBUILDMAT.2022.128008
203. Teles F, Martins G, Antunes F (2022) Fire retardancy in nanocomposites by using nanomaterial additives. J Anal Appl Pyrolysis 163:105466. https://doi.org/10.1016/J.JAAP.2022.105466
204. Teng Z, Ye C, Zheng C et al (2019) Layer-by-layer self-assembly of Ce-doped titanate nanotube-based hybrid coating on bamboo fibre materials for flame retardancy and thermostability. Polym Degrad Stab 168:108950. https://doi.org/10.1016/J.POLYMDEGRADSTAB.2019.108950
205. Trotta F, Da Silva S, Massironi A et al (2023) Silver bionanocomposites as active food packaging: recent advances & future trends tackling the food waste crisis. Polymers 15:4243. https://doi.org/10.3390/POLYM15214243
206. Uflyand IE, Irzhak VI (2021) Recent advances in the study of structure and properties of fiber composites with an epoxy matrix. J Polym Res 28:1–18. https://doi.org/10.1007/S10965-021-02783-9/METRICS
207. Uz Zaman S, Shahid S, Shaker K et al (2022) Development and characterization of chemical and fire resistant jute/unsaturated polyester composites. J Text Inst 113:484–493. https://doi.org/10.1080/00405000.2021.1889131
208. Velmurugan G, Chohan JS, Muhammed Abraar SA et al (2023) Investigation of nano SiO_2 filler loading on mechanical and flammability properties of jute-based hybrid polypropylene composites. SILICON 15:7247–7263. https://doi.org/10.1007/S12633-023-02578-1/METRICS
209. Venu G, Jayan JS, Saritha A, Joseph K (2022) Thermal decomposition behavior and flame retardancy of bioepoxies, their blends and composites: a comprehensive review. Eur Polym J 162:110904. https://doi.org/10.1016/J.EURPOLYMJ.2021.110904
210. Versino F, López OV, García MA (2021) Green biocomposites for packaging applications, pp 1–30. https://doi.org/10.1007/978-981-33-4091-6_1
211. Vidakis N, Petousis M, Michailidis N et al (2023) Optimized PCL/CNF bio-nanocomposites for medical bio-plotted applications: rheological, structural, and thermomechanical aspects. Bioprinting 36:e00311. https://doi.org/10.1016/J.BPRINT.2023.E00311
212. Wang D, Peng H, Wu Y et al (2020) Bioinspired lamellar barriers for significantly improving the flame-retardant properties of nanocellulose composites. ACS Sustain Chem Eng 8:4331–4336. https://doi.org/10.1021/ACSSUSCHEMENG.9B07745/ASSET/IMAGES/LARGE/SC9B07745_0004.JPEG
213. Wang D, Peng H, Yu B et al (2020) Biomimetic structural cellulose nanofiber aerogels with exceptional mechanical, flame-retardant and thermal-insulating properties. Chem Eng J 389:124449. https://doi.org/10.1016/J.CEJ.2020.124449
214. Wang P, He S, Sun G et al (2021) Favorable pleiotropic loci for fiber yield and quality in upland cotton (*Gossypium hirsutum*). Sci Reports 11(1):1–12. https://doi.org/10.1038/s41598-021-95629-9

215. Wang S, Zhang L, Semple K et al (2020) Development of biodegradable flame-retardant bamboo charcoal composites, Part II: Thermal degradation, gas phase, and elemental analyses. Polymers 12:2238. https://doi.org/10.3390/POLYM12102238
216. Wang Z, Feng W, Ban J et al (2023) Sisal-fiber-reinforced polypropylene flame-retardant composites: preparation and properties. Polymers 15:893. https://doi.org/10.3390/POLYM15040893
217. Wu M, Huang Y, Zhang T et al (2020) Cellulose nanofibril-based flame retardant and its application to paper. ACS Sustain Chem Eng 8:10222–10229. https://doi.org/10.1021/ACSSUSCHEMENG.0C02892/ASSET/IMAGES/LARGE/SC0C02892_0003.JPEG
218. Xu J, Niu Y, Xie Z et al (2023) Synergistic flame retardant effect of carbon nanohorns and ammonium polyphosphate as a novel flame retardant system for cotton fabrics. Chem Eng J 451:138566. https://doi.org/10.1016/J.CEJ.2022.138566
219. Xu Q, Mensah RA, Jin C, Jiang L (2022) A critical review of the methods and applications of microscale combustion calorimetry for material flammability assessment. J Therm Anal Calorim 147:6001–6013. https://doi.org/10.1007/S10973-021-10963-4/METRICS
220. Yadav M, Mathur A (2021) Bamboo as a sustainable material in the construction industry: an overview. Mater Today Proc 43:2872–2876. https://doi.org/10.1016/J.MATPR.2021.01.125
221. Yan M, Pan Y, Cheng X et al (2021) "Robust-Soft" anisotropic nanofibrillated cellulose aerogels with superior mechanical, flame-retardant, and thermal insulating properties. ACS Appl Mater Interfaces 13:27458–27470. https://doi.org/10.1021/ACSAMI.1C05334/ASSET/IMAGES/LARGE/AM1C05334_0011.JPEG
222. Yang XM, Qiu S, Yusuf A et al (2023) Recent advances in flame retardant and mechanical properties of polylactic acid: a review. Int J Biol Macromol 243:125050. https://doi.org/10.1016/J.IJBIOMAC.2023.125050
223. Yu J, Yang D, He Q et al (2022) Strong, durable and fire-resistant glass fiber-reinforced bamboo scrimber. Ind Crops Prod 181:114783. https://doi.org/10.1016/J.INDCROP.2022.114783
224. Zhang L, Chai W, Li W et al (2021) Intumescent-grafted bamboo charcoal: a natural nontoxic fire-retardant filler for polylactic acid (PLA) composites. ACS Omega 6:26990–27006. https://doi.org/10.1021/ACSOMEGA.1C03393/ASSET/IMAGES/LARGE/AO1C03393_0016.JPEG
225. Zhang Y, Liu Y, Wen H, Wang Q (2021) Durable flame retardant and mechanism of bamboo fabric through grafting based on arginine. Cellulose 28:8187–8204. https://doi.org/10.1007/S10570-021-04056-6/METRICS
226. Zhang Y, Yu HY, Dong Y et al (2023) Facile strategy to design a cellulose nanocrystal-based nanocomposite fire retardant with strong smoke suppression efficiency. ACS Sustain Chem Eng 11:12983–12991. https://doi.org/10.1021/ACSSUSCHEMENG.3C02458/SUPPL_FILE/SC3C02458_SI_003.MP4
227. Zheng C, Wen S, Teng Z et al (2019) Formation of H_2 $Ti_2O_5{\cdot}H_2O$ nanotube-based hybrid coating on bamboo fibre materials through layer-by-layer self-assembly method for an improved flame retardant performance. Cellulose 26:2729–2741. https://doi.org/10.1007/S10570-019-02252-Z/METRICS
228. Zhou X, Fu Q, Zhang Z et al (2021) Efficient flame-retardant hybrid coatings on wood plastic composites by layer-by-layer assembly. J Clean Prod 321:128949. https://doi.org/10.1016/J.JCLEPRO.2021.128949
229. Zhou Y, Fan M, Chen L (2016) Interface and bonding mechanisms of plant fibre composites: an overview. Compos Part B Eng 101:31–45. https://doi.org/10.1016/J.COMPOSITESB.2016.06.055
230. Zwawi M (2021) A review on natural fiber bio-composites, surface modifications and applications. Molecules 26:404. https://doi.org/10.3390/MOLECULES26020404

Fire-Retardant Properties of the Styrene-Butadiene Rubber-Based Biocomposites and Bionanocomposites

Nurjannah Salim, Siti Noorbaini Sarmin, and Siti Noorhidayah Mustapha

Abstract This chapter provides a comprehensive review of recent advances in flame retardant of styrene-butadiene rubber (SBR), with a particular focus on biocomposites and bionanocomposites. Due to the complexity of materials with increasing demand, SBR is emerging as a promising matrix suitable for its versatile properties. The incorporation of natural fibers, fillers, and nanofillers into SBR matrices has shown the ability to enhance flame retardant properties by maintaining mechanical properties. Furthermore, this chapter will emphasize synergistic effects, fire-retardant additives, and thermal stability processes. A discussion of the future views is also included in this chapter. The purpose of this discussion is to provide direction for future attempts to develop products that are long-lasting and have improved fire safety qualities.

Keywords Fire-retardant · Styrene-butadiene · Rubber · Biocomposite · Bionanocomposite

N. Salim (✉) · S. N. Mustapha
Faculty of Industrial Sciences and Technology, Universiti Malaysia Pahang Al-Sultan Abdullah, Kuantan, Pahang, Malaysia
e-mail: njannah@umpsa.edu.my

S. N. Sarmin
Department of Wood Industry, Faculty of Applied Sciences, Universiti Teknologi MARA Pahang, Bandar Pusat Jengka, Pahang, Malaysia

C. Muthukumar et al. (eds.), *Flammability Performance of Biocomposites and Bionanocomposites*, Composites Science and Technology,
https://doi.org/10.1007/978-981-95-2389-4_11

1 Introduction to Styrene-Butadiene Rubber (SBR) Biocomposites

1.1 Overview of Styrene-Butadiene Rubber (SBR)

The copolymerization of 75% butadiene and 25% styrene leads to the manufacture of styrene-butadiene rubber (SBR), a type of rubber that is widely used. This type of rubber is produced through the employment of free radical initiators. The production of styrene-butadiene rubber (SBR) from styrene and butadiene can be accomplished through the use of either the emulsion or solution polymerization technique. The manufacturing of styrene-butadiene rubber materials (SBRs) can be accomplished by several different methods. One of these methods involves the polymerization of styrene and butadiene in a solution using an alkyllithium initiator, such as butylthium, and a hydrocarbon solvent, such as hexane or cyclohexane [1].

For example, in manufacturing SBR rubber seals, the rubber compound must be forced through a die during the manufacturing process using an extruder. A combination of heat and pressure is utilized in the extruder to form the surface of the die into the appropriate shape. To treat and reinforce the sample, vulcanization processes such as steam lines, salt baths, or autoclave treatment can be done [2]. These procedures are chosen based on the dimensions and length of the sample. When the SBR is equipped with a calendar loader that can be precisely positioned, it has the capability of flattening the sample into sheets. This method allows for the sheets to be readily divided into narrow strips, which can then be used for a variety of purposes [3].

The qualities of the rubber are affected by the amount of the ratio of styrene to butadiene mix in the rubber. An increase in the ratio of styrene to butadiene leads to an increase in the polymer's level of hardness. When it comes to rubber compositions, SBR is generally favored over natural rubber (NR) because of its lower viscosity grade, which eliminates the need for pre-cycling on the part of the manufacturer. Rubber seals, textiles, and gaskets are all good candidates for this material because it is not only inexpensive but also offers a multitude of benefits, making it an attractive choice. Due to its great resistance to corrosion and strength, it is often used in applications that require synthetic substitutes for natural rubber [4]. Additionally, in comparison to other synthetic compounds of a similar kind, NR possesses higher abrasion resistance, while SBR demonstrates exceptional flexibility. As a result of their high level of durability, seals that are created using SBR are particularly well-suited for usage in circumstances where there is abrasion on car seats and tires [5]. SBR demonstrates exceptional flexibility in comparison to other basic materials. SBR seals are strong and ideal for applications requiring friction and skid resistance, such as tires and vehicle belts.

SBR and NR share many qualitative similarities but are inferior in areas like heat retention, tack, and foam tensile strength. The effectiveness of SBR for a certain application depends on the specific requirements and circumstances, despite its several benefits such as abrasion resistance, compression recovery, and tensile strength [2]. There are instances where NR or alternative synthetic rubber compounds would be more appropriate. SBR rubber has a working temperature range of − 25 to 100 °C, which is somewhat better than the temperature resistance of natural rubber, which ranges from − 20 to 70 °C [6]. However, if a material with exceptional tensile strength and compression recovery is not necessary, this temperature range can restrict certain sealing options. It is not appropriate for high-temperature applications due to its tendency to soften or deteriorate at elevated temperatures. While SBR is often regarded as superior to NR for outdoor uses, it exhibits much lower resistance to weathering and UV radiation compared to other synthetic rubber compounds. Extended exposure to the elements can cause degradation, cracking, or color fading in styrene-butadiene rubber. However, the addition of resins and reinforcing fillers effectively enhances these properties to an acceptable level [7].

1.2 Reinforcing or Non-reinforcing Additives in SBR

Rubber engineering involves the transformation of raw rubber materials into finished rubber products through the implementation of engineering principles. Basic rubbers (natural or synthetic), rubber compounds (including curing agent, accelerator, activator, co-activator, and additives), and fillers (reinforcing or non-reinforcing) are opted for during the procedure [8]. In recent decades, there has been extensive research on filler procedures for elastomers. Multiple experiments have been carried out throughout this period to ascertain the mechanism by which the addition of fillers strengthens an elastomer. Fillers are small particles added to polymeric materials, such as rubber, plastics, composites, and concrete, to improve certain qualities or reduce the need for more expensive binder components [9]. Figure 1 shows recent studies to improve the sustainability and biodegradability of synthetic rubber by incorporating raw materials from biological sources.

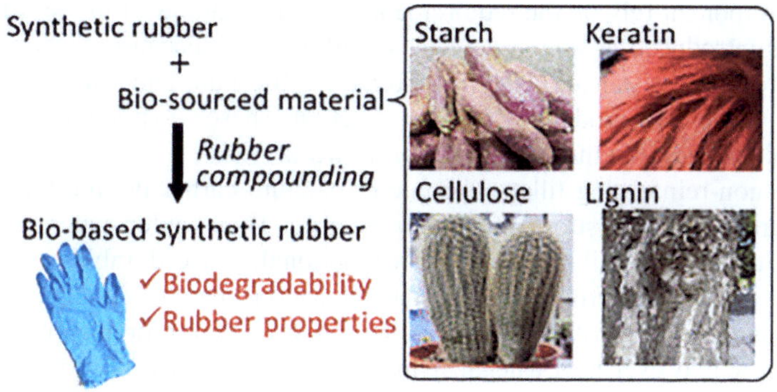

Fig. 1 Recent studies to improve the sustainability and biodegradability of synthetic rubber by incorporating raw materials from biological sources [10]

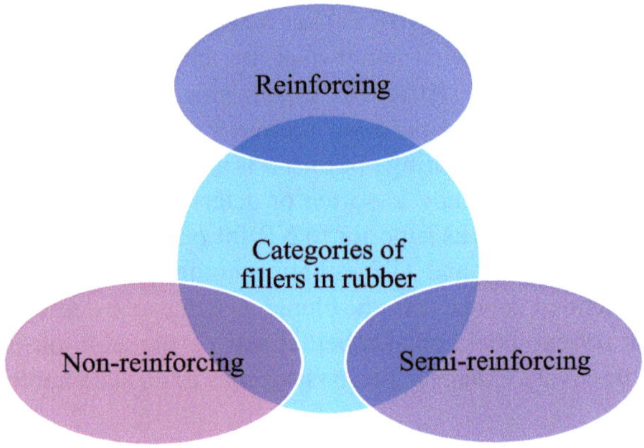

Fig. 2 Classification of fillers as being utilized in rubber composites

Unreinforced rubber is unsuitable for industrial purposes and must be supplemented with fillers to create composite constructions. Figure 2 illustrates the functions of fillers that are frequently included in rubber composites.

Reinforcing fillers are employed to enhance the mechanical, thermal, electrical, and chemical characteristics of composites. Non-reinforcing fillers are employed to reduce costs; however, they can also diminish the mechanical qualities. The primary types of reinforcing fillers are carbon black, precipitated silica, fumed silica, calcium silicate, aluminum silicate, activated calcium carbonate, hard clays, and zinc oxide, among others [5]. Non-reinforcing fillers encompass many substances such as soft clay, white barite, talc, brown asphalt, pulverized coal dust, and powdered vulcanized rubber (crumb). Due to its non-uniform molecular structure, SBR does not form crystals, hence necessitating the use of reinforcing filler to get desirable physical characteristics. Table 1 illustrates the type of fillers that are used in rubber composites.

Carbon black and silica are the primary reinforcing fillers utilized in the manufacturing of SBR, with different proportions and loadings. The utilization and ratio of each component rely on the requirements and applications of the resultant SBR. Silica was introduced as a complement to carbon black because of its capacity to decrease rolling resistance, hence extending the lifespan, and enhancing the fuel efficiency of the SBR product by reducing friction. These two popular fillers exhibit comparable behavior in the polymer-filler network.

Unlike non-reinforcing fillers like talc or calcium carbonate, reinforcing fillers do not merely reduce cost or occupy empty space. They confer material properties that are crucial to the ultimate product's functionality and durability. The size of the filler is a crucial determinant of the mechanical properties of composites. Fillers that have particles larger than 1 μm may not enhance the reinforcing qualities of the elastomer and may even have a negative impact on them [1]. The reinforcement effect is primarily observed when the size is below 100 nm, and the structural characteristics in this size range become more significant than the mere particle size. A smaller filler size exhibits a larger surface area, hence increasing the likelihood of binding to the rubber. Conversely, it is more challenging to distribute evenly within polymer

Table 1 Classification of filler types used in rubber composite

Filler type	Example materials	Function	References
Carbon-based	Carbon black, carbon nanotubes (CNTs), graphene	Enhance mechanical strength, abrasion resistance	Wilke et al. [11], Kumar et al. [12], Wang et al. [13]
Silica-based	Precipitated silica, fumed silica	Improve tensile strength, reduce rolling resistance	Song et al. [14], Huang et al. [15]
Carbon–silica hybrids	Carbon–silica dual-phase filler (CSDPF)	Combine benefits of CB and silica, better dispersion	Gao et al. [16]
Inorganic minerals	Calcium carbonate, talc, clay	Cost reduction, control of density and hardness	Chang et al. [17]
Organic/biomass	Lignin, starch, cellulose powder	Renewable source, moderate reinforcement	Alam et al. [18]
Flame retardant	Aluminum hydroxide, magnesium hydroxide	Enhance fire resistance and thermal stability	Lai et al. [19]
Conductive	High-loaded carbon black, CNTs, metal powders	Improve electrical and thermal conductivity	Roy et al. [20]
Nanomaterials	Nanoclay, nanocellulose, graphene oxide, halloysite	Improve mechanical, barrier, and thermal properties	Wang et al. [21]
Agro/industrial waste	Fly ash, rice husk ash, oil palm ash, kenaf core powder	Cost-effective, sustainable filler source	Torres et al. [22]

matrices. Consequently, composites that are filled with smaller fillers have superior mechanical capabilities compared to those that contain bigger particles, due to their effective dispersion. On the other hand, the crucial interaction relates to the filler-base polymer. During the processing mill, a part of the rubber adheres to the reinforcing fillers, making it resistant to typical rubber solvents that are used for extraction.

1.3 SBR Biocomposites/Bionanocomposites

Synthetic rubbers could undergo a crosslinking reaction, which makes them stronger. However, to further enhance their strength, they also require the addition of fillers. NBR stands out as the most intriguing base material for biocomposite applications among all synthetic rubbers, primarily because of the existence of polar nitrile groups (C≡N) in its polymer structure. SBR, in conjunction with NBR, is the synthetic matrix most frequently employed in rubber biocomposite applications. Tire manufacturing is the primary application of SBR due to its exceptional abrasion resistance, thermal stability, and resistance to fracture formation [7].

The surging demand for synthetic rubber significantly accelerates the depletion rate of fossil fuels and intensifies the negative environmental impact of their overexploitation. Furthermore, the disposal of rubber and plastic products has been a significant concern for several decades, resulting in harm to the environment and a growing number of landfills [8]. The integration of components derived from biological sources has the potential to create sustainable synthetic rubber products. They possess the characteristics of being renewable, economically affordable, and, of utmost significance, capable of undergoing biodegradation.

Recent research has transformed many raw materials, including industrial waste lignin, dead leaves, birchwood, corn starch and corn stover, *Paulownia elongata*, coconut shell [4], abandoned wood pallets [1], rice bran, and rice husk [6], into biochar. This biochar is then utilized as fillers in rubber composites. Instead, researchers have investigated the use of protein sources including soy protein isolate or concentrate and keratin from tannery effluent as fillers in elastomer composites. The protein sources were directly added without undergoing any thermochemical transformations. Furthermore, the utilization of chicken feathers as feedstock has been successfully transformed into biochar and tested as a filler in thermoplastic polymers. Prior research has primarily concentrated on the conversion of lignocellulosic feedstocks, which have a low nitrogen concentration, into biochar. These studies have evaluated the potential of biochar as a reinforcing agent in rubber blends. Amine groups are sometimes used in rubber and tires to enhance the interaction between filler and rubber, such as carbon nanotubes (CNTs) [3].

The use of environmentally friendly materials has had a significant impact on integrated design. Biocomposites containing natural fibers provide a viable alternative for improving the overall properties of the polymer. Various natural fibers have been added to the SBR matrix to enhance the mechanical properties or to increase the crystallinity of the SBR.

2 Fire-Retardant Mechanisms in Styrene-Butadiene Rubber Biocomposites

In recent years, SBR biocomposites have gained a lot of attention due to their widespread use in a wide range of industries. Flexible elastomers are attractive to manufacturers because they have unique properties such as flexibility, strength, and affordability, which is low. These qualities make it an attractive material for manufacturing. On the other hand, the disadvantage of SBR biocomposites is that they are flammable, posing serious safety concerns in a wide range of industries, as shown in Fig. 3 [19, 23, 24].

Fig. 3 Disadvantages of rubber products [19]

The natural polymers in SBR biocomposites provide high thermal conductivity, which means that they can burn under heat or in flames, and therefore, it may be necessary to use more durable and resistant fireplace preservatives. Enhancing the overall safety standard performance of SBR biocomposites and reducing their flammability is a potential strategy that has emerged as a solution to this challenge posed by flame retardants [24, 25].

Understanding the mechanisms that affect fire retardancy in SBR biocomposites is crucial for developing powerful fire-retardant techniques. Researchers can enhance the fire resistance of SBR biocomposites and ensure their applicability for a whole lot of sectors with the aid of know-how the interactions among diverse additives and their synergistic consequences [25].

There are numerous types of fire-retardant additives applied in SBR biocomposites, including phosphorus-based totally compounds, halogen-based totally compounds, inorganic fillers, and nanostructured substances. Fire-retardant additives are materials incorporated into SBR biocomposites to improve their fire resistance and sluggish down or prevent combustion. These components can act through various mechanisms, which include [25–27]:

1. Gas-phase mechanisms: These additives launch inert gases or dilute the combustible gases while heated, thereby reducing the concentration of flammable vapors, and slowing down the combustion method.
2. Condensed section mechanisms: These additives promote the formation of a shielding char layer on the surface of the composite, appearing as a barrier to warmth and mass switch, for this reason insulating the underlying fabric from similar degradation.
3. Cooling mechanisms: Some additives go through endothermic reactions while heated, absorbing warmth, and lowering the temperature of the composite below the vital point for sustained combustion.

2.1 Fire-Retardant Additives

Fire-retardant additives are materials integrated into SBR biocomposites to inhibit or postpone the combustion process. Various varieties of fireplace-retardant additives are applied, each with specific mechanisms of action [24, 26, 27].

2.1.1 Phosphorus-Based Additives

Phosphorus-based components, together with ammonium polyphosphate (APP) and red phosphorus, are characteristic on the whole through char formation, gasoline-section flame inhibition, and radical scavenging. When exposed to heat, phosphorus compounds undergo chemical alterations, leading to the formation of a shielding char layer on the material's surface. This char layer acts as a physical barrier, hindering heat and oxygen diffusion and thereby retarding the unfolding of flames. Additionally, phosphorus compounds can interrupt combustion reactions with the aid of scavenging unfastened radicals, which are essential for sustaining the fire [27].

2.1.2 Halogen-Based Additives

Halogen-based totally additives, which include brominated and chlorinated compounds, exert their fire-retardant impact ordinarily through gasoline-phase flame quenching and radical scavenging. When subjected to warmness, halogen compounds launch halogen radicals, which react with flame radicals, suppressing the combustion process. Moreover, halogen compounds can disrupt radical chain reactions, further inhibiting flame propagation [26, 27].

2.1.3 Inorganic Fillers

Inorganic fillers, such as aluminum hydroxide (ATH) and magnesium hydroxide (MDH), are performed by way of liberating water vapor and soaking up warmness in the course of combustion. This endothermic procedure results in a cooling impact, lowering the temperature of the fabric and slowing down the ignition and combustion techniques. Additionally, the decomposition of inorganic fillers can contribute to the formation of a strong char layer, enhancing fireplace resistance [27].

2.1.4 Nanostructured Additives

Nanostructured additives, which include layered double hydroxides (LDHs) and carbon nanotubes (CNTs), provide unique homes that lead them to be powerful fire retardants in SBR biocomposites. LDHs have high floor vicinity and ion trade abilities, which enable them to soak up flammable gases and inhibit flame propagation. CNTs, alternatively, act as physical boundaries, preventing the diffusion of heat and gases and lowering the material's flammability [27].

2.2 *Synergistic Effects and Interactions*

Synergistic consequences and interactions between extraordinary fire-retardant additives play an essential role in improving the overall fire resistance of SBR biocomposites. These interactions can occur through numerous mechanisms, resulting in synergistic effects that surpass the character contributions of each additive [24].

One commonplace synergistic mechanism involves the enhancement of char formation. Certain combos of components, such as phosphorus-based compounds with inorganic fillers, can synergistically enhance the formation of a stable char layer throughout combustion. The presence of phosphorus compounds enables char formation, while inorganic fillers offer structural support, resulting in an improved and protective char layer [28].

Gas-phase flame inhibition is another area wherein synergistic results may be observed. Halogen-primarily based components, when mixed with phosphorus-containing compounds, can synergistically suppress flames in the gas phase. Halogen radicals launched from halogen compounds react with flame radicals, while phosphorus compounds scavenge loose radicals, leading to an extra effective inhibition of combustion reactions [24].

Radical scavenging and chain-breaking mechanisms can also show off synergistic outcomes. Additive combinations might also possess complementary radical scavenging properties, leading to the interruption of radical chain reactions and the suppression of flame propagation. This synergistic effect complements the overall heat retardant efficiency of the additives, making the SBR biocomposites extra immune to ignition and combustion [27, 29].

Furthermore, certain additive mixtures can improve the thermal stability of SBR biocomposites, lowering the likelihood of ignition and improving fire resistance. This enhancement in thermal balance can also result from interactions among additives that inhibit decomposition reactions and sell char formation at elevated temperatures. By synergistically improving the fabric's thermal stability, those additive mixtures make contributions to standard fireplace retardancy [27].

3 Recent Advances and Future Perspectives

3.1 Recent Advances in SBR Biocomposites for Fire Retardancy

Styrene-butadiene rubber (SBR) is a synthetic rubber known for its exceptional mechanical qualities, durability, and chemical resistance, making it preferred in many applications. Despite all its advantages, SBR itself is not inherently a fire retardant. Fire retardant is an important safety property that is essential in many SBR applications, including in electrical, electronic, transportation, and construction. SBR is commonly utilized as carpet backings and low-cost back coating in public contracting carpeting, as well as in rubber-insulated cable, insulating foam, conveyor belting, roofing, and hoses in the electrical, electronic, and transportation industries where fire prevention is crucial [30]. In addition to the crucial role of fire-retardant characteristics in preventing rapid combustion, it is well-established that approximately 80% of fire-related casualties are caused by toxic gases and smoke inhalation. Fatal incidents occurred due to asphyxia caused by these conditions. Rubber products are known to contain carbon black, which leads to substantial smoke emissions when they undergo thermal degradation [31]. It is important to emphasize that the smog resulting from burning SBR is affected by soot particles suspended in the air which will be the carrier for cancerogenic and mutagenic compounds that are dangerous for human life [32].

Fire-retardant elements function by decreasing the flammability of materials through physical obstruction or by triggering a chemical response that slows the fire. To enhance the fire-retardant qualities of SBR rubber, a common method is to incorporate a fire-retardant chemical into the SBR system. An additive of the flame retardants can be made through physical mixing which does not bind to polymer chemically, or by chemical reactions known as a reactive flame retardant in which once it combines it will be part of the chemical structure of the SBR. There are many examples of flame retardant that are used depending on the synergy between the filler and polymer matrices component including halogen-based, phosphorus-based, melamine-based, metal hydroxide-based, silicone-based, and phosphate-based flame retardants [33]. The halogen fire-retardant system in SBR is quite effective. These systems produce combustion byproducts with unfavorable traits such as corrosiveness, toxicity of smoke, and other emissions resulting from burning SBR as in Fig. 4 [30, 34]. Various halogen-free fire-retardant additives exist on the market, functioning through different methods based on their chemical composition. Nevertheless, their efficacy and adaptability are said to be significantly inferior than halogen-based compounds [30].

Natural lignocellulosic filler is one of the latest approaches in improving the fire retardancy of SBR-carbon black composite. When natural lignocellulosic filler is added to SBR rubber composites in a synergistic system with non-halogen flame retardant chemicals, it considerably decreases the rate of thermal decomposition and the heat generated during combustion [36]. Utilizing a lignocellulose filler with an

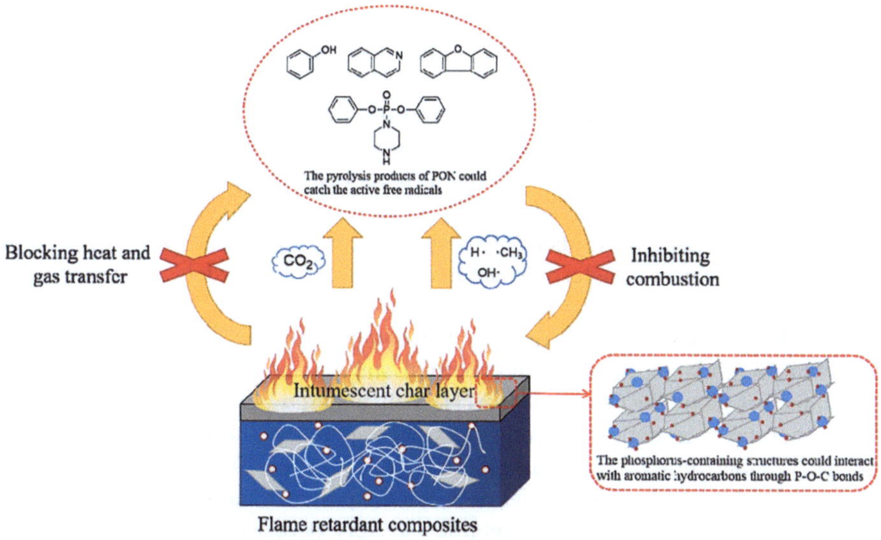

Fig. 4 Schematic diagram of flame retardants burning [35]

organized fibril structure, along with melamine polyphosphate, aluminum hydroxide, and zinc borate, creates an insulating barrier that restricts the exchange of mass and energy between the flame and the sample. This leads to a decrease in smoke production and enhances the fire-retardant characteristics of SBR. Besides that, different types of clays are also used as one of the non-halogen fire-retardant agents for SBR. One of the most used clays is organic montmorillonite (OMT) nanoclay [37]. Incorporating a small amount of organomodified nanoclay into the polymer matrix creates a protective layer during combustion [38]. The clay addition serves as a barrier that limits the transmission of heat into the substance, as well as the release of flammable decomposition products and the penetration of oxygen into the material [23]. Zinc hydroxystannate (ZHS) is also an alternative to halogen-free flame retardant. However, ZHS does not have flame retardant properties on its own and needs to be combined with chlorinated paraffin (CP) in SBR rubber.

Besides the use of non-halogen fire-retardant agents, one of the promising alternatives to halogen-based systems would be an intumescent system [23, 39] in which the substance swells as a result of heat exposure. When heated, the material swells on the surface to form a protective char layer, which acts as an insulating barrier that shields the underlying polymer matrix from the flame. The fire-retardant coating is thermally stable and provides continuity to reduce heat and mass transfer between the flame and the polymer, resulting in flame extinguishment. Ammonium polyphosphate ($[NH_4PO_3]_n$–APP) is a well-known component of intumescent fire retardants. This is due to it allowing the formation of a continuous crosslinked vitreous phase called ultraphosphate during the thermal degradation, which acted as a shielding coating and fire-retardant component.

In the twenty-first century, there has been tremendous growth in the development and use of ceramifiable composites due to increased demands related to fire prevention regulations for public property and high-rise buildings [40]. When exposed to high temperatures or fire, ceramifiable composites transform from a polymer-matrix mineral dispersion into a porous and continuous mineral barrier char shield, initiating the production of a ceramic structure on the composite's surface. The layer formed restricts the spread of flammable volatile substances to the burning region and the oxide to the main composites, hence lowering the combustion rate [41]. The ceramic structure restricts heat transfer into the material's core, hence delaying the thermal degradation of the SBR rubber and the creation of fuel [42, 43].

This char barrier not only reduces the combustibility of the composites but also exhibits good mechanical and thermal stability of the SBR. For example, aluminum trihydroxide ($Al(OH)_3$—ATH) will undergo endothermic dehydration during thermal degradation and release water vapor with the in situ formation of thermally stable ceramic materials (alumina-Al_2O_3) [44]. This ceramic layer acts as a shield to heat and mass transfer between the polymer and the flame. However, the effectiveness of ATH as a fire retardant is very poor which needs generally > 50 wt% to reach a suitable fire-retardant property. Thus, in 2005, Castrovinci [30] et al. combined the theory of intumescent and ceramifiable systems by combining APP-ATH to allow better performance of fire retardants on SBR rubber. It was found that 15 wt% APP and 1 wt% ATH show a good fire-retardant property of SBR.

3.2 Future Prospects and Challenges

SBR rubber is very widely used in many applications, including automotive tires, conveyor belts, footwear, and construction materials, due to its desirable properties such as high elasticity, high abrasion resistance, and low cost. According to the Styrene-Butadiene Rubber Global Market Report 2024, the SBR global market grew about 6.3% from 2023 to 2024 and it is expected to see strong growth in the next few years [17]. However, inherent flammability is the major drawback of SBR rubber. Thus, enhancing the fire retardancy of SBR is crucial, aiming to mitigate the fire risk and ensure safety in various sectors. The challenge of improving the fire-retardant properties begins as early as the selection of the chemicals during SBR vulcanization. It is because sulfur and peroxide also provide different viscoelastic and thermal behavior which finally affects the fire-retardant modification optimization.

As discussed in previous sections, there are a lot of fire-retardant agents that could be added to improve its fire-retardancy property. However, the challenges include maintaining the desirable mechanical properties, achieving cost-effectiveness, and ensuring environmental compatibility. Halogen-based fire-retardant filler was proven the best option in improving the fire retardancy of SBR rubber with maintaining its other mechanical properties. For example, a combination of decabromodiphenyl oxide (DBBO) and antimony trioxide (Sb_2O_3) is widely used as the flame retardant system for SBR compounds [34]. In addition to cost-effectiveness, thermal, and mechanical performances, halogen content is responsible for its low flammability characteristics. However, due to the toxicity and the smoke during combustion which also be the soot for the dangerous gaseous trapped during the fire, researchers have started to explore the potential of non-halogen filler and natural filler such as lignocellulosic and risk-husk nanofiller [45] to improve the fire retardancy of SBR rubber.

The challenge for non-halogen and biobased filler is low flammability properties as compared to halogen-based filler. Thus, an appropriate combination of more than one reinforcing material is always needed to support the flammability-resistant property. However, the use of more than one material needs a proper analysis of its synergy between the materials and also with the SBR matrix to allow good fire-retardant and mechanical properties of the product. Thus, the synergy between the filler's combination and the SBR matrix will also be a challenge. The fire-retardant filler will either function by blocking the fire physically or by initiating a chemical reaction that stops the fire. Due to the limitation of the non-halogen fire-retardant agent to function as fire fire-blocking agent, an intumescent and ceramifiable system is an alternative. An intumescent system allows the material substance to swell as the result of heat exposure, while ceramifiable system initiates the process of formation of a ceramic structure on the surface of the composite by changing their structure from polymer-matrix mineral dispersion into porous and continuous mineral and provides a barrier that functioned as char shield during heat exposure.

However, the disadvantage of ceramifiable composites of SBR is the weaker rheological properties [41]. As a solution, the addition of silicone oil as a plasticizer is necessary in a ceramizable composite composed of SBR as polymer matrix and mica as mineral filler. Overall, SBR rubber modification to improve the fire-retardant properties has a challenge in maintaining its original properties and reducing its dependency on halogen-based and chemical-based materials in the future. It is expected that fire-retardant SBR lies in the integration of advanced materials, engineering principles, and sustainable practices in the future to achieve enhanced fire safety without compromising performance or environmental responsibility. By overcoming current challenges and leveraging emerging technologies, researchers can pave the way for the widespread adoption of fire-retardant SBR in diverse applications, thereby reducing fire hazards and ensuring the well-being of society.

References

1. Lee JH, Kim J, Jeong MH, Ahn KH, Lee HL, Youn HJ (2023) Visualization of styrene-butadiene rubber (SBR) latex and large-scale analysis of the microstructure of lithium-ion battery (LIB) anodes. J Power Sources 557:232552
2. Cao R, Yang J, Li G, Niu M (2023) Effect of sulfuric acid on cementitious composites containing ethylene-vinyl acetate and styrene-butadiene rubber. J Build Eng 80
3. Xing C, Li M, Liu L, Yang R (2023) Long-term aging behavior of plastic/styrene butadiene rubber (SBR) composite modified bitumen. Materials 16(3)

4. Deredas K, Kępczak N, Urbaniak M (2021) Influence of doping with styrene-butadiene rubber on dynamic and mechanical properties of polymer concrete. Compos Struct 268
5. Idrees M, Saeed F, Amin A, Hussain T (2021) Improvement in compressive strength of styrene-butadiene-rubber (SBR) modified mortars by using powder form and nanoparticles. J Build Eng
6. Calarnou L, Traïkia M, Leremboure M, Therias S, Gardette JL, Bussière PO, Malosse L, Dronet S, Besse-Hoggan P, Eyheraguibel B (2023) Study of sequential abiotic and biotic degradation of styrene butadiene rubber. Sci Total Environ
7. Wang LE, Luo Z, Yang L, Wang H, Zhong J (2020) Effect of styrene content on mechanical and rheological behavior of styrene butadiene rubber. Mater Res Express 8(1)
8. Wu Q, Liu L, Meng J, Liu S, Hui J, Wang X, Xu L, Zhang L (2023) Microstructure and performance of green tire tread based on epoxidized solution polymerized styrene butadiene rubber and epoxidized natural rubber. Ind Eng Chem Res 62(13):11
9. Zhang T, Huang H, Li W, Chang X, Cao J, Hua L (2020) Vulcanization modeling and mechanism for improved tribological performance of styrene-butadiene rubber at the atomic scale. Tribol Lett 68:11
10. Boon ZH, Teo YY, Ang DT-C (2022) Recent development of biodegradable synthetic rubbers and bio-based rubbers using sustainable materials from biological sources. RSC Adv 12(52):34028–34052
11. Wilke LA et al (2023) Detailed understanding of the carbon black–polymer interface in filled rubber composites. Carbon 201:520–528
12. Kumar Singaravel D, Sharma S, Kumar P (2022) Recent progress in experimental and molecular dynamics study of carbon nanotube reinforced rubber composites: a review. Polym-Plastics Technol Mater 61(16):1792–1825
13. Wang J et al (2024) Graphene-based hybrid fillers for rubber composites. Molecules 29(5):1009
14. Song SH (2021) Study on silica-based rubber composites with epoxidized natural rubber and solution styrene butadiene rubber. Polym Polym Compos 29(9):1422–1429
15. Huang L et al (2022) Understanding the reinforcement effect of fumed silica on silicone rubber: bound rubber and its entanglement network. Macromolecules 56(1):323–334
16. Gao X et al (2022) Study on properties of carbon-coated silica prepared by polymer pyrolysis reinforced rubber composites. Polym Testing 110:107583
17. Chang BP et al (2021) Bioresourced fillers for rubber composite sustainability: current development and future opportunities. Green Chem 23(15):5337–5378
18. Alam MN et al (2023) Naturally abundant silica-kaolinite mixed minerals as an outstanding reinforcing filler for the advancement of natural rubber composites. J Polym Res 30(2):59
19. Lai L et al (2023) Recent advances for flame retardant rubber composites: mini-review. Adv Ind Eng Polym Res 6(2):156–164
20. Li Y-T et al (2022) Processing, thermal conductivity and flame retardant properties of silicone rubber filled with different geometries of thermally conductive fillers: a comparative study. Compos B Eng 238:109907
21. Wang Y et al (2022) Preparation and reinforcement performance of RGO-CNTs-SiO_2 three-phase filler for rubber composites. Compos Sci Technol 228:109633
22. Torres LF et al (2021) Torrefied agro-industrial residue as filler in natural rubber compounds. J Appl Polym Sci 138(28):50684
23. Khobragade PS et al (2016) Flame retarding performance of elastomeric nanocomposites: a review. Polym Degrad Stab 130:194–244
24. Wan L et al (2020) Flame retardation of natural rubber: strategy and recent progress. Polymers 12(2):429
25. Rahmah M et al (2016) Antioxidant and flame retardant additive effect on the thermal degradation and flammability of SBR filled EPS/UPR composite. Int J Res Adv Eng Technol 2(2):17–20
26. Lu S-Y, Hamerton I (2002) Recent developments in the chemistry of halogen-free flame retardant polymers. Prog Polym Sci 27(8):1661–1712

27. Mohd Sabee MM et al (2022) Flame retardant coatings: additives, binders, and fillers. Polymers 14. https://doi.org/10.3390/polym14142911
28. Wang N et al (2017) Effect of tea saponin-based intumescent flame retardant on thermal stability, mechanical property and flame retardancy of natural rubber composites. J Therm Anal Calorim 128:1133–1142
29. Muzakir S et al (2018) Fabrication of hydrophobic compressed oil palm trunk surface by sol-gel process. IOP Conf Ser Mater Sci Eng
30. Castrovinci A et al (2005) Ammonium polyphosphate–aluminum trihydroxide antagonism in fire retarded butadiene–styrene block copolymer. Eur Polymer J 41(9):2023–2033
31. You J-H et al (1994) Polycyclic aromatic hydrocarbons (PAHs) and mutagenicity of soot particulates in air emissions from two-stage incineration of polystyrene. J Hazard Mater 36(1):1–17
32. Valavanidis A et al (2008) Persistent free radicals, heavy metals and PAHs generated in particulate soot emissions and residue ash from controlled combustion of common types of plastic. J Hazard Mater 156(1–3):277–284
33. Fontaine G, Gallos A, Bourbigot S (2014) Role of montmorillonite for enhancing fire retardancy of intumescent PLA. Fire Saf. Sci 11:808–820
34. Ismawi DHA, Harper JF, Ansarifar A (2008) Influence of flame retardant additives on the flammability behaviour of natural rubber (NR). J Rubber Res 11:223–236
35. Wang X et al (2023) A novel halogen-free flame-retardant fabrication for the study of smoke suppression and flame retardancy of polystyrene. Polymer 283:126240
36. Rybiński P et al (2023) The effect of the lignocellulosic filler on the reduction of fire hazard of styrene-butadiene rubber composites, including the reduction of smoke, PCDD/F, PAH emissions and toxicity during its thermal decomposition. Polym Testing 118:107900
37. Simionescu TM, Minea AA, dos Reis PNB (2019) Fire properties of acrylonitrile butadiene styrene enhanced with organic montmorillonite and exolit fire retardant. Appl Sci 9(24):5433
38. Zhang Y et al (2014) Nano-scale blocking mechanism of MMT and its effects on the properties of polyisocyanate-modified soybean protein adhesive. Ind Crops Prod 57:35–42
39. Lomakin SM, Zaikov GE (1996) New type of ecologically safe flame retardant based on polymer char former. Polym Degrad Stab 51(3):343–350
40. Anyszka R et al (2016) Thermal stability and flammability of styrene-butadiene rubber-based (SBR) ceramifiable composites. Materials 9(7):604
41. Imiela M et al (2023) Effect of silicone oil on properties and performance of ceramizable styrene-butadiene rubber-based composites. Polymers 15(15):3204
42. Gong X-H et al (2017) Improved self-supporting property of ceramifying silicone rubber composites by forming crystalline phase at high temperatures. J Alloy Compd 706:322–329
43. Imiela M et al (2019) Synergistic effect of mica, glass frit, and melamine cyanurate for improving fire resistance of styrene-butadiene rubber composites destined for ceramizable coatings. Coatings 9(3):170
44. Camino G et al (2005) Effect of magnesium hydroxide on fire retardance mechanism of intumescent EVA composition. In: Fire retardancy of polymers: new applications of mineral fillers, pp 248–263
45. Attia NF, Saleh BK (2020) Novel synthesis of renewable and green flame-retardant, antibacterial and reinforcement material for styrene–butadiene rubber nanocomposites. J Therm Anal Calorim 139(3):1817–1827

Flammability Characteristics of the Silicone Rubber-Based Biocomposites and Bionanocomposites

V. M. Gobinath, V. Parthasarathy, P. Senthil Kumar, and T. Arivazhagan

Abstract The materials with the chemical composition of higher levels of carbon, hydrogen, and oxygen are often more combustible than those with the lower levels of these elements. The flammability qualities of SR biocomposites are greatly influenced by the loading of nanofillers and the composition of the fillers. The flammability properties of the materials and the measurement of flammability by various testing methods are discussed in detail. The flammability measurement techniques like oxygen index test, smoke density test, and test for fire propagation are highlighted in this chapter. The ignition temperature, smoke production rate, and release of any flammable gas or vapor during combustion must be taken into account while assessing the flammability of the materials. The flammability properties and the flammability measurement using various testing methods are discussed in detail. The chapter also outlines the flammability measurement techniques like the UL94 test, oxygen index test, smoke density test and fire propagation test with the aim of providing insightful information on a material's flammability and fire behavior. The reinforcing effect of natural fillers like montmorillonite (MMT), sepiolite, and silica on the flame retardancy of silicone rubber biocomposites is also summarized.

V. M. Gobinath
Department of Mechanical Engineering, Rajalakshmi Institute of Technology, Chennai, India

V. Parthasarathy (✉)
Department of Physics, Rajalakshmi Institute of Technology, Chennai, Tamil Nadu, India
e-mail: parthu0406@gmail.com

P. Senthil Kumar
Centre for Pollution Control and Environmental Engineering, School of Engineering and Technology, Pondicherry University, Kalapet, Puducherry, India

T. Arivazhagan
Department of Physics, Sri Sairam Engineering College, Chennai, Tamil Nadu, India

C. Muthukumar et al. (eds.), *Flammability Performance of Biocomposites and Bionanocomposites*, Composites Science and Technology,
https://doi.org/10.1007/978-981-95-2389-4_12

1 Introduction

Silicone rubber (SR) is a flexible substance that has many uses in a variety of sectors due to its special qualities. When stress is applied, it can bend significantly and then revert to its previous shape upon the removal of the applied stress. SR possesses excellent temperature resistance and outstanding low-temperature flexibility, and the service temperature of SR is in the range of −70 to 250 °C [28]. Atoms of silicon, oxygen, carbon, and hydrogen are organized in a polymer framework to form silicone rubber. It is appropriate for various applications due to its good thermal stability, electrical insulating qualities, low toxicity, and tolerance to extreme temperatures. SR is classified into various types based on the applicable environments, differences in monomer and vulcanization conditions, etc., (Fig. 1. Emerging materials that combine the benefits of natural fibers or nanoparticles with a polymer matrix are called biocomposites and bionanocomposites [1, 21]. Compared to conventional composites, these materials have better mechanical qualities, biodegradability, and environmental sustainability. The bionanocomposites integrate nanoparticles like clay, and cellulose nanocrystals into the polymer matrix [11, 23], whereas biocomposites are made of natural fibers like hemp, flax, or bamboo embedded in a polymer matrix [2].

For several reasons, it is crucial to investigate the flammability properties of silicone rubber-based biocomposites and bionanocomposites. First of all, it guarantees the security and dependability of these materials in various applications, particularly in areas where fire risks are a concern, including the automotive, aerospace, and building sectors. Designing fire-resistant products for efficient fire prevention and

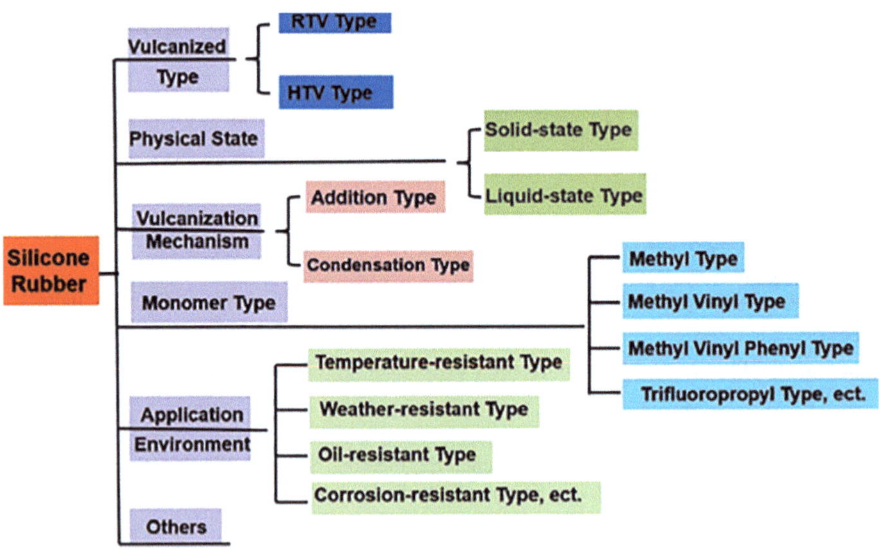

Fig. 1 Classification of SR [28]

protection requires an understanding of how these materials react to fire. Secondly, it is easier to follow fire safety rules and regulations by knowing the flammability traits. To assess the flammability features of materials, regulatory organizations like the National Fire Protection Association (NFPA) and Underwriters Laboratories (UL) have set standards and tests [5]. These standards may be used by the manufacturers to guarantee their goods fulfill the necessary safety requirements for particular applications. Thirdly, research into the flammability of biocomposites and bionanocomposites based on silicone rubber aids in the creation of flame-retardant products. Researchers can create new additives, coatings, or processing methods to improve the flame retardancy of these materials by comprehending the principles underlying combustion and the variables that affect fire behavior. The creation of safer and more environmentally friendly goods may result from this understanding [18]. The reinforcing polymers with traditional flame retardants such as halogenated substances and heavy metals are recognized as harmful to both human health and the environment [26]. It is possible to create safer and more environmentally friendly materials by investigating alternate flame retardant systems and comprehending how well they operate in silicone rubber-based biocomposites and bionanocomposites.

Silicone rubber is a flexible substance with special qualities, such as high thermal stability, superior electrical insulation, and resistance to aging and deterioration that make it appropriate for a range of applications [13, 16]. Environmental sustainability and improved mechanical qualities are the benefits of biocomposites and bionanocomposites. It is crucial to investigate the flammability properties of silicone rubber-based biocomposites and bionanocomposites to ensure safety, abide by laws, create flame retardant techniques, and evaluate the environmental effects. Researchers and producers can endeavor to develop goods that are safer and more sustainable by learning more about how these materials behave in fire circumstances. Silicone rubber possesses unique features due to the presence of alternative silicon and oxygen atoms in its backbone. Due to its flexibility, resilience, and biocompatibility, silicone rubber is used in various industries, including construction, automotive, electronics, and healthcare. The flammability properties of SR bionanocomposites are outlined in this article, along with the effects of nano clays on their flammability behavior, and the relationship between the content of nanofillers and flammability characteristics.

2 Reinforcing Effect of Fillers on the Physicochemical Properties of Flameretardant Materials

Natural fillers can be derived from a variety of sources, such as plants, animals, and minerals, and are utilized as reinforcing agents to fabricate flame retardant biocomposites [25]. Natural fillers include starch, chitosan, cellulose fibers from hemp and bamboo, and lignocellulosic fibres from wood flour are frequently employed as reinforcing agents to improve the performance of the flame retardant biocomposite materials. These fillers include benefits like affordability, regenerative nature, and biodegradability. The glass fibers [19], carbon fibers [24], and carbon nanotubes [17] are few examples of synthetic fillers, and these fillers have excellent mechanical and thermal qualities because of their high strength. While reinforcing polymeric materials with these fillers, they create a network-like structure and thereby improve the tensile, flexural, impact resistance, and mechanical properties of the materials. The type of fillers used and their loading percentage have a big impact on how well the composites function mechanically. Fillers serve a critical role in increasing the characteristics of flame retardant materials. The fillers' main functions are described below.

2.1 Thermal Properties

Fillers can improve the composites' thermal stability and heat resistance. The ability of the composites to dissipate heat is enhanced by fillers with high thermal conductivity, such as nanoclays and carbon-based nanoparticles. Applications where thermal management is crucial can benefit greatly from these characteristics.

2.2 Barrier Qualities

Fillers can enhance the composites' barrier qualities, such as moisture resistance and gas permeability. For instance, the gas permeability of silicone rubber composites can be greatly reduced by adding nanoclays or graphene oxide, making them ideal for applications needing gas barrier qualities.

2.3 Flammability Properties and Testing Methods

Flammability describes a substance's capacity to catch fire and maintain when exposed to an ignition source, such as heat, flame, or sparks. As it directly links to the safety and protection of people and property, it is a crucial feature to take into account in a variety of industries, including construction, automotive, textiles, and electronics. It is easier to identify all possible fire hazards and enables the creation of efficient fire protection measures by understanding the flammability of the substance. The risk of fires and their possible effects, such as property damage, injuries, or even loss of life, can be decreased by manufacturers making informed decisions regarding the selection of suitable materials for particular applications. The flammability of materials is influenced by several factors. Effective fire risk assessment and mitigation require an understanding of these factors. Here are a few crucial ones:

2.4 Chemical Make-Up

The kinds of elements and compounds that make up a substance have a significant impact on flammability. For instance, materials with higher levels of carbon, hydrogen, and oxygen atoms are often more combustible than those with lower levels of these elements. Each material has a specified ignition temperature at which it can ignite. This temperature, also referred to as the ignition temperature, differs for various materials. Lower ignition temperature materials are more likely to catch fire and keep it going.

Another significant aspect is the rate at which a substance produces heat when it burns. Materials that produce a lot of heat during combustion can hasten the spread of a fire and increase the severity of its effects. The amount of a material's surface that is exposed to an ignition source can influence how quickly it ignites. The greater surface area offers more opportunities for heat transfer and may hasten the spread of a fire. When heated or ignited, some materials release flammable gases or vapors. These gases or vapors may raise a material's overall flammability and may heighten the intensity of a fire.

3 Flammability Measurement Techniques

The flammability of materials has been measured by carrying out the standard flammability tests. These tests include the cone calorimeter, vertical burning, and limiting oxygen index tests. These techniques give information about a material's fire resistance as well as how it responds to various fire scenarios. Among the often employed testing techniques are:

3.1 Cone Calorimetry

In this technique, a small sample of the material is exposed to the regulated radiant heat while various quantities, including the heat release rate, the ignition time, the mass loss rate, and the smoke and gas release rate are measured [7]. Cone calorimetry offers insightful information on a material's flammability and fire behaviour. Examples are the UL 94 test for plastic materials' flammability and the ASTM D6413 test for textiles' flame resistance. These tests measure a material's flammability in a vertical configuration using a flame.

3.2 Limiting Oxygen Index Test (LOI)

The oxygen index test gauges the amount of oxygen that must be present for a substance to burn. It shows how relatively flammable the substance is in an oxygen-rich environment. The relative flammability of plastics and composite materials can be measured by employing the LOI test [27].

3.3 Smoke Density Test

These tests gauge how much smoke a substance produces when it burns. Smoke density tests assist in evaluating the dangers that could result from inhaling smoke during a fire. The smoke density (D_s) can be determined by using the following equation [15]:

$$D_s = \frac{V}{A \times L} \log \frac{100}{T} \tag{1}$$

where, A—an area of sample exposure (m^2), optical path length (m), T—the rate of transmittance (%), and V—some box volume (m^3)

3.4 Fire Propagation Test

These tests gauge a material's capacity to fend off fire. The Steiner tunnel test, also known as ASTM E84 [20], is one instance where materials used for interior finishes are tested for their ability to spread flame and create smoke.

4 Overview of Key Parameters Measured in Flammability Tests

Several important factors are examined during flammability testing to evaluate the material's fire resistance and behavior. These variables offer quantifiable information that can be utilized to evaluate various building materials and make fire safety decisions. The ignition temperature, flame spread rate, mass loss heat release rate, smoke production, and emission of toxic gases are some of the important parameters, and they provide valuable information about a material's behavior in a fire scenario and also assist in the development of fire safety regulations and standards [30].

When exposed to an ignition source, a substance must have enough time to ignite. The rate at which a flame travels across a surface is known as the flame spread rate

[12]. The rate at which heat is released during material burning, measured in units of time. The place at which a substance loses mass as a result of burning is taken into account for flammability assessment. The volume of smoke produced during combustion. The measurement of dangerous gases such as carbon monoxide or hydrogen cyanide, released during the combustion of the material. All these parameters are taken into consideration by assessing the material's flammability.

5 Flame Retardancy Strategies of Flameretardant Materials

Silicone rubber-based biocomposites and bionanocomposites can be made as flame-resistive materials by adding different additives. These materials are highly resistive to ignition and can also stop the spread of flames. The incorporated additives work in several ways to achieve flame retardancy. The characteristics of different fire retardants and their fire resistive mechanism are discussed in this section. When exposed to heat, halogenated flame retardants, including brominated or chlorinated chemicals, release halogen radicals [4]. By scavenging free radicals and preventing the creation of combustible gases, these radicals halt combustion. Phosphorus-based flame retardants have a number of different ways by which they work in preventing the spread of flame [10]. They can create a shielding char layer on the material's surface that acts as a barrier against heat and oxygen by restricting the flow of combustible gases. They can also encourage the development of condensed phase processes, which minimizes the emission of combustible volatiles. Nanoparticles, such as carbon nanotubes, nanoclays, or nanosilica, can improve silicone rubber composites' flame retardancy. By acting as physical barriers, these nanoparticles stop the transmission of heat and fuel vapor. They may also encourage the development of a dense char layer, which slows the spread of flames.

6 Role of Flame Retardant Additives in Improving Flammability Resistance

The ignition time, burning rate, and flame spread can be decreased by adding different additives to improve the fire safety qualities of flame retardant materials. Flame retardants postpone the material's ignition by raising the temperature needed to start combustion. This delay increases overall safety by giving more time for evacuation or fire suppression. The rate at which flames spread across a material's surface can be slowed down by the use of flame retardants. This characteristic restricts the spread of the fire, making it simpler to contain or put out. Flame retardants can help reduce smoke and toxic gas emissions produced by combustion. To safeguard public health and prevent additional risks in the event of a fire, this decrease is essential. The mechanical and thermal properties of silicone rubber can be affected by the addition of natural/synthetic additives. To guarantee that the necessary balance between flame retardancy and material performance is achieved, it is crucial to take these impacts into account. Some flame retardants, especially those that contain halogens, may have a deleterious impact on the composites' mechanical characteristics. They may result in reduced elongation at break, stiffness, or brittleness. However, certain nanoparticles and additives based on phosphorus have demonstrated less negative impacts on mechanical characteristics. These additives can improve a material's thermal stability by raising the decomposition temperature or by encouraging the development of stable char layers. Because of its increased thermal stability, the material can resist higher temperatures without suffering serious damage.

7 Flame Retardant Studies of Silicone Rubber Biocomposites

The silicone rubber was reinforced with the hybrid fillers of glass fiber (GF)/montmorillonite (MMT) nanoclay/aluminium hydroxide (ATH) to investigate the flame retardancy of the fabricated SR/GF-MMT nanoclay-ATH bionanocomposite by comparing with that of SR/ATH and SR/ATH-MMT biocomposites [22]. The flammability analysis was carried out at a heat flux of 35 kW/m^2 using laboratory cone calorimeter to assess the flame resistive parameters such as peak heat release rate (HRR), total heat release (THR), average rate of heat release (ARHR), smoke production rate (SPR), total smoke production (TSP) and total oxygen consumed (TOC) for the fabricated SR-based bionanocomposite. It was concluded that there was a decrement in the THR, HRR, ARHE, TOC, peak SPR, and TSP values of silicone rubber with the addition of ATH, ATH-MMT and GF-MMT-ATH filler, resulting in an improved fire resistance as compared to pristine SR and SR/ATH-GF biocomposite (Table 1).

Table 1 Fire resistive parameters of SR composites obtained from cone calorimetric analysis [22]

Samples	MARHE (kW/m^2)	Peak HRR (kW/m^2)	Peak SPR (m^2/s)	THR (MJ/m^2)	TOC (g)	TSP (m^2)
SR	140.69	316.44	0.108	49.32	34.22	11.48
SR/ATH	61.29	223.76	0.085	17.82	12.83	6.02
SR/ATH/MMT	46.62	208.55	0.030	21.02	14.61	1.49
SR/ATH/GF	33.11	166.49	0.018	16.41	11.35	1.03
SR/ATH/MMT/GF	24.93	144.63	0.015	12.49	8.68	0.48

Among the SR biocomposities, SR/GF-MMT-ATH bionancompositties exhibited remarkably lowest fire resistive parameters with a better fire resistance making it suitable for electrical insulation applications. GF is a non-flammable reinforcing agent and acts as a fire barrier. ATH is a natural compound that releases water vapour upon decomposition due to the presence of three hydrated water molecules, which slows down the combustion process of polymers with the formation of char reside of aluminum oxide. MMT also acted as a natural fire retardant during the combustion process without releasing toxic gases. MMT creates protonic catalytic sites upon flame exposure that aid in creating a stronger interaction with the char residue. Therefore, the formation of compact char residue due to the combined effect of ATH and MMT acts as a heat barrier against the combustible gas, resulting in the enhanced flame resistance for SR composite reinforced with hybrid fillers of GF/ATH/MMT. The surface morphology of pristine SR was greatly altered by the incorporated fillers, and their physical interaction with the SR matrix was affirmed by the SEM micrographs of the SR composites (Fig. 2).

The surface morphology of the char residue of the SR and its composites was investigated using an SEM micrograph to comprehend the mechanism of fire resistance. The char residue of pristine SR seems to be fluffy and fragile (Fig. 3). However, the surface morphology of the char residue of SR/ATH biocomposite and SR/AHT-MMT bionanocomposite reveals a denser surface with cracks and few weak points. The surface morphology of the char residue of SR/ATH-GF-MMT bionanocomposite is

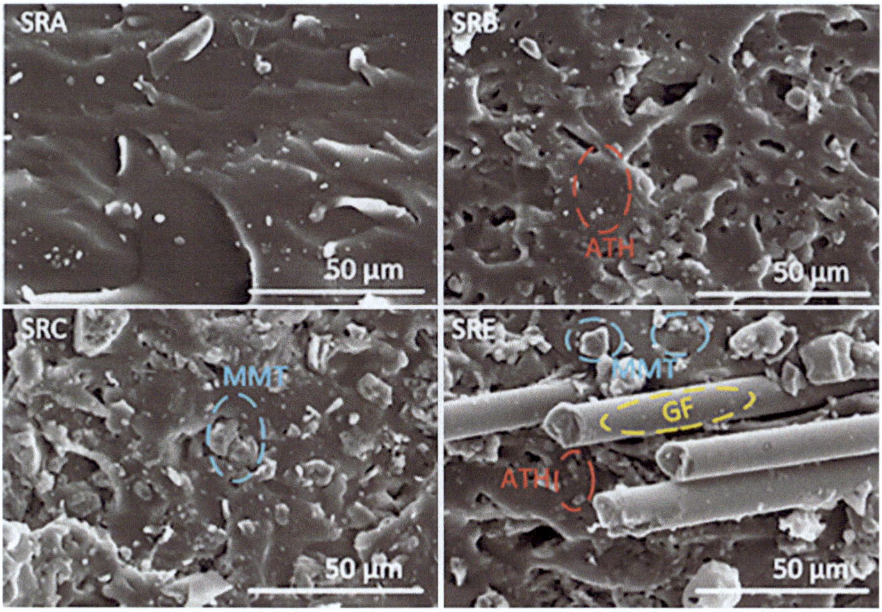

Fig. 2 SEM micrographs of **a** SR (SRA), **b** SR/ATH (SRB), **c** SR/ATH/MMT (SRC) and **d** SR/ATH/MMT/GF (SRE) biocomposites [22]

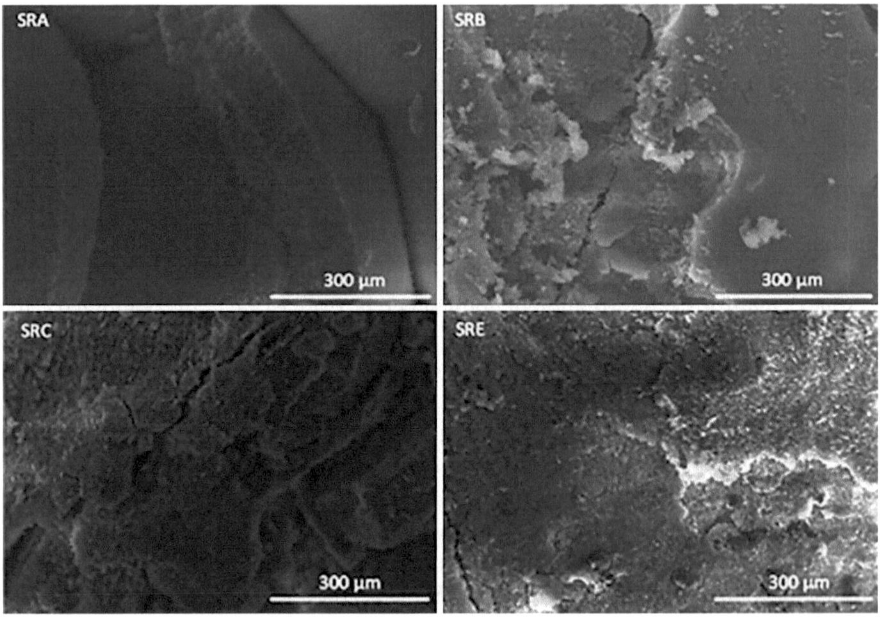

Fig. 3 SEM micrographs of char residues of **a** SR (SRA), **b** SR/ATH (SRB), **c** SR/ATH/MMT (SRC) and **d** SR/ATH/MMT/GF (SRE) biocomposites [22]

observed to be compact, rough, and continuous due to the interlocking of GF threads with the compact char resulting in an improved flame resistance as compared to other samples.

Sepiolite, a natural silicate mineral improves the flame retardant property of silicone rubber as a reinforcing agent. Silicone rubber foam was reinforced with microencapsulated sepiolite and zinc borate (ZB) to improve its smoke suppression and flame retardancy [14]. A sample dimension of $100 \times 5 \times 3$ mm^3 was used to measure the limiting oxygen index (LOI) of SR foam and its biocomposites with the help of a JF-3 oxygen index meter in accordance with the GB/T 2406-2009 standard. The dimension of the samples was preferred as $25 \times 25 \times 6$ mm^3 to carry out smoke density analysis in accordance with the GB/T 8323v-2008 standard using a JCY-2 smoke density tester. The SR Foam reinforced with 6 wt% ZB and 9 wt% MSEP showed a higher LOI value of 30.9% than pure SR foam (27.8%). Figure 4 illustrates the HRR and THR curves of SR foam and its biocomposites with ZB/MSEP.

A decrement in the HRR and PHR values of the SR foam/ZB/MSEP biocomposites was observed by the researchers with the increase in ZB content while comparing the HRR and PHR values of pristine SR foam (Table 2). Moreover, the lowest HRR and PHRR values (37.10 MJ/m^2 and 87.7 kW/m^2) were achieved with SR foam/ 6wt% ZB/3wt% MSEP biocomposites owing to the formation of compact metal oxide layer on the surface of the biocomposites. During the combustion of the SR foam biocomposites, the produced water vapors also absorbed lots of heat resulting

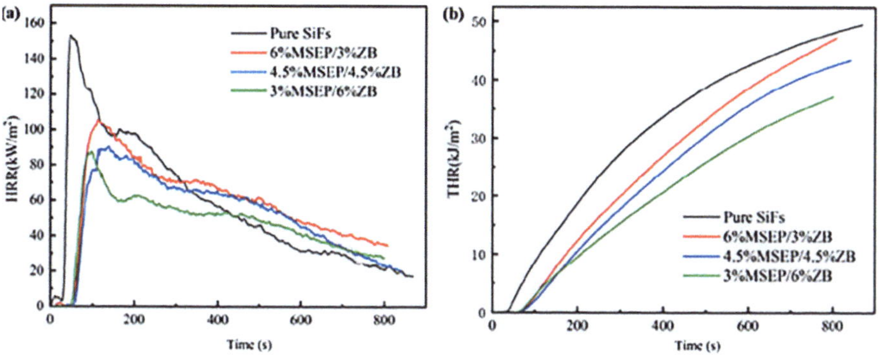

Fig. 4 A **a** HRR and THR curves of SR form and IIS biocomposites [14]

in low heat release. The fire growth index (FGI) and fire performance index (FPI) can also be used to investigate the fire safety of a material. A material with a lower FGI or higher FPI possesses good fire retardant efficiency.

There was a gradual decline in the total smoke production (TSP) and smoke production rate (SPR) of SR foam biocomposites due to the reinforcement of different wt% ZB/MSEP when compared to that of SR foam (Fig. 5). The values of TSP and Peak SPR were estimated as 11.09 and 0.045 for the SR foam/6wt%ZB/3wt%MSEP biocomposite, and it was also reported as lowest among the fabricated SR foam biocomposites (Table 2).

The synergistic effect of the incorporated ZB/MSEP led to the solid formation and the deposition of dense carbon layers during combustion, which could isolate air and heat resulting in improved smoke suppression for SR foam biocomposites compared to pristine SR foam.

The vulcanized silicone rubber (VSR) at room temperature finds applications in many fields owing to its good chemical stability, low-temperature toughness,

Table 2 Flame resistive parameters of SR foam biocomposites with ZB/MSEP [14]

Sample	TTI (s)	THR (MJ/m^2)	PHRR (kW/m^2)	TSP (m^2/m^2)	PSPR (m^2s)	FGI (kW/m^2s)	FPI ($m^2.s/kW$)
Pure SR foam	17	49.48	153.1	17.30	0.083	3.12	0.111
SR foam/ 3wt%ZB/ 6wt%MSEP	25	47.18	105.8	14.99	0.066	0.93	0.236
SR foam/ 4.5wt%ZB/ 4.5wt%MSEP	28	43.45	90.6	13.30	0.056	0.65	0.309
SR foam/ 6wt%ZB/ 3wt%MSEP	28	37.10	87.7	11.09	0.045	0.90	0.319

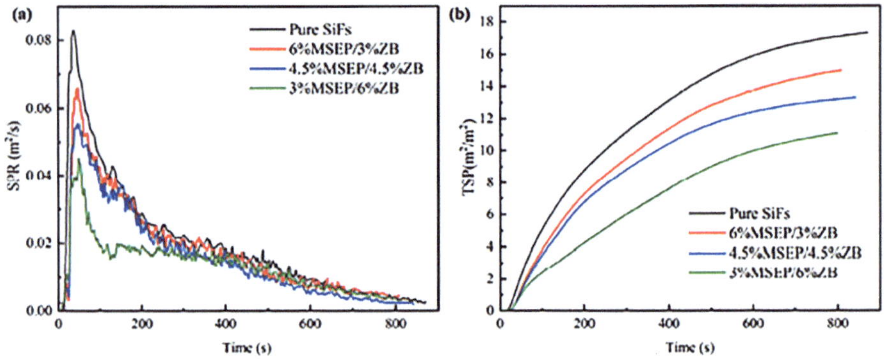

Fig. 5 **a** SPR and TSP curves of SR form and is biocomposites [14]

electrical insulation, low toxicity, and excellent thermal stability [6, 8]. However, it is flammable due to the presence of combustible C and H atoms in its molecular chains and it can be ignited at low temperatures, which probably restrict its applications in electrical and electronic appliances, aerospace, and sealing products [9]. The flame retardation of VSR can be improved by incorporating the fillers like MMT, silica, magnesium dehydrate and aluminum trihydrate. An ideal way of improving the flame retardancy of VSR at low cost is the incorporation of surface modified-MMT (OMMT) into VSR, which enables the fabrication of VSR biocomposite with improved thermal, mechanical, and flame retardant properties. The study on the reinforcement of VSR with different fillers such as OMMT, fumed silica, and OMMT/fumed silica was reported to improve flame retardant characteristics of the VSR composite in the literature [3]. The VSR composites filled with three different fillers were subjected to the UL 94 rating test. The time taken by the specimens for color change, catching of fire, burning onset and fume generation was recorded and compared with the fabricated VSR composites. The flame retardant properties of the VSR composites reinforced with OMMT, fumed silica and OMMT/fumed silica were evaluated from UL 94 test and listed in Table 3. The VSR composite containing fumed silica showed the lowest ignited rate with poor spreading of fire, transfer of heat, degree of damage diameter of flame, and smell production among the prepared samples. The VSR/OMMT biocomposite had better fire retardant properties compared to VSR/OMMT/fumed silica biocomposites and pristine VSR. This may be due to the formation of an intumescent char layer during the combustion of VSR composites by the introduced fillers, which leads to the release of low heat resulting in improved smoke suppression and flame retardancy compared to pristine VSR.

SR alone cannot survive in a high-temperature environment as it lacks in forming a strong char layer. The incorporation of inorganic fillers such as layered clay, metal oxides, carbonaceous fillers, and silica into the SR matrix assists in improving ablation resistance by forming a strong char layer at high temperatures, besides improving flammability, wear abrasion, and mechanical properties of SR. [29] reinforced SR

Table 3 Flame retardant properties of VSR composites [3]

Samples	Fumes	After fire (sec)	Transform of heat	Smell	Degree of damage diameter of flame	Color	Ignited rate
Pure VSR	++	7	High	++	3.5 cm	Black	Fast
VSR/ silica	+	2	Lowest	+	1 cm	Brown light	Slowest
VSR/ OMMT	+	3	Lower	+	2.5 cm	Brown dark	Slower
VSR/ OMMT/ fumed Silica	+	5	Low	+	3 cm	Black	Slow

with MMT and silicon carbide whisker (SCW) to improve its ablation resistance. SCW is used as reinforcing fillers in polymer, metal, and ceramic due to its high thermal stability, thermal shock resistance, and excellent oxidative stability. The SR/ MMT/SCW hybrid biocomposite was fabricated by incorporating SCW as secondary fillers along with MMT to improve the ablation resistance of SR. The advantage of using MMT with SCW is that the formation of a dense and compact covering layer by MMT while bonding with silica, and this process delays the diffusion of heat and oxygen into the inner layers of the fabricated composites, which remarkably improves the ablation resistance. The mixture of SR, MMT, and SCW along with the kynol fibers, fumed SiO_2 (added to improve mechanical), and curing agent of 2,5-bis(tert-butyl peroxy)-2,5-dimethyl hexane (DBPMH) was compression-molded under 10 MPa at 165 °C for 15 min to fabricate SR/MMT/SCW composites. The SR and SR/MMT/SCW biocomposites were subjected to an oxyacetylene torch test to assess their ablation resistance. The samples were irradiated by the oxyacetylene torch with an inner diameter of 2 mm at a distance of 10 mm for 20 s to study their ablation resistance against flame. An infrared thermometer was used to monitor the surface temperature which was around 2200 °C. The schematic diagram of the oxyacetylene torch test (ablation test) is illustrated in Fig. 6a and b. The mass ablation and linear rates were estimated by using the following relations:

$$R_m = \frac{m_1 - m_2}{t} \tag{2}$$

$$R_l = \frac{d_1 - d_2}{t} \tag{3}$$

where t–testing time; m_1, d_1 and m_2, d_2 are the mass and thickness of the sample before and after the ablation test. After testing, the char layer was removed from the sample to measure thickness d_2.

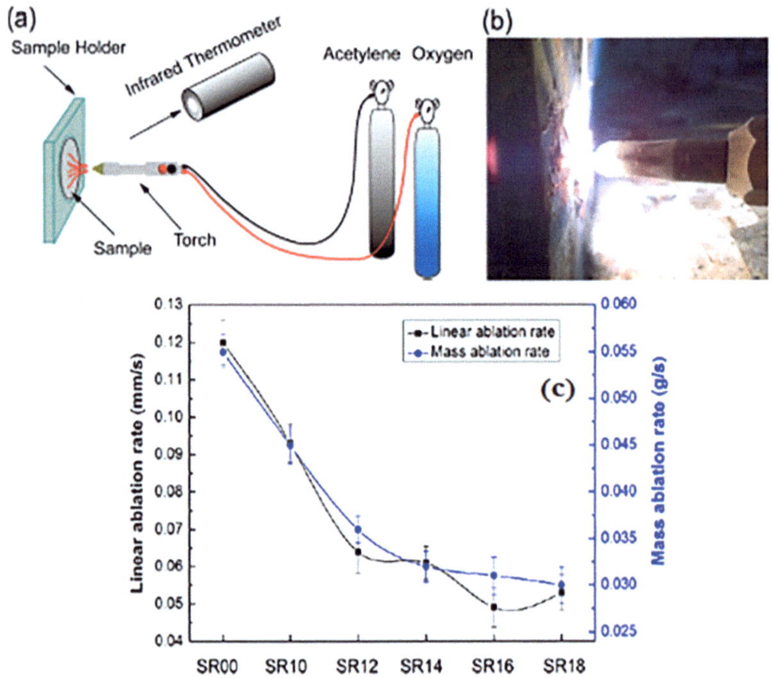

Fig. 6 **a** Schematic illustration of ablation test and **b** sample under the test and **c** ablation results of SR composites [29]

The mass and linear ablation rates of the SR composites were evaluated from the oxyacetylene torch test to comprehend their ablation resistance ability. The results of the ablation test for the fabricated SR biocomposites are shown in Fig. 6c. The ablation resistance of SR improved slightly with the addition of MMT and this was understood from the decrement of mass and linear ablation rates by 18.2% and 22.5%, respectively over the pristine SR. However, the mass and linear ablation rates of SR/MMT/SCW were decreased by 43.6% and 59.2%, respectively as compared to that of pristine SR due to the addition of SCW as secondary fillers. The reason for the better ablation resistance of SR/MMT/SCW composite was due to its more compact and stable char layer forming ability, which acted as a heat shield by protecting an inner region of the material. Meantime, the char layer of SR and SR/MMT biocomposite seemed to be porous and fragile.

The photographs of the ablated SR composites are illustrated in Fig. 7a and b. The macro cracks and broad ablated center were seen from the surface of pristine SR due to its poor ablation resistance. However, the SR/MMT/SCW composite exhibited a surface with no macro cracks and considerably reduced ablated center due to its greater ablation resistance. The ablation process of SR composites is illustrated schematically in Fig. 7a–c

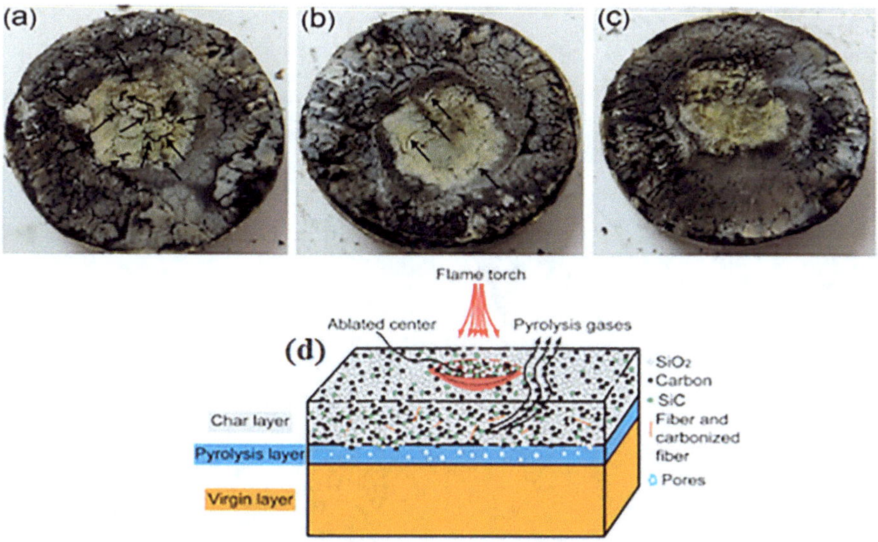

Fig. 7 **a-c** Photographs of the ablated SR composites and **d** schematic diagram of the ablation process [29]

The morphology of the deposited char layer on the SR composites after the oxyacetylene test was investigated by analyzing the SEM micrographs to study the effect of MMT and SCW fillers on the ablation resistance of the SR composites (Fig. 8a–d). The existence of cracks and pores was observed from the SEM micrographs of the char layer of pristine SR due to the accumulation of pyrolysis gases within the inner region of the material. It is concluded that the char layer is not strong enough to withstand different thermal stress acting along the longitudinal direction which leads to the formation of cracks in the case of pristine SR. The SEM micrograph of SR/MMT biocomposites was seen to be continuous with the absence of cracks, and the size of the created pores was larger as compared to that of SR (Fig. 8b). The formation of the viscous molten MMT film at high temperatures protects the char layer and the inner material against flame by acting as a heat shield that prevents the diffusion of heat and oxygen into the specimen. However, the char layer of SR/MMT/SCW biocomposite was harder and denser due to the incorporated oxidation inhibitor (SCW) that consumed oxygen resulting in better ablation resistance at higher temperatures as compared to that of other samples (Fig. 8c).

8 Conclusion and Future Perspectives

Studies have demonstrated that silicone rubber bionanocomposites are much more fire resistant while incorporating flame retardant additives. These flame retardant additives work by making materials less flammable and preventing the spread of fire.

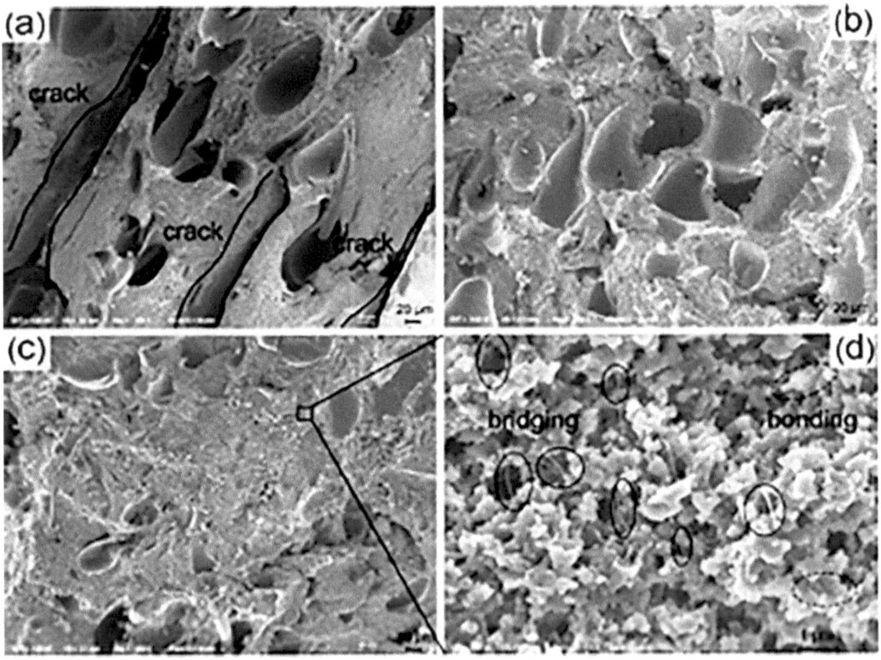

Fig. 8 SEM micrographs of the char layer of the SR composites (cross-sectional image) after ablation test **a** Pristine SR, **b** SR/MMT and **c** SR/MMT/SCW and **d** selected region of SR/MMT/SCW biocomposites (magnified) [29]

Through processes like the release of flame-quenching gases, the development of protective char layers, or the diluting of flammable gases, they are able to do this. Natural fillers such as montmorillonite (MMT), sepiolite, and silica have been found to improve the flame retardancy of the resulting biocomposites. By acting as barriers, these natural fillers slow down the rate of heat release and stop fires from spreading by forming the char layers that protect the underlying materials from more combustion. The flame retardancy of SR biocomposites is typically improved by increasing the natural filler loading. However, higher loadings of fillers may have a deleterious impact on the composites' mechanical capabilities. Silicone rubber bionanocomposites have demonstrated enhanced flammability performance in comparison to conventional silicone rubber composites. The incorporation of MMT into silicone rubber enhanced the material's fire resistance by lowering the rate of heat release and smoke formation. It is quite challenging to improve flame resistance of the SR biocomposites by preserving their inherent mechanical and thermal properties. The selection of suitable natural fillers and the optimization of filler loadings need to be focused on improving the flame resistance of biocomposites by preserving the other desired properties. Continuous research is essential for creating safer and more environmentally friendly materials, enhancing fire safety, and satisfying the world's rising need for cutting-edge solutions. Further investigation on the optimization of flame retardant systems by taking compatibility, processing, toxicity, and environmental impact into account will provide a safer and more sustainable future.

References

1. Ali MS, Al-Shukri AA, Maghami MR, Gomes C (2021) Nano and bio-composites and their applications: a review. In: IOP conference series: materials science and engineering. IOP Publishing, vol 1067(1):012093
2. Atmakuri A, Palevicius A, Janusas G, Eimontas J (2022) Investigation of hemp and flax fiber-reinforced EcoPoxy matrix biocomposites: morphological, mechanical, and hydrophilic properties. Polymers 14(21):4530
3. Attia MK (2020) Improving thermal stability and electrical insulation of gamma-irradiated silicone rubber nanocomposites. J Vinyl Add Tech 26(3):354–361
4. Bocchini S, Camino G (2010) Halogen-containing flame retardants. Fire Retard Polym Mater 2:75–106
5. Charbonnet JA, Weber R, Blum A (2020) Flammability standards for furniture, building insulation and electronics: benefit and risk. Emerg Contam 6:432–441
6. Chen D, Yi S, Wu W, Zhong Y, Liao J, Huang C, Shi W (2010) Synthesis and characterization of novel room temperature vulcanized (RTV) silicone rubbers using Vinyl-POSS derivatives as cross linking agents. Polymer 51(17):3867–3878
7. El Gazi M, Sonnier R, Giraud S, Batistella M, Basak S, Dumazert L, Hajj R, El Hage R (2021) Fire behavior of thermally thin materials in cone calorimeter. Polymers 13(8):1297
8. Gonzalez-Perez G, Burillo G, Ogawa T, Avalos-Borja M (2012) Grafting of styrene and 2-vinylnaphthalene onto silicone rubber to improve radiation resistance. Polym Degrad Stab 97(8):1495–1503
9. Hamdani S, Longuet C, Perrin D, Lopez-Cuesta JM, Ganachaud F (2009) Flame retardancy of silicone-based materials. Polym Degrad Stab 94(4):465–495
10. Hörold S (2014) Phosphorus-based and intumescent flame retardants. In: Polymer Green Flame Retardants. Elsevier, pp 221–254
11. Jawaid M, Qaiss A, Bouhfid R (2016) Nanoclay reinforced polymer composites
12. Johnston JM, Wheatley MJ, Wooster MJ, Paugam R, Davies GM, DeBoer KA (2018) Flame-front rate of spread estimates for moderate scale experimental fires are strongly influenced by measurement approach. Fire 1(1):16
13. Kaneko T, Ito S, Minakawa T, Hirai N, Ohki Y (2019) Degradation mechanisms of silicone rubber under different aging conditions. Polym Degrad Stab 168:108936
14. Kang F, Tu J, Zhao H, Bai Z, Zhang T (2023) Flame retardancy and smoke suppression of silicone rubber foam with microencapsulated sepiolite and zinc borate. Polymers 15(13):2927
15. Liu SH, Ke CY, Chiang CL (2021) Thermal stability, smoke density, and flame retardance of ecotype bio-based flame retardant agricultural waste bagasse/epoxy composites. Polymers 13(17):2977
16. Lothongkam C, Siebler D, Heidmann G, Plath R, Gockenbach E (2014) The influence of thermal aging on AC dielectric strength of transparent silicone rubbers for HV insulation. In: Proceedings of 2014 international symposium on electrical insulating materials. IEEE, pp 346–349
17. Mittal G, Dhand V, Rhee KY, Park SJ, Lee WR (2015) A review on carbon nanotubes and graphene as fillers in reinforced polymer nanocomposites. J Ind Eng Chem 21:11–25
18. Mohd Sabee MMS, Itam Z, Beddu S, Zahari NM, Mohd Kamal NL, Mohamad D, Zulkepli NA, Shafiq MD, Abdul Hamid ZA (2022) Flame retardant coatings: additives, binders, and fillers. Polymers 14(14):2911
19. Morampudi P, Namala KK, Gajjela YK, Barath M, Prudhvi G (2021) Review on glass fiber reinforced polymer composites. Mater Today: Proc 43:314–319

20. Morgan AB (2023) Revisiting the "Steiner Tunnel" test method covered under ASTM E84: an overview and analysis of the test method. J Fire Sci 41(6):288–320
21. Nandhini J, Karthikeyan E, Rajeshkumar S (2024) Eco-friendly bio-nanocomposites: pioneering sustainable biomedical advancements in engineering. Discover Nano 19(1):86
22. Nazir MT, Khalid A, Wang C, Kabir I, Yeoh GH, Phung BT (2024) Enhancing flame and electrical surface discharge resistance in silicone rubber composite insulation through aluminium hydroxide, clay, and glass fibre additives. Adv Compos Hybrid Mater 7(2):57
23. Nkuna CN, Mhike W, Ojijo V, Shongwe MB, John MJ (2024) Biocomposites and bionanocomposites from poly (lactide) and cellulosic materials–a review. Cellulose, pp 1–24
24. Sayam A, Rahman AM, Rahman MS, Smriti SA, Ahmed F, Rabbi MF, Hossain M, Faruque MO (2022) A review on carbon fiber-reinforced hierarchical composites: mechanical performance, manufacturing process, structural applications and allied challenges. Carbon Letters 32(5):1173–1205
25. Shah AUR, Prabhakar MN, Song JI (2017) Current advances in the fire retardancy of natural fiber and bio-based composites–a review. Int J Precis Eng Manuf-Green Technol 4:247–262
26. Sharma V, Agarwal S, Mathur A, Singhal S, Wadhwa S (2023) Advancements in nanomaterial based flame-retardants for polymers: a comprehensive overview. J Ind Eng Chem
27. Shen J, Liang J, Lin X, Lin H, Yu J, Wang S (2021) The flame-retardant mechanisms and preparation of polymer composites and their potential application in construction engineering. Polymers 14(1):82
28. Shit SC, Shah P (2013) A review on silicone rubber. Natl Acad Sci Lett 36(4):355–365
29. Zhang G, Wang F, Huang Z, Dai J, Shi M (2016) Improved ablation resistance of silicone rubber composites by introducing montmorillonite and silicon carbide whisker. Materials 9(9):723
30. Zhisheng X, Long Y, Yong L (2014) Study on correlations between the flammability and dynamic smoke properties of four decorative materials. Procedia Eng 84:498–505

Synergistic Effects of Flammability Performance of the Hybrid Composites Reinforced with the Various Natural Fibers

Swagata Dutta, Avishek Chanda, and Vikram Yadama

Abstract Natural fiber-reinforced polymeric composites (NFRPCs) possess excellent potential for effective applications in many applications for various industries. There are significant advantages of using these materials over synthetic fibers, which apart from being biodegradable and environment-friendly also include excellent weight-to-stiffness and strength-to-stiffness ratios, low densities and light-weight structures, abundant availability, and others. Some of the limitations do include strength compared to their synthetic polymers, an aspect that has been widely addressed through hybridization. Additionally, natural fibers have traditionally faced issues regarding their flammability performance, limiting their applications where heat exposures are on the higher end. Natural fiber-reinforced hybrid composites also have similar issues, and therefore, it is imperative to observe and study the thermal properties and decomposition of the constituent fibers, polymers, and the derived composite under heat and fire. Another advantage of hybridization is also the ability to increase the fire performance of NFRPCs. The current chapter will primarily focus on the thermal behavior of the constituent of hybrid NFRPCs and the fire reaction and resistance properties of their composites, which can be used in both structural and non-structural applications within the built environment.

Keywords Natural fiber composite · Hybrid natural fiber composite · Thermal properties · Fire reaction properties · Fire resistance properties · Plant fibers · Animal fibers · Mineral fibers

---

S. Dutta
Institute of Appropriate Technology (IAT), Bangladesh University of Engineering and Technology (BUET), Dhaka, Bangladesh

A. Chanda (✉) · V. Yadama
Department of Civil and Environmental Engineering, Composite Materials and Engineering Center, Washington State University, Pullman, WA, USA
e-mail: avishek.chanda@wsu.edu

C. Muthukumar et al. (eds.), *Flammability Performance of Biocomposites and Bionanocomposites*, Composites Science and Technology,
https://doi.org/10.1007/978-981-95-2389-4_13

1 Introduction

Composite materials have been widely used in a variety of applications, with the usage still expanding rapidly and a current worldwide market of higher than US$100 billion [1]. Composite materials currently constitute one of the broadest and the most important classes of engineered materials, second to steel. Many materials can be easily classified as composite materials, where the material is made from two or more constituents. This is particularly true for biological materials. In general, a composite constitutes of a strong and stiff component, which is often found in elongated form and are also called fibers, embedded in a complaint and softer component, also known as matrix. One such example is wood, where fibrous chains of cellulose and hemicellulose are embedded in the lignin matrix [2, 3]. Another example being bone and teeth, both of which are primarily composed of inorganic crystals embedded in a collagen matrix [4, 5]. Composites are anisotropic materials, resulting in varying mechanical properties in different directions. This is primarily due to the fibrous reinforcements resulting in increased strength of the material along the direction in which the fibers are mostly aligned. In synthetic composites, this direction can be tailored, whereas in natural composites, like wood, the material is generally stronger in the direction of the fiber tracheid that is usually aligned parallel to the trunk's axis or in other words in the transverse directions [1].

The fibers form the main load carrying component in the composites, where a pure polymer does not constitute the required mechanical properties for the desired application. This results in enhanced mechanical properties of the fiber-reinforced polymer composites (FRPCs), making them suitable for a variety of applications [6]. Traditionally, FRPCs have been made from various synthetic fibers, namely glass, carbon, aramid, Kevlar, and others. The primary advantage is the ability to control the direction of the fibers and thus, tailoring the strength of the FRPCs as desired for the particular application. High strengths and stiffnesses can be easily achievable with long fatigue lives, with the possibility of imparting higher corrosion resistance, better thermal properties, improved dimensional stability, higher wear resistance, and greater conductivity [7]. However, there are certain disadvantages of using synthetic FRPCs, which include high cost, poor recyclability, high density, and non-biodegradability. The application field of synthetic FRPCs is facing significant issues with end-of-life recycling of the materials, with one such critical example being the aerospace industry. A huge number of investments are being made by the governments of many countries to develop recycling techniques of carbon and glass fiber composites after their end of life. Due to these issues and the current inclination toward green and biodegradable technologies, natural fibers have gained substantial importance with increased applications in various industries as the viable alternative to synthetic FRPCs [8–13].

Studies have even proven natural FRPCs (NFRPCs) to be competitive in material properties compared to glass FRPCs [14], while studies from many decades on NFRPCs have proven their applicability in various applications. Apart from the environmental benefits, NFRPCs have low pollutant emissions, high specific strength and stiffness values, are made from renewable sources, and have better energy recovery capabilities [15, 16]. These have resulted in their huge demand in industries where along with environmental benefits weight reduction is also critical, namely automotive and aerospace. Another advantage of NFRPCs over their synthetic counterparts is their affinity to hydrophilic polymers that is enhanced through chemical treatment of the fibers, that directly influences the specific mechanical properties of the composites, as desired for the intended application [17, 18]. Studies [19] have also proven that the substantially high moisture absorption experienced in NFRPCs can be reduced through alkali treatments, resulting in comparable dimensional stabilities. It is important to note that as per recorded estimations, almost half of the energy required to harvest and process the world's vegetation is lost, with only a small part of it returning to the soil as melioration agents and the other by-products having no economic value [8]. Therefore, utilizing these agricultural and forest wastes and by-products is critical for carbon sequestration and ensuring a habitable environment for future generations. Natural fibers can be broadly classified into mineral, plant, and animal fibers, with further subdivisions that are detailed in Fig. 1. The current chapter will primarily focus on the flammability performance of plant and animal fibers-reinforced polymer composites. Despite of the various advantages, NFRPCs have a critical limitation for engineering applications, which is their flammability performance or fire resistance, especially in the plant fibers, where the presence of cellulose and hydrocarbons leads to extremely high flammability in the fabricated composites. Polymers tend to decompose between the temperature range of 300 °C and 500 °C, forming gases and condensed phases that result in liquid formation, combustible gas generation, and smoke production. These can be really hazardous in many applications, specifically in areas of high thermal exposure [20].

Natural fibers are known to generally increase the hazardous aspects with substantially higher amount of heat and smoke production that can potentially instigate poor visibility and fire spread, respectively, resulting in serious loss of properties and human risks [21]. Additionally, the softening of the polymer composites due to fire,

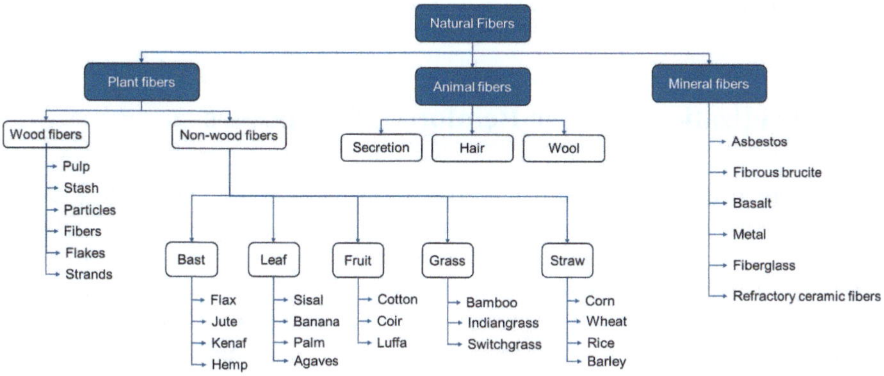

Fig. 1 Classification of natural fibers

resulting in degraded creep properties, can drastically impact the composite structures' load bearing capacity and structural integrity [22]. Therefore, it is absolutely critical to understand the limitations and vulnerability of NFRPCs under fire for asserting their applicability in various fields with growing stringent fire regulations and standards. Federal Aviation Regulations have the most stringent requirements for any composite to be acceptable for interior aircraft applications, with it being able to meet the vertical burn test, smoke density, and heat release requirements [23]. On the other hand, international building codes [24] have specified the requirement of both cone calorimeter properties and full-scale ASTM E119 [25] or equivalent test requirements to understand the applicability of a particular composite material in structural and non-structural aspects. The test details will be further explained later in the chapter.

In the present scenario, as can be observed with the various standard requirements in every sector, fire-retardant treatments have become critical to enhance the fire properties and resisting abilities of the NFRPCs. Studies have shown that the burning properties can be effectively prohibited in the NFRPCs through fire-retardant fillers and treatments, although this often results in degraded mechanical properties [26, 27]. Therefore, it is critical to find a balance. Another method, which has seen a significant rise in research since 2013, is hybrid composite that has the ability to balance the mechanical and fire performances with two or more fillers [28]. China has been leading the research with India and US holding the second and third spots, respectively [29]. The primary objective behind these materials is to overcome the limitations of single fiber/filler-reinforced polymer composites to achieve better performance than the parent composite as per the requirements of the desired applications. The concept also involves blending more than one polymer and the same filler with varying sizes or dimensions, which is also known an auto-hybridization. A more detailed explanation will be provided in the later half of the chapter. The primary focus of the current chapter is to highlight the fire performance of different hybrid NFRCs to illustrate the advantages and limitations of hybridization when it comes to the flammability performance of the composites while focusing only on plant and animal fibers.

2 Hybridization of Fiber-Reinforced Composites

Hybrid composites are generally defined as those materials that are made with two or more fillers in a matrix, where the matrix can be steel, ceramic, or polymer. The current chapter deals only with polymeric matrix hybrid composites, where the definition of hybrid composites also includes a material that comprises of two or more polymers along with at least one or more types of fillers. Another new term, known as auto-hybrid, is being used for composites that are made of the same filler but with different particle sizes [29]. Hybrid composites are developed to enhance the capabilities of composite materials for high-end uses. Three key elements impact the performance and quality of hybrid composites. The final properties of the hybrid composites are primarily dependent on three aspects:

1. Materials used: The constituent materials, matrix, and filler, which are primarily selected as per the application has a significant influence on the final properties. Studies have shown that jute/glass hybrid composites had higher compressive strengths compared to bamboo/glass hybrid composites, which actually showed higher tensile strengths [30]. Therefore, the application is critical to understand the ideal constitution of the composite. Mansor et al. [31] employed the Analytical Hierarchy Process to establish the ideal natural fiber combination with glass fiber for application in automotive brake levers. Out of the 13 natural fibers studies, kenaf was established to be the top scorer in the study.
2. Preparation method: The method of preparation used to fabricate the hybrid composite comes next, which is again dependent on the filler(s) and matrix used. As should be understandable, designing a new product should consider the environment that it needs to be used in, based on which the respective conditions need to be met by the product. These conditions include indoor or outdoor usage, usage in high temperature or heat conditions, high moisture conditions, direct or indirect fire exposure, and extreme weather conditions. This enables studies to compare the new product with the commercially available ones that are currently used for the specific application. Often this leads to developing a balance between sustainability and performance, requiring synthetic fibers to be used to achieve the required performance in the final composite. Moisture transfer is another critical aspect, where the natural fibers often have issues, requiring the addition of hydrophobic synthetic fibers to reduce moisture uptake of the final composite. On the other hand, synthetic fibers are generally denser, resulting in the need for natural fibers to reduce the weight of the overall composite, along with making the composites environment-friendly. Thus, the critical aspects to consider while choosing the fillers include weight, mechanical performance, recyclability, density, biodegradability, moisture affinity, cost, production cost, compatibility with the current manufacturing systems and recyclability techniques used in the industry, and testing methods.
3. Interaction: The matrix and filler interaction forms the third most important criterion for choosing the constituents of the composite. Chemical treatments are commonly used in the current industries along with coupling agents that can enhance the fiber-matrix interaction. This directly influences the performance of the final hybrid composite. Generally, the desired properties of the composite, assuming no physical or chemical interactions, can be predicted using the rule of mixture equation [32]. The general equation is:

$$P_{HC} = \sum P_{C_n} V_{C_n} \tag{1}$$

where P_{HC} is a specific property of the hybrid composite (HC), P_{C_n} is that property of the individual constituents, and V_{C_n} is the volume fraction of that constituent in the composite. The value of n can be 2 or more, depending on the constituents of the composite. The summation of all the volume fractions should always be unity.

2.1 Filler Selection

The current work primarily focuses on the utilization of natural fibers in fabricating hybrid composites. Over the last decade, the use of natural fibers has increased significantly, due to the increasing need for decarbonization and meeting the requirements of the Paris Treaty of 2015 [33]. Petroleum-based products are extremely hazardous to the environment and have been causing substantial landfill issues in many countries around the world. However, there are some limitations while utilizing the NFs, which include low thermal stability, high flammability, variation in mechanical properties, high moisture absorption, and others. Therefore, fillers have gained significant importance with a wide range of applications. The main goal was to replace the synthetic fibers with NF alternatives making the NFRCs biodegradable and economical. The advantages of natural fibers can be listed as:

- Recyclable.
- Excellent strength-to-weight and stiffness-to-weight ratios.
- Zero carbon footprint.
- High specific mechanical properties.
- Equipment friendly and non-abrasive.
- Produces non-harmful gases during production and processing, with negligible skin irritation.
- Contains good acoustic and thermal properties.

However, there are certain disadvantages as well, which include, high moisture absorption, low dimensional stability, high flammability, lower strength, anisotropic behavior, limitations in processing temperatures, odor generation during degradation, sensitivity to UV, prone to microbial and fungus attacks, and variable properties and price depending on location and market. The majority of these issues can be tackled using additional fillers or through chemical treatment of the fibers. The current work will focus primarily on NFRCs and their flammability aspect due to the presence of more than one filler or reinforcement. The chemical composition of the constituent fibers and fillers will thus play the most critical role in influencing the fire reaction and fire resistance properties of the hybrid NFRPCs.

2.2 *Resin Selection*

Resin used in the composite helps to ensure the required appearance, durability, and environmental tolerances are met for the intended application. Another critical aspect to consider is the processing technique of the composite, depending on which two types of resins are used, thermosets and thermoplastics. Thermosets have certain advantages over thermoplastics, including moldability, recyclability, re-processibility, and post-curing processability including thermoforming. Thermoplastics are generally obtained in the solid form in the room temperature and usually involve heating to convert the solid pellets into a viscous form for fiber impregnation and encapsulation. The most common techniques include melt mixing [34–36], compression molding [37–40], solution casting [41], and single or double screw extrusion [42–44]. The widely used thermoplastic composites include polypropylene (PP) [45], polyethylene (PE) [46], polystyrene (PS) [47], polyamide (PA) [48], and polyvinyl chloride (PVC) [49]. Thermosets are generally used for advanced composite applications due to their higher strengths, stiffnesses, dimensional stability, and ability to incorporate higher amounts of fiber compared to thermoplastics [29]. However, they are generally costly and are not recyclable. These include aerospace [50], automotive [51], and wind turbine blades [52]. The most commonly used fabrication techniques for thermoset polymer-based composites include resin transfer molding (RTM) [53], vacuum assisted resin transfer molding (VARTM) [12, 54], sheet molding compound [55], hand lay-up [56], and pultrusion [57]. The primary focus of this chapter being fire performance of hybrid composites, the thermal properties of the individual constituents will be discussed before discussing those of the composites.

3 Thermal Characteristics of Natural Fibers and Polymers

3.1 *Plant-Based Fibers (Cellulosic)*

Cellulose forms the primary fiber part in plant fibers along with hemicellulose, lignin (the matrix in which the fibers are embedded), wax, pectin, and traces of secondary metabolites and inorganic ash [58]. Therefore, the plant fibers are themselves a hybrid composite made by nature, often referred to as natural composites [11, 59]. Cellulose is highly crystalline and consists of glucose, whereas hemicellulose being a polysaccharide acts as the cementing agent between the micro-cellulose fibrils, constituting the main structural part of the plant fibers [27]. Under heat, cellulose decomposes between 50 and 340 °C at various steps as: (a) water desorption, (b) dehydrocellulose formation (cross-linking of cellulose chains), (c) char and volatiles (decomposition of dehydrocellulose), (d) levoglucosan formation, and (e) tar, gases, flammable and non-flammable volatiles, and char (decomposition of levoglucosan)

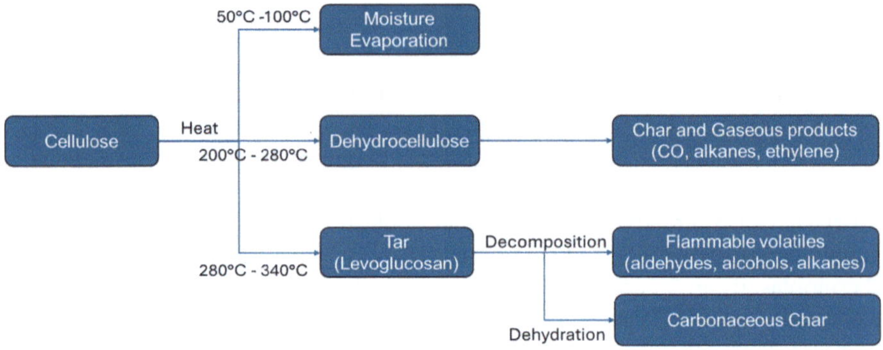

Fig. 2 Schematic representation of the degradation process of cellulose. Modified from [15, 61, 65, 66]

[60, 61]. The endothermic and exothermic steps involved during cellulose degradation are provided in Fig. 2. Hemicellulose experiences thermal decomposition in the temperature range of 200 and 260 °C, resulting in the release of incombustible gases like CO_2 [15]. Lignin has various complex bonds resulting in a wider range of thermal decomposition temperature, 160–400 °C. The weaker bonds decompose at lower temperatures, whereas the carbon–carbon and ether bonds, which form the stronger bonds in the aromatic ring, require higher decomposition temperatures [62]. Studies have also proven that lignin has a significant influence on the degradation of the other constituents due to its nature of promoting dehydration and char formation. Therefore, lignin-rich bast fibers, namely hemp and flax, have better fire resistance and lower heat release rate compared to leafy fibers, namely abaca and cabuya [63]. Impurities also play a significant role in the fire performance of the fibers, where Galaska et al. [64] proved that the influence of lignin is negligible if sodium salts are present on the fiber. The fibril orientation also plays a critical part in the eventual fire performance of the fibers, with highly oriented ones having better performance with reduced pyrolysis and diminished flammability [21].

3.2 *Animal-Based Fibers (Protein)*

Animal fibers primarily constitute amino acid groups, which directly influence the thermal decomposition and fire performance of the fibers. One such example is wool fiber, which consists of eighteen α-amino acids, primarily classified as lysine, cystine, aspartic, and glutamic acids [15]. A pair of cystine residues forms a disulfide bond that plays a critical role in determining the stability of the wool fibers against heat, water, cold, light, mechanical distortion, and biological attacks [67]. Under heat, moisture evaporation initiates at 100 °C, followed by an endothermic reaction at 200 °C. This results in the release of hydrogen sulfide (H_2S). The cleavages formed

Table 1 Thermal properties of commonly used animal fibers [15]

Fiber	Glass transition temp (T_g—°C)	Melting temp (T_M—°C)	Pyrolysis temp (T_P—°C)	Ignition temp (T_I—°C)	LOI (%)
Wool	NA	Doesn't melt	245	600	25.2
Silk	NA	Doesn't melt	320	600	22–23
Cotton	NA	Doesn't melt	350	350	18.4

at the disulfide bonds result in char formation above 230 °C. Additionally, cross-linking and dehydration reactions experienced by the sulfide containing amino acids at temperatures higher than 250 °C further facilitate char formation, limiting drip and fiber melting [67]. Moreover, sulfur containing compounds, namely carbonyl sulfide and carbon disulfide are released at around 247 °C. Finally, wood catches fire at around 600 °C, after the peak mass loss rate was reached at 500 °C, which is higher than any cellulosic fibers [68].

Silk is also known to have good fire resistance, with both wool and silk having a higher limit oxygen index (LOI) compared to other fibers. Since LOI estimate the minimum oxygen requirement for sustaining combustion after ignition, the presence of higher amounts of nitrogen in both wool (15–16%) and silk (15–18%) interferes with the oxidation process [66]. Silk can form char during combustion, which is influenced greatly by the cross-linking and dehydration of the hydroxyl groups [69]. Table 1 illustrates the thermal properties of some of the common animal fibers. The combustion behavior and pyrolysis temperatures of natural fibers are generally influenced by their source, and these also directly influence the fire performance of the composite materials.

3.3 *Mineral Fibers*

Mineral fibers are generally naturally occurring fibers that are produced from minerals found in the earth. These include asbestos, glass, aluminum, silicon carbide, basalt, boron carbide, and others. Asbestos is a term used to define both serpentine and amphibole minerals that naturally occur as fibers in the environment [70]. The general chemical formula for asbestos is $WX_2Y_5Si_8O_{22}(OH)_2$, where the substitution of *X* and *Y* results in different types of asbestos fibers. Asbestos fibers can be further classified into four most commonly used types, chrysotile, tremolite, crocidolite, and amosite [71]. Other types of asbestos include actinolite and anthophyllite. These can be broadly classified into two groups, serpentine, and amphiboles, as illustrated in Fig. 3.

Serpentines are generally less stable under heat, with initial endothermic peaks appearing around or greater than 500 °C, where dihydroxylation of chrysotile takes place [70]. This marks the beginning of sample degradation, with a complete degradation at about 1100 °C. The pyrolysis range varies between 500 and 750 °C, with

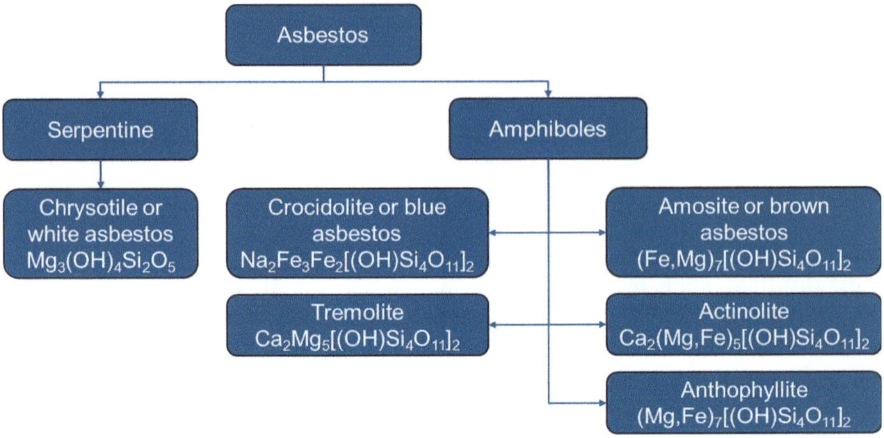

Fig. 3 Classification of various types of asbestos, adopted and modified from [71]

complete re-crystallization to forsterite and enstatite at around 1100 °C, resulting in an entirely altered fiber morphology due to the transformation into silicate phases. Amphiboles have higher thermal stability, with the first endothermic event, or initiation of mass loss being experienced at around 950 °C, with a shorter pyrolysis range, between 950 and 1050 °C. Complete re-crystallization is again observed at 1100 °C, when heated for 1 h, transforming into non-toxic silicate phases [70].

Basalt forms another type of mineral fiber that is acquired from basalt rocks that are formed through the solidification of volcanic lava [72]. In order to achieve the fibers, shattered basalt rocks are initially melted, with the melting temperature being 1400 °C [73]. The molten material is then drawn into filaments of desired cross-sections to achieve the fibers without any additives [74]. These basalt fibers are attractive options for reinforcing polymer matrices due to their potential low cost, along with their superior strengths and stiffness, which are often equal to or higher than glass fibers [72]. The disadvantages of basalt fibers and the subsequent composites include brittleness, which with higher stiffnesses although stronger can result in catastrophic and sudden failures [75]. One of the influential constituents of basalt is silica oxide (SiO_2), the presence of which helps categorizing the rocks into acidic basalts with > 46% SiO_2, mildly acidic basalts with 43–46% SiO_2, and alkaline basalts with $\leq$ 42% of SiO_2 [76]. The thermal stability of basalt fibers is exceptional with no major noticeable mass loss till 800 °C, resulting in mass residue during thermogravimetric analysis of about 98% [77]. Moreover, basalt fibers are also known to be excellent electric insulators that are biologically inactive and environmentally friendly [78].

Glass fibers form another type of mineral fiber, which is currently the most used form and has a wide variety of applications. The possibility of using glass fibers has been known to mankind for centuries, but the use started picking pace in the mid-nineteenth century [79]. In 1920, the first commercial use of glass fibers was to replace metal fibers in air filters, which were produced in Newark, Ohio, USA.

Fiberglass has dominated the market for many decades in filtration and insulation systems, with polymers currently replacing the majority of the applications from the last decade. The second major application of glass fibers lies in reinforcing polymers, due to their relatively higher tensile strength and elastic modulus compared to other commonly available mineral fibers. Their melting point is generally between 1400 and 1600 °C [80]. The atomic structure of glass has no long-range orders, resulting in its solidification in an amorphous arrangement. The concept of glass transition temperature, widely used in the polymer and composite industries, has been adopted from the solidification of glass, which is a function of the kinetics involved in the relaxation process [81]. All glass types experience significant viscosity change with temperatures and the final frozen structure is cooling rate dependent. Therefore, the transition from the liquid to the solid phase occurs over a range of temperatures, also known as the glass transition temperature range. Glass is often known as a vitreous substance, which is similar to a frozen liquid or ice, where the permanence is similar to a crystalline solid and the randomness is similar to a liquid [82]. Silica oxide or SiO_2 also forms one of the major constituents of glass fibers, which have a similar processing technique as that of basalt fibers. The thermal stability of the glass fibers varies significantly based on the constituents, production technique, and additives used. The LOI also ranges between 0.3 and 0.65 [83]. In general, glass fibers also have great thermal stability with almost negligible mass loss during TGA analysis till 575 °C [83].

Many other types of mineral fibers are used for various applications in the industries, including metal fibers. However, the aforementioned fibers are majorly used in the current time, with the utilization of metal fibers reducing significantly, and those of the others being very limited.

3.4 Thermoset and Thermoplastic Polymeric Resins

Polymeric matrices can be generally subdivided into thermosets and thermoplastic polymers. The behavior under heat of these two types of polymers is significantly different and is widely governed by the differences in their structures [66]. Thermoset polymers are composed of three-dimensional cross-linked molecular structures, which do not allow forming or flowing under heat, and experience direct degradation [84]. Thermoplastic polymers, on the contrary, have a linear chain structure and undergo reversible cross-linking reactions during the curing phase. Therefore, under heat, the thermoplastic polymers first melt and flow before the onset of degradation [85]. In general, burning any type of polymer is accompanied by heating, decomposition, ignition, and combustion [27]. Both thermoset and thermoplastic polymers experience molecular modification with the exposure of heat, at lower temperatures, experiencing further decomposition with the breakage in the molecular bonds and release of volatiles during pyrolysis. Bond scission with flammable products results in free radicals, which eventually lead to the ignition of the polymers in the presence of oxygen. The combustion cycle can be easily established in the presence

Table 2 Thermal properties of commonly used polymers

Polymer type	Glass transition temp (T_g—°C)	Melting temp (T_M—°C)	Pyrolysis temp (T_P—°C)	Ignition temp (T_I—°C)	LOI (%)
Nylon 6 [86, 87]	50	215	431	450	20.75
Polypropylene [88]	− 20	165	469	580	18
Polyethylene [89]	− 120	120	340	350	18
Polyester [90]	85	255	452	480	20.75
Epoxy [91, 92]	60	No melting	360–430	550	23
Phenolic [93]	300	No melting	440–520	614	25
Polycarbonate [94]	147	295	320–440	530	27
Polyvinyl chloride [95]	82	100	260–500	391	42
Polyimide [63, 96]	300	250	500–630	550	43
Polystyrene [63, 97]	100	240	390–450	285	18
ABS [63, 98]	105	190	350–460	349	19

Note ABS—Acrylonitrile Butadiene Styrene

of enough energy and combustible volatiles [84]. Certain polymers have inherent fire resistance properties, namely phenolic and polyimide polymers, which yield char during combustion. Phenolic resins initially release water on exposure to heat at lower temperatures due to phenol-phenol condensation. This is followed by the oxidation of the methylene groups leading to carbonyl linkage. Finally, under higher heats, the polymer undergoes further decomposition liberating CO, CO_2, and other volatiles, which ultimately leads to char formation [15]. The intrinsic fire properties of certain common polymers are detailed in Table 2.

4 Fire Reaction Properties of Hybrid NFRCs

In recent years, various applications have seen the introduction of NFRCs [21]. These natural fibers, such as jute, hemp, and flax, have gained attention because they have the potential to enhance the mechanical properties and reduce the environmental impacts of composites. The combination of different natural fibers with synthetic resins can lead to hybrid composites with improved fire performance characteristics. This section explores the fire reaction characteristics of fiber-reinforced hybrid polymer composites. Various fire reaction properties such as time-to-ignition, heat release rate, mass loss, extinction flammability index, thermal stability index, limiting oxygen index, smoke density, smoke toxicity, and surface spread of flame are discussed. Moreover, the inherent fire reaction properties along with fire-retardant fillers, testing methods, and standards are discussed in particular for hybrid natural fiber composites.

There are different fire reaction (alternatively known as reaction-to-fire) properties. The time-to-ignition (alternatively known as ignition time) is a very important property. It is the time it takes for a material to ignite when exposed to a flame or heat source. Ignition stands as a critical fire reaction property for combustible substances since it marks the commencement of flaming combustion. Composites are frequently made from organic resins, such as polyesters, vinyl esters, and epoxies and have a propensity to ignite shortly after encountering intense heat from a fire. Once ignited, composites typically produce substantial, high-temperature flames, facilitating the swift propagation of fire. Consequently, ignition serves as a significant characteristic for assessing the fire risk associated with composite materials. A longer ignition time implies a higher resistance to fire and better safety in potential fire incidents. The determination of composites' time-to-ignition involves experimental fire testing methodologies like cone calorimetry and the ISO ignitibility test. This parameter can be assessed under two conditions: spontaneous ignition and piloted ignition. Spontaneous ignition is known as the phenomenon when flames spontaneously arise in the flammable vapor/air mixture surrounding the heated composite surface. On the other hand, piloted ignition is triggered within the flammable vapor/air mixture by the means of an external ignition source or pilot, such as an electrical spark or flame. Furthermore, heat release rate (alternatively known as the rate of heat release) is the single most important property of the fire reaction properties. Heat release refers to the thermal energy generated per unit surface area when the combustible by-products ignite and burn near a material exposed to fire or heat flux. The peak heat release rate (HRR) typically happens rapidly following ignition and provides a strong indication of a material's maximum flammability. On the other hand, the average HRR represents the cumulative heat released throughout the combustion period and is regarded as the most dependable measure of a material's heat contribution to a sustained fire. The heat release rate (HRR) emerges as the pinnacle fire reaction property due to the fact that the heat liberated during combustion can supply the extra thermal energy essential for fire progression and expansion. Among all fire reaction properties, none exerts as significant an impact on the fire dynamics of different composites as the HRR.

Various other reaction properties, including surface spread of flame, smoke production, and CO emission, hinge upon or correlate with the HRR. For hybrid composites, a lower heat release rate is desirable, as it indicates slower fire propagation and can provide more time for evacuation or fire suppression in case of a fire event. Another fire reaction property that is closely related to the time-to-ignition is the ignition temperature [15]. The ignition temperature is the minimum temperature at which a material can ignite and sustain combustion. In addition, smoke production during combustion is a critical fire reaction property. Hybrid composites with natural fibers may exhibit different smoke production characteristics compared to traditional natural fiber composites. Lower smoke production is preferable, as it can improve visibility and reduce the health hazards associated with smoke inhalation during a fire event. The ability of a material to form a protective char layer during combustion is beneficial for enhancing fire resistance and limiting smoke production. Natural fiber-reinforced hybrid composites may demonstrate improved char formation, which can act as a barrier to heat transfer and reduce the spread of flames. The emission of toxic gases during combustion is a concern for fire safety. Hybrid composites with natural fibers may influence the types and quantities of toxic gases produced during a fire. Minimizing toxic gas emissions is crucial for ensuring the safety of occupants and emergency responders in the event of a fire. One important fire reaction property to consider in hybrid composites is flame spread, which measures the rate and extent of flame propagation over the material's surface. Factors such as the type of natural fibers used, the composition of the resin, and the arrangement of fibers within the composite all contribute to the flame spread behavior.

In order to understand the flammability performance of hybrid composites reinforced with natural fibers, it is important to consider the characteristics of these fibers and their compatibility with the matrix material [21, 99, 100]. It is crucial to examine the adhesion between the natural fibers and matrix, as well as the thermal stability and moisture absorption properties of the natural fibers. Furthermore, the incorporation of fire-retardant additives, such as intumescent ammonium polyphosphate-based compounds, can further enhance the flammability performance of these hybrid composites. Overall, the incorporation of natural fibers in the development of hybrid composites offers a potential solution for improving the flammability performance of materials. These materials have advantages such as low environmental impact and cost, making them a desirable option in industries where reduced flammability is crucial. With deeper investigation into the research on hybrid composites reinforced with natural fibers, it becomes evident that the synergistic effects of these materials offer a pathway to address flammability concerns in various industries [15]. Furthermore, the development of hybrid composites has demonstrated the potential to enhance the flammability performance of natural fiber-reinforced materials. Through the combination of natural fibers and synthetic resins, hybrid composites have shown promise in improving the flammability and performance of materials [101]. The utilization of natural fibers in hybrid composites has shown promising results in improving the flammability performance of materials [102, 103].

It is essential to conduct further studies to optimize the combination of natural fibers and synthetic resins to maximize the flammability performance of the resulting hybrid composites. Understanding the interplay between different natural fibers and their corresponding resin matrices will be critical in tailoring the material properties to meet specific flammability requirements. Additionally, exploring novel methods for enhancing the fire resistance of these composites, such as incorporating nanomaterials or functionalized natural fibers, holds promise for further improvement in flammability performance [15, 102, 104]. In the upcoming sub-sections, specific case studies and experimental investigations that highlight the application of hybrid composites in real-world scenarios will be highlighted, showcasing their enhanced flammability performance and potential for mitigating fire hazards [29, 105]. The economic and environmental benefits associated with the widespread adoption of natural fiber-reinforced hybrid composites in various industries will also be discussed. The current section explores the fire reaction characteristics of fiber-reinforced hybrid polymer composites. Various fire reaction properties are discussed. Moreover, the inherent fire reaction properties along with fire-retardant fillers, testing methods, and standards are discussed in particular for hybrid natural fiber composites. These hybrid composites offer a unique solution to addressing flammability concerns by combining the inherent fire-resistant properties of natural fibers with the enhanced performance of synthetic resins.

4.1 Inherent Fire Reaction Properties

In the field of engineering materials, natural fiber/polymer composites have attracted considerable attention due to their environmental friendliness and potential for low-impact sourcing. These composites are commonly reinforced with plant or mineral fibers, which present various challenges when exposed to fire hazards. When considering the fire reaction properties of hybrid composites reinforced with natural fibers, several key aspects need to be evaluated. These include the materials' ignitability, flame spread, heat release rate, smoke production, and toxic gas emissions. Understanding these properties is crucial for assessing the safety and potential applications of composite materials. Ignitability refers to the ease with which a material can be ignited. It is essential to assess the ignition resistance of hybrid composites to ensure their safe use in various environments. Flame spread describes the speed at which fire propagates across the material's surface. Controlling flame spread is vital for minimizing the fire hazard associated with the composites. These basic fire reaction properties are fundamental for evaluating the safety and performance of hybrid composites reinforced with natural fibers. Additionally, natural fiber-reinforced hybrid composites have shown promise in enhancing the flammability performance of materials while reducing environmental impact [63, 106, 107]. The use of jute, hemp, and flax fibers in combination with synthetic resins has been found to improve the flammability, mechanical, and thermal properties of composites [108, 109]. Additionally, the synergistic effects of these natural fibers have resulted in enhanced fire resistance and reduced smoke and toxic gas emissions. Research has indicated that the incorporation of natural fibers can lead to the formation of insulating char layers when the composite is exposed to fire. These char layers act as a barrier, slowing down the combustion process and protecting the underlying material. Moreover, the presence of natural fibers can promote the release of water vapor and other fire-retardant gases, further contributing to the suppression of flames. The development of hybrid composites reinforced with various natural fibers presents promising opportunities for addressing flammability concerns in diverse applications, including the automotive, construction, and aerospace industries. By exploring the unique synergies between natural fibers and synthetic resins, we can unlock the potential for advanced materials with improved fire safety properties.

Hybrid polymer composites reinforced with both synthetic and natural fibers are being increasingly utilized to improve material properties, decrease moisture uptake, afford cost benefits, mitigate environmental impact, and reduce energy consumption and carbon emissions. Natural fibers, being hygroscopic due to surface hydroxyl groups, tend to absorb water, which can compromise their performance. To address this, while limiting the use of harsh and non-eco-friendly chemical treatments, a fusion of hydrophilic natural fibers with hydrophobic synthetic fibers is an attractive solution. Through hybridization, water absorption is minimized, and the overall material quality is enhanced. Significant research has been conducted on hybrid composites combining both synthetic and natural fibers. Introducing synthetic glass fibers into polyester composites reinforced with natural jute fibers was found to increase both the tensile and bending strength [106]. This addition also curbed the composite's water absorption tendency. Comparable results were presented by other researchers who combined flax and carbon fibers, as well as jute with carbon fibers in woven laminates. Kureemun et al. [110] indicated that increasing the carbon fiber content bolstered tensile strength, however, substituting jute with carbon fibers can also considerably reinforce the material.

There have been numerous studies on the fire reaction properties of natural fiber-based composites as well as hybrid composites [111–119]. The ease of ignition is commonly assessed through the parameter known as time-to-ignition, representing the shortest duration needed to initiate ignition and sustained flaming of a combustible material under external heat exposure. This ignition time is influenced by the availability of oxygen, the chemical as well as the thermo-physical characteristics of both the fiber and matrix, and temperature [22]. The presence of fiber reinforcement can have an impact on the time it takes for a composite to ignite. Both glass and carbon fibers do not react to fire until the heat flux surpasses the range of 100–125 kW/m^2; yet they can still play a role in the ignition cycle. Research shows that enhancing the volume of fibers in a glass/polyester composite from zero percent (pure resin) to approximately seventy percent significantly lengthens the time until ignition occurs. By increasing the fiber volume, the quantity of resin, which is more prone to catching fire, is reduced, and thus, there are fewer flammable substances emitted that could ignite (Fig. 4).

Fibers are typically enveloped by a slight coating composed of various organic substances, including those that form films, reduce static, and provide lubrication, with the collective weight of these agents comprising approximately 1–2%. During a fire, the breakdown of this layer can emit combustible gases, potentially shortening the time it takes for ignition to occur. While the organic film enveloping the fibers is typically quite thin, about 0.1 μm in thickness, it generates fewer volatile substances compared to the resin matrix. Nonetheless, specific reinforcement forms, particularly those using chopped strand mats, are held together by a sizable quantity of organic emulsion binder, which, when exposed to fire, can release gases that contribute to reducing the time required for ignition. Glass and carbon fibers maintain stability in fires that reach temperatures up to approximately 1000–1200 °C. However, during production, fibers are treated with sizing, emulsion binders, and other organic materials to enhance their adhesion, binding, anti-static qualities, and resistance to wear, which will break down at high temperatures, as mentioned before. It has been observed that composites with organic fibers exhibit shorter ignition times compared to those reinforced with non-flammable fibers like glass. A study illustrated in Fig. 5 [22] indicates that vinyl ester composites incorporating woven glass fibers demonstrate longer ignition times than those incorporating aramid or polyethylene fibers, particularly at two different levels of heat flux. The presence of organic

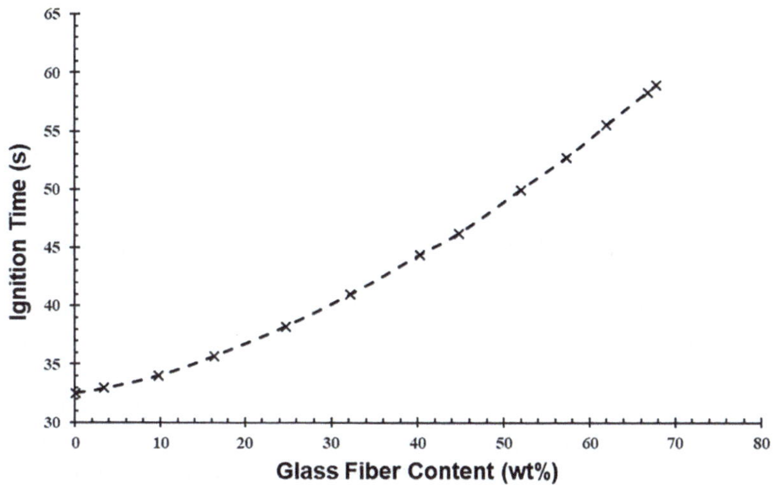

Fig. 4 Impact of fiber content on the ignition time of a glass/polyester composite [22]. Reproduced with permission from Mouritz from his book—"Fire Properties of Polymer Composite Materials" referenced in [22]

fibers, especially polyethylene, tends to facilitate quicker ignition. In the event of a fire, polyethylene fibers, and to a more moderate degree, aramid fibers, break down alongside the polymer matrix, accelerating the release of volatile substances, which in turn shortens the time required for ignition.

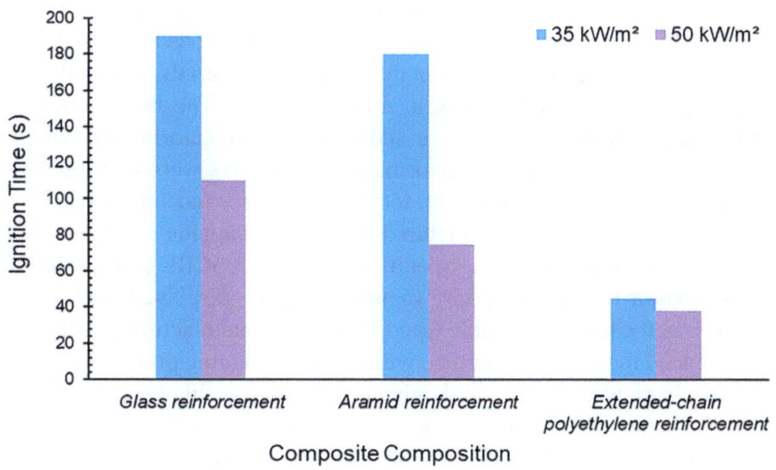

Fig. 5 Effect of heat flux on the ignition times of vinyl ester-based laminates containing glass, aramid or extended chain polyethylene fibers [22]. Reproduced with permission from Mouritz from his book—"Fire Properties of Polymer Composite Materials" referenced in [22]

The ignition behavior of polymer matrix also plays a pivotal role in the fire behavior of hybrid composites. It is commonly recognized that phenolic laminates have a notably longer ignition delay compared to other materials. This highlight in fire response behavior of phenolic composites is attributed to their high thermal decomposition threshold, strong inclination to form char upon decomposition, and the minimal production of combustible gases by the phenolic resin matrix [63, 118–121]. The time it takes for different fiberglass composites with various polymer bases to catch fire was studied by Sorathia et al. [118]. The composite made with a cyanate ester polymer took the longest to ignite, at approximately 326 s, while the one with a vinyl ester polymer had the shortest ignition time, about 50 s. Therefore, it can be seen that the inherent fire properties of fibers and polymers play a dominant role in the ignition behavior of hybrid composites.

Furthermore, the rate at which heat is released during burning (heat release rate or HRR) is critical as it can supply the extra energy needed to fuel the expansion and propagation of a fire. No other property impacts the fire performance of composites as significantly as HRR. Moreover, several other reaction characteristics, including how quickly a flame spreads across a surface, the production of smoke, and the emission of carbon monoxide, are influenced, or correlated with HRR. Heat release refers to the thermal energy generated per surface area when combustible breakdown products catch fire near a material exposed to fire or heat. Typically, the highest rate of heat release happens over a brief period shortly following ignition, and this peak is often viewed as a reliable gauge of a material's ultimate flammability. The average heat release rate, which is the total heat emitted over the duration of combustion divided by that time period, is regarded as the most dependable indicator of a material's contribution to the intensity of an ongoing fire. A study by Sałasińska et al. [117] explores the production and evaluation of hybrid layered composites tailored for mechanical properties and fire resistance. Different manufacturing methods were used, including hand lay-up, vacuum bagging, and resin infusion. The materials used in the composites were glass, aramid, carbon, basalt, and flax fibers. The burning behavior of the manufactured composites was assessed using a cone calorimeter, which simulated fire conditions. Critical fire performance parameters were measured, such as peak heat release rate, time of ignition, total heat release, and smoke emission. The study found that composites manufactured by vacuum bagging exhibited the lowest flammability and smoke emission. Specifically, the V-AGCBF composite (aramid-glass-carbon-basalt-flax fabricated by vacuum bagging) displayed the lowest HRR, indicating it was the least flammable (see Fig. 6). The manufacturing method significantly influenced these fire properties, with vacuum bagging producing composites with high mechanical properties and enhanced fire safety, likely due to lower resin content in the outer layer, which is the most flammable component. Moreover, a significant finding related to ignition time was that laminates prepared by the hand lay-up process ignited quicker compared to those made by other methods, suggesting that the manufacturing process influences the ignition behavior of the composites. The longest ignition time was recorded for the V-AGCBF composite, which means

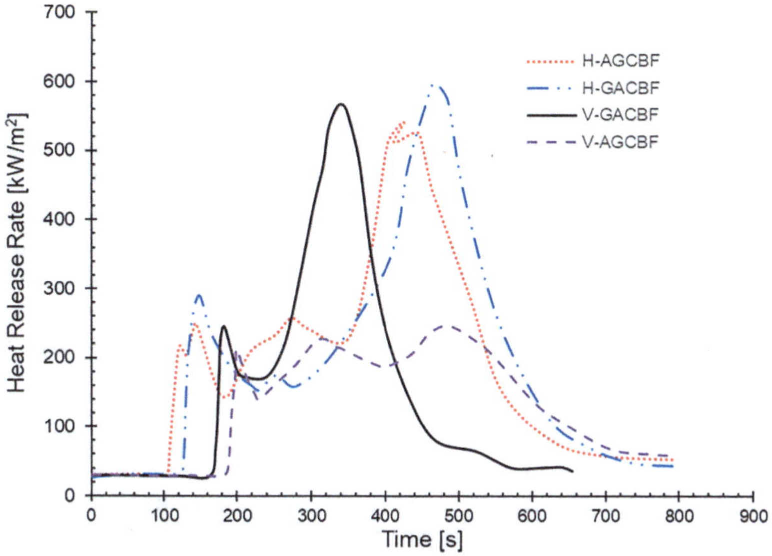

Fig. 6 Heat release rate of the epoxy-based composites [117]. Some parts of the graph are reproduced with permission from Sałasińska et al. from their article—"The Effect of Manufacture Process on Mechanical Properties and Burning Behavior of Epoxy-Based Hybrid Composites" referenced in [117]

this material was slower to ignite. This is an indicator of better fire resistance, especially considering that the V-AGCBF composite also exhibited a low peak HRR, aligning with a tendency to resist fire spread and intensity.

Moreover, the fiber content in composite materials can significantly influence the amount of heat they release when on fire. Research by Kozłowski et al. [63] and Mouritz et al. [116] has demonstrated that a higher concentration of fibers can decrease the heat emanating from a combusting composite. For instance, a glass/polyester composite depicted a marked decrease in the peak heat release rate as fiber content increased, owing to the reduced presence of the polymer matrix contributing to the heat during its thermal breakdown. Furthermore, the particular type of fabric used for reinforcement has been observed to impact the heat release characteristics of composites. In a separate study, Mouritz et al. [116] explored the relationship between heat release rate and different other fire reaction properties of fiber-reinforced polymer composites. The study used the oxygen consumption cone calorimeter to measure HRR and other fire reaction properties, including time-to-ignition, smoke density, carbon monoxide yield, carbon dioxide yield, mass loss rate, and total mass loss, at different heat fluxes. They found that for composites reinforced with non-combustible fibers, there is a linear relationship between HRR and the measured fire reaction properties. However, when composites are reinforced with combustible fibers, the HRR is only related to carbon monoxide yield, mass loss rate, and smoke density. The study underscores the importance of HRR in determining the fire hazard of materials; it is directly linked to the fire's temperature and spread and influences other critical factors like smoke density and carbon monoxide production, which are crucial for human survivability in a fire. The composition of the composites, including the type of fibers used, plays a significant role in the HRR and thus affects the overall fire behavior of the material.

Moreover, the work by Ghazzawi et al. [115] investigates how fiber length and fiber type impact the fire performance of a polymeric composite with a moderately high char yield, specifically focusing on Polycarbonate. The authors explore the behaviors of glass, basalt, and carbon fiber-reinforced PC composites with fibers of varying lengths (6, 12, 20 mm), created through compression molding using pre-compounded PC pellets. The fire properties of the PC composites were tested using cone calorimetry with various heat fluxes and additional tests like flame spread tests and thermal conductivity measurements. The study examines the in-plane and out-of-plane thermal conductivity over a range of temperatures and measures the flame spread rate and critical heat flux necessary for flame spread. The authors found that fiber length plays a significant role in the inherent fire performance when high thermal conductivity fibers are used. The results also indicate that the effect of fiber length can vary under different conditions and that longer fibers can sometimes negatively impact fire performance. Moreover, Dutta et al. [112, 113] investigated how different orientations of flax fiber within polypropylene composites affect their fire reaction properties. The study involved both experimental and numerical analysis under varying heat flux conditions using a cone calorimeter, which measures the response of materials to controlled fire conditions. The authors examined a range of fire reaction properties, including ignition time, ignition temperature, burn time, heat release rate, and critical heat flux. The study demonstrates that the orientation of flax fibers within the composite significantly influences the fire behavior of the material. Different sample angles were tested to observe how orientation impacts the heating process and subsequent ignition and combustion. The authors found that orientation can affect HRR by altering the convective and radiation heat transfer to the material.

Natural fibers are primarily composed of cellulose, hemicelluloses, and lignin, where the cellulose and hemicellulose endow the material with inherent flammability, making fire safety a critical concern [15]. As their thermal properties and combustion behavior form the key factors in their applications, especially in industries with stringent fire safety requirements, extensive research has been directed toward enhancing flame retardancy. Approaches to improving this characteristic include modifications to the polymer matrix, surface functionalization of the fiber, or the application of fire-protective coatings. Through such methods, the rate of heat release can be reduced, the ignition time can be delayed, or the generation of smoke and toxic gases can be minimized. There are two primary types of flame retardants used within these composites: additive and reactive. Additive flame retardants are mixed physically during the manufacturing process, while reactive types chemically bond with the polymer structure, creating intrinsic fire resistance. For natural fiber composites, various reports have highlighted the effective roles of these retardants, including mineral fillers and intumescent systems. The latter are particularly notable for their ability to form an insulating char layer when exposed to high temperatures, shielding the material from continued combustion. Yet, despite the advancements in flame-retardant technologies, achieving the desired level of fire performance for natural fiber composites remains challenging. Of particular note are the vertical burn experiments, which reveal that many composites fail to withstand a mere 10 s of direct flame application, signifying the pressing need for continued innovation in this field. The interplay between the components of natural fiber composites—fibers and polymer matrix—affects the overall fire behavior, thus complicating the development of flame-retardant solutions. By dissecting the interactions and effects of various flame-retardant strategies, ongoing research aims to surmount these challenges and pave the way for safer and more reliable use of natural fiber composites in fire-sensitive applications.

Furthermore, natural fibers, such as kenaf and wool, possess inherent fire reaction properties that can contribute to the flammability performance of hybrid composites [21]. Kenaf fibers, for example, have a high cellulose content, which can increase flammability due to the presence of levoglucosan. However, their high lignin content can lead to char formation after initial ignition, providing a thermal barrier. Xiang Ao and colleagues [99] have discussed in their extensive review that the inherent flammability of natural fibers, due to their organic composition, poses significant challenges in fire safety for natural fiber/polymer composites. They emphasized that natural fibers in composites are generally considered to be sources of fuel for fire due to their organic composition, mainly cellulose, hemicelluloses, and lignin. This makes the fire safety of natural fiber/polymer composites an important issue. Moreover, the flammability of natural fibers is influenced by their cellulose and lignin content. High levels of cellulose can increase flammability, whereas lignin content can lead to char formation, which provides a thermal barrier during combustion. To enhance the fire safety of NFCs, approaches include modifying the polymer matrix, functionalizing the fiber surface, or applying fire-protective surface finishes. These approaches can reduce flammability and improve fire resistance [99]. Furthermore, the work by Bhattacharyya et al. [21] emphasizes that the cellulose content in natural fibers can increase flammability, whereas lignin contributes to char formation offering a thermal barrier. These studies underscore the critical need for effective flame retardancy strategies in the use of natural fiber composites. In another study conducted by Kim et al. [15], the authors discussed that flame retardants can be additive or reactive. Additives are incorporated physically during manufacturing, while reactive types chemically modify the polymer structure. Mineral fillers and hybrids are commonly used as additive flame retardants. Moreover, intumescent flame-retardant systems have been developed as a more effective alternative to mineral-based additives. They form a char layer when exposed to fire, insulating, and protecting the underlying composite material. The research by Khuntia and Biswas [122] examines the flammability, thermal stability, and mechanical properties of coir fiber-reinforced polypropylene composites, emphasizing the potential benefits of using eco-friendly materials in composite fabrication. Using a combination of techniques such as UL-94 flammability tests, thermogravimetric analysis, and Fourier Transform Infrared Spectroscopy, the study explores how the treated coir fibers can enhance the fire resistance and mechanical strength of polymer composites. The findings suggest that incorporating coir fibers into polypropylene can improve its flammability resistance, which carries significant implications for developing safer and more sustainable materials in various industries.

The ignition behavior of hybrid composites reinforced with natural fibers plays a pivotal role in determining their overall flammability performance. In contrast to conventional composites, hybrid composites offer a unique solution by integrating the inherent fire-resistant properties of natural fibers with the superior performance of synthetic resins. This combination results in a material that exhibits promising ignition behavior under varying fire conditions. When natural fibers, such as kenaf and wool, are incorporated into hybrid composites, their inherent fire reaction properties significantly influence the ignition behavior of the resulting material. For instance, kenaf fibers, with their high cellulose content, can contribute to increased flammability due to the presence of levoglucosan. However, the high lignin content in these fibers can lead to char formation upon ignition, providing a crucial thermal barrier. Understanding and manipulating the ignition behavior of natural fibers in hybrid composites is essential for optimizing the overall flammability performance of these materials. Researchers and industry professionals have been exploring innovative approaches to enhance the ignition behavior of hybrid composites, focusing on modifying the polymer matrix, functionalizing the fiber surface, and applying advanced fire-protective coatings. These strategies aim to not only mitigate the flammability risks associated with natural fibers but also improve the ignition characteristics of the hybrid composites, thereby addressing the pressing need for effective flame retardancy in the use of natural fiber composites.

Understanding the ignition behavior of natural fibers in hybrid composites is crucial for developing effective flame retardancy strategies. The interplay between the inherent fire reaction properties of natural fibers and the synthetic resins in hybrid composites determines their ignition behavior under different fire conditions [107]. By comprehensively studying and manipulating the ignition behavior, researchers and industry professionals can optimize the overall flammability performance of these materials. Modifying the polymer matrix, functionalizing the fiber surface, and applying advanced fire-protective coatings are innovative approaches that aim to enhance the ignition behavior of hybrid composites. These strategies not only help mitigate the flammability risks associated with natural fibers but also improve the ignition characteristics of the composites. This is essential for addressing the pressing need for effective flame retardancy in the use of natural fiber composites in various industries [109]. The ongoing research in this field is focused on surmounting the challenges associated with the ignition behavior of natural fibers and developing safer and more reliable materials for fire-sensitive applications..

Smoke production involves assessing the amount, composition, and behavior of smoke emitted during combustion. Measuring the smoke production includes evaluating parameters such as smoke density, smoke toxicity, and the mass of smoke released during combustion. kim et al. [114] investigated the impact of horizontal versus vertical orientations on the fire behavior of flax fiber-reinforced composites. The main aim was to understand how orientation affects fire reaction properties such as ignition timing, heat release rate, and smoke production. In relation to smoke and heat production, the study found that the orientation of the flax fiber-reinforced composites significantly influenced the fire properties. For example, composites in a vertical orientation showed delayed ignition and produced less heat and smoke compared to horizontally oriented composites (see Fig. 7). This finding is primarily due to a thinner pyrolysis layer and a shorter burn time in vertical alignment. Moreover, while flame retardants did reduce the composites' flammability, the vertical orientation and resulting irregular char promoted greater heat and smoke production than expected. The research revealed that the way flax fiber-reinforced composites are oriented can significantly alter their fire behavior in terms of heat and smoke production, which are critical factors for fire safety.

In a study by Mazur et al. [111], the author focused on examining the mechanical properties, as well as the fire and smoke behavior of the resulting composites with varying fiber content. As the fiber content increases (5/5, 7/7, and 10/10 wt.%), the paper indicates that the addition of fibers to the polymer matrix leads to a decrease in the rate of heat emission during a fire. For composites with both 10 wt.% basalt and 10 wt.% carbon fibers, the maximum average rate of heat emission decreased by approximately 37% compared to the unmodified polymer. This suggests better fire resistance as compared to neat polyamide 6. The study also reported that the presence of fibers within the polyamide 6 matrix could potentially contribute to a reduction in smoke generation. This is because the fiber additives enhance the composite's thermal stability and possibly promote char formation, both of which are mechanisms that can reduce visible smoke and combustion products during burning. The research also suggested that incorporating small amounts of both basalt fibers and carbon fibers significantly enhances the strength and elasticity of polyamide 6

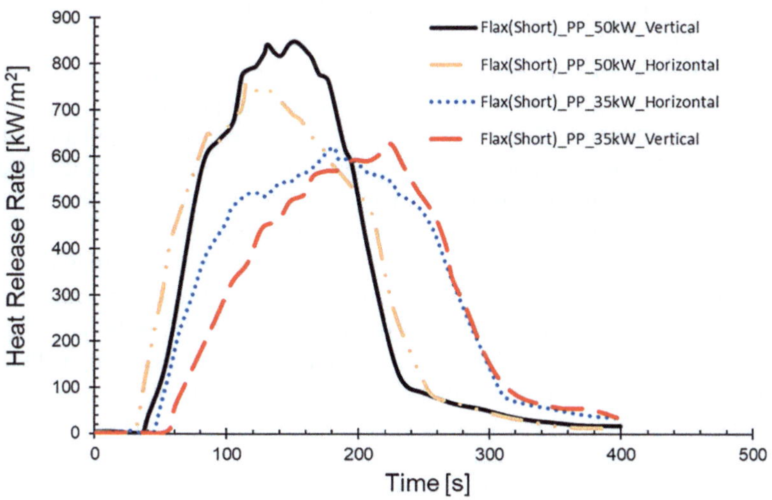

Fig. 7 Heat release rate curves of short flax-PP composites under horizontal and vertical orientations. Some parts of the graph are reproduced with permission from Kim et al. from their article—"Heat and smoke production of flax fiber-reinforced composites under horizontal and vertical orientations" referenced in [114]

composites. The most effective results were achieved with a balanced addition of 7 wt.% of each type of fiber, leveraging their synergistic effects: CFs with their fine diameters improve micro-crystalline surface connections, while the larger and longer BFs bridge more separated areas. This leads to notably better flexural strength for composites with a total fiber content of 20 wt.% compared to pure PA6 samples. Additionally, integrating a mix of BFs and CFs, which do not absorb water, greatly mitigates water absorption's impact on the composite's mechanical features. This is because the fibers are well integrated into the polymer matrix, reducing water penetration at the fiber boundaries. This resistance to water sorption can potentially be enhanced by using adhesion promoters or modifying the BF surfaces. The study also showed that even a modest amount of CFs and BFs could significantly lower heat release and gas emissions during combustion.

4.2 Fire-Retardant Fillers

The demand for fire-retardant materials has grown significantly, especially in industries that require high safety standards. The need for developing effective flame retardancy strategies for natural fiber composites has led to extensive research in the field of fire-retardant fillers for hybrid NFRCs. The incorporation of fire-retardant fillers in hybrid natural fiber-reinforced composites has garnered attention as a viable approach to enhance their flammability performance. Fire-retardant fillers, when integrated into the composite matrix, contribute to improving the ignition behavior and overall fire resistance of the material. These fillers play a crucial role in mitigating the flammability risks associated with natural fibers and synthetic resins, thereby addressing the pressing need for effective flame retardancy in various industries. Fire-retardant fillers are specialized materials added to polymers to enhance their resistance to combustion. These fillers can be generally categorized based on their chemistry and function. Some common types include metal hydroxides, hydrates, and carbonates, which are valued for their endothermic decomposition that cools the material and dilutes flammable gases. Active inorganic fillers like aluminum trihydroxide are often used due to their thermal insulating properties and smoke suppression effects. Other fillers include phosphorus and halogen additives, each with unique mechanisms for reducing polymer flammability. Phosphorus compounds can promote char formation, while halogenated fillers act by interfering with the free radical propagation during the combustion process. Various types of fire-retardant categories and their characteristics are given in Table 3.

Inorganic non-reactive substances, known as fillers, are incorporated into polymers during the later stages of their manufacturing to enhance the fire resistance of the end product. Their affordability, straightforward integration process, and effectiveness in increasing fire resistance make them a pragmatic choice in materials engineering. Fillers generally are not utilized in isolation but are combined with other fire retardants, such as compounds containing halogens or phosphorus, to maximize resistance to fire.

Table 3 Various fire retardants

Flame retardants (FR)	Additives	Merits	Demerits	Polymer
Halogenated (organ bromine, organochlorine)	Brominated—FR [99]	Dissolves in polymers Has a tendency to reduce the glass transition temperature (Tg) Provides ignition resistance and exhibits self-extinguishing properties	Significant deterioration in mechanical properties Emits large amounts of smoke and carbon monoxide (CO) Generates corrosive gases	Epoxy VE
Phosphorus (single molecules)	Intumescent additives [108]	Depending on the polymer's chemical structure, it will act as an active flame retardant in both the vapor phase and condensed phase	Certain phosphorus-based compounds act as plasticizers for the matrix	PP PA PC PVC
	DOPO [63, 108]	Combination of vapor-phase and condensed-phase activity. Excellent hydrolytic stability	Challenging to process Increased heat results in a reduction of mechanical properties Higher vapor-phase activity leads to greater smoke and corrosive gas emissions Difficult to synthesize Costly	Epoxy
Mineral fillers	Metal hydroxides, carbonates [15, 63]	Capable of meeting most fire safety standards Non-flammable gases help dilute the total fuel during combustion Produces low levels of smoke Highly effective in fire resistance	Smoke production and fire performance Has notable drawbacks Restricted effectiveness in fire situations Provides effective resistance in the initial stages of a fire but is less effective later on Not suitable as a fire retardant	Epoxy

(continued)

Table 3 (continued)

Flame retardants (FR)	Additives	Merits	Demerits	Polymer
Intumescent flame retardants	Carbon-based char, Pentaerythritol [63, 109]	Safeguards areas exposed to flames Provides strong ignition resistance and self-extinguishing properties Reduces heat release Ensures effective time-to-temperature performance	Cannot be mixed directly Can only be applied as a surface treatment Absorbs water, raising concerns about the coating's durability Water absorption leads to corrosion and weight-related problems	PP PA PS
Polymer nanocomposites	[103, 116]	Reduces the rate of heat release Imparts anti-drip properties to thermoplastics	Some have worse ignition resistance Polymer viscosity causes challenges in fabrication Chances for voids and dry spots are usual	Epoxy VE PE Ph

Note PP Polypropylene, *PA* Polyamide, *PS* Polystyrene, *PE* Polyester, *VE* Vinyl Esters, *PVC* Polyvinyl chloride, *Ph* Phenolics, *PC* Polycarbonate

The mechanisms by which fillers contribute to flame retardancy in the condensed phase include:

1. Diluting combustible components within the material.
2. Acting as a heat sink to reduce the composite's temperature.
3. Undergoing endothermic decomposition, which produces water or other non-flammable substances with high heat capacities, effectively acting as a cooling agent.
4. Encouraging endothermic breakdown processes which lessen the heat release.
5. Promoting higher aromatic content during breakdown, leading to the formation of a carbon-rich char layer that insulates the composite from heat and curtails the escape of flammable gases.

The interaction between the filler and the polymer is essential; a mismatch can result in significant losses in the material's mechanical strength and durability. Fillers can also affect the composite's physical properties by increasing its viscosity and affecting the setting time of the polymer, therefore, complicating the processing. Over time, exposure to moisture can lead to the fillers degrading through hydrolysis, reducing their effectiveness as flame retardants. Aluminum trihydroxide is a predominant fire-retardant filler utilized in polymers and their composites. It is effective in both the solid and vapor phases of burning, and when used in significant quantities, it is particularly successful in reducing flames mainly by decomposing and releasing water, and in some instances, smoke [123]. A notable advantage of this decomposition is the absence of harmful or corrosive emissions. However, the use of aluminum trihydroxide is restricted to polymers that do not require processing or curing at temperatures higher than its own decomposition point of 220 °C. Magnesium hydroxide operates in a similar manner to aluminum trihydroxide, with multiple fire retardancy mechanisms taking place during combustion. Like aluminum trihydroxide, magnesium-based compounds need to be used in substantial amounts—ranging from 30 to 60% by weight—to significantly enhance fire resistance. This high loading of fillers serves to dilute the polymer's combustible content, resulting in lowered flammability. On the other hand, the fire resistance mechanism associated with carbon nanofillers relies on forming protective layers made of inorganic and organic residues atop the burning composite. Fillers like expandable graphite swell in response to heat, forming an expansive insulating layer on the polymer composite's surface to prevent flames from spreading. For effective protection, the residue's structure and surface must be compact and uninterrupted. Thus, the efficient distribution of nanoparticles within the polymer is crucial for these fillers to enhance fire resistance effectively, as illustrated in Fig. 8.

In addition, compounds containing bromine or chlorine are known to be potent fire retardants, particularly during the gas phase portion of combustion. Their effectiveness arises from the breakdown of carbon-halogen bonds, which form part of their primary mechanism. These halogen-based materials interrupt combustion by disrupting the reactions in the gas phase that contribute to the fire's intensity. As a result, when these substances decompose under heat, they release halogen elements that quench the combustion of volatile organic compounds, thus reducing the overall temperature. However, the use of these halogenated elements in polymers and composites has triggered significant concern due to their potential to emit harmful smoke that includes toxic, acidic, and corrosive gases, creating severe risks to health and the environment. Prompted by these risks, environmental advocacy groups have been active in urging governmental bodies to either completely prohibit or place

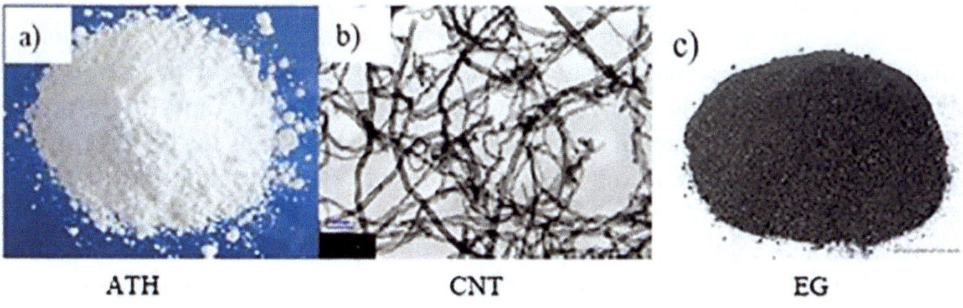

Fig. 8 Various types of fire-retardant fillers include: **a** Aluminum trihydroxide (ATH); **b** carbon nanotubes (CNT); **c** expandable graphite (EG) 124. Permission granted by Aruna Dilshantha Lelwela Subasinghe from his Ph.D. thesis referenced in [124]

stringent restrictions on halogen utilization. Also, phosphorus serves as an effective flame retardant by functioning in either the gas or condensed phases, influenced by the host polymer's composition and stability. In many thermoplastic and non-oxygen containing thermoset polymers, the gas phase mechanism is prevalent. This process involves the emission of phosphorus radicals at high temperatures, which requires volatilization to occur below 350–400 °C to be effective and prevent the polymer itself from breaking down. Additionally, a significant mechanism in the gas phase includes a covering effect over the polymer's heated surface. Conversely, when phosphorus compounds are integrated into polymers that contain oxygen or hydroxyl groups, their flame-retardant action is predominantly observed in the condensed phase. Here, phosphorus encourages the development of a protective layer of char, which limits the release of combustible volatile compounds during a fire. Furthermore, in certain thermoplastics, phosphorus can promote quicker heat dispersal by facilitating the melting and dripping processes [124].

Furthermore, the term "intumescence" originates from the Latin word "intumescere," meaning "to swell". Dramatist John Webster (1580–1624) applied the term in two senses: to signify growth and escalation in size in the presence of heat, and to describe the bubbling effect that indicates expansion. In practice, when a material is exposed to temperatures above a certain threshold, it begins to swell and then broaden out. This behavior results in the creation of a bubbly, expanded layer of char on the surface that serves as a barrier, shielding the material beneath from intense heat or direct flame contact. Polymer fire retardancy based on intumescence is classified within the condensed-phase mechanism. Intumescent systems interfere with the self-sustaining combustion of a polymer during the initial phase, when thermal degradation and the release of gaseous fuels begin. The intumescence response involves both charring and expansion at the polymer's surface as it burns. The foam-like carbonaceous char that forms becomes less dense as the temperature rises and acts as an insulating layer, offering protection to the material beneath from heat or flame [123], as depicted in the referenced Fig. 9. The formation of a char layer creates a protective shield that impedes the exchange of heat and mass between the gas and solid phases. Intumescent systems that perform efficiently not only enhance the fire-resistant properties of polymers but also influence factors such as the overall cost of material development.

Intumescent flame-retardant systems have proven to be an effective solution for enhancing the fire resistance of hybrid composites reinforced with natural fibers [21]. The ignition behavior of hybrid natural fiber-reinforced composites represents a complex interplay of material properties, influenced heavily by the hybridization of fibers. As outlined in the research by Xiang Ao et al. [99] hybrid natural fiber composites often exhibit varied ignition characteristics due to the combination of fibers with different thermal stabilities and combustion properties. For instance, the integration of natural fibers with higher lignin content can influence the time-to-ignition and subsequent burning rates, as lignin acts as a natural flame retardant. However, hybrid composites that combine natural fibers with synthetic counterparts can present a contrasting behavior, where the synthetic fibers may reduce the ignition resistance due to their thermoplastic nature. Crucially, the matrix used within the composite

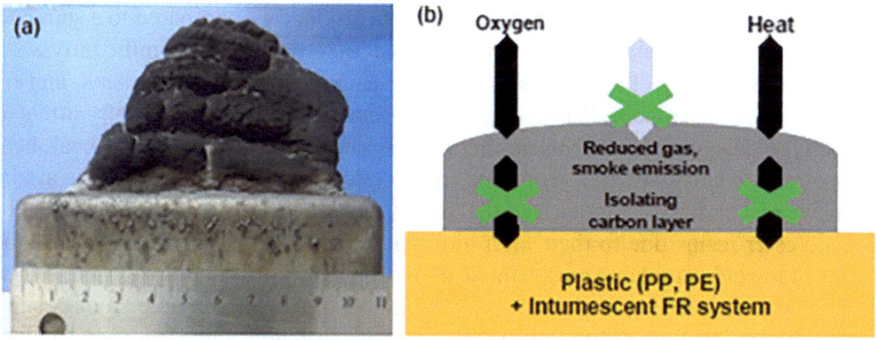

Fig. 9 **a** Char formed from PP-IFR following the cone calorimeter test and **b** a schematic illustration of the intumescent char generated from IFR and polymer. Permission granted by Swagata Dutta from her Ph.D. thesis referenced in [123]

also has major contributions, as thermosetting matrices may have different thermal degradation profiles compared to thermoplastics. Consequently, the overall ignition response of hybrid NF composites is a product of the synergistic effects of their constituents, and understanding this is essential for tailoring their fire performance to meet specific safety standards and regulatory requirements.

Prabhakar et al. [100] present a methodology to increase the flame retardancy and thermal properties of hybrid NF composites through exploitation of surface treatments and flame-retardant additives. The approach described in the article involves modifying flax fabric with ammonium polyphosphate to create flame-retardant fibers and additional flame retardants, including chitosan, ammonium polyphosphate (APP), and modified calcium sulfate (CS), are incorporated into the vinyl ester matrix. This combination not only enhances the thermal stability of the material but also improves its fire resistance, achieving a V-0 rank in the UL-94 test, and indicating self-extinguishing characteristics. The dual strategy of modifying the reinforcement fibers and integrating flame retardants into the matrix contributes to higher thermal resistance, lower peak heat release rate, and reduced total heat release and smoke production during combustion. Dreyer and Motoc [105] investigated the reaction-to-fire properties, specifically looking at a blend of cyanate ester/epoxy polymer with natural (flax) and synthetic (carbon or basalt) fibers. The study's focus was on assessing the intrinsic properties of the base polymer matrix, consisting of a blend of cyanate ester and epoxy resins, and their interaction with natural (flax) and synthetic (carbon and basalt) fiber reinforcements. The matrix was formulated without additional flame retardants to evaluate the inherent reaction-to-fire characteristics of the combination of resin system and fiber types. The epoxy resin selected was used because of its high thermoforming stability, and it was employed alongside flax fiber reinforcements for comparative purposes against the hybrid composites' behavior in fire conditions. The study used cone calorimetry to measure the fire response of these materials under an incident heat flux. Results showed that the cyanate ester/epoxy-based composites have better fire resistance—with lower rates of heat release, total heat emitted, and CO2 produced—compared to a standard epoxy benchmark. Synthetic fibers improved the fire growth index significantly, with approximately 70 and 103% improvement for carbon fiber reinforcements, and 63 and 99% for basalt fibers. While the ignition times were not substantially affected by the type of synthetic fiber used, there were notable differences in the peak heat release rates when switching from one resin system to another. In the same study, extensive research was summarized that highlighted the intrinsic flame resistance of cyanate ester resins due to their high nitrogen content and network structure. This results in lower heat release rates, smoke density, and production of toxic gases when compared to standard epoxy resins, which tend to have unacceptable smoke emissions or emit highly toxic and corrosive gases when halogenated epoxy resins are involved in a fire. The study incorporated carbon fiber and basalt fiber reinforcements and discussed their significant effects on fire performance and char residue formation. Although the structural optimization of these reinforcements was not the focus, further studies are suggested to optimize their layout to improve fire resistance further. It was also observed that during combustion, carbon fibers contribute to creating temperature-resistant boundary layers, enhancing fire resistance.

IFR systems, which prominently feature APP, leverage the combinatorial effects of phosphorus and nitrogen during ignition. This dynamic interaction releases nitrogen and ammonia, gases known to thin out combustible vapors. Concurrently, IFR leads to the creation of an expanding protective layer, depicted in one figure, acting as a reliable block against the flow of heat, combustibles, and gases from the source material to the flames, showcased in Fig. 6. Such IFR technology is increasingly adopted in enhancing the flame-retardant properties of composites infused with natural fibers. Through the integration process, IFR particles are mixed with other components. Significantly, the APP's phosphoric acid can transform the hydroxyl groups in natural fibers into a phosphorus ester linkage. This phosphorylation on cellulose hydroxyl groups can induce dehydration reactions, favoring the formation of char and water, and reducing the production of levoglucosan.

Utilizing diverse intumescent APP variants, a study by Subasinghe and Bhattacharyya [125] evaluated the fire performance of kenaf-polypropylene composite materials. Their investigation, which incorporated morphological assessments, determined that the use of a twin-screw extruder in the material synthesis process effectively dispersed both the kenaf fibers and the APP particles (see Fig. 10). This well-distributed IFR presence was linked to reductions in both prolonged burning and peak heat release rates, as indicated by a particular figure. Post-fire examination, employing the cone calorimeter test, revealed that the interplay between the fibers and IFR during combustion resulted in char formation, proving beneficial as an insulating barrier for the material beneath. Additionally, these composites exhibited enhancements in tensile and flexural rigidity, as depicted in another figure. Nonetheless, while evaluating these formulations using the vertical burning test, the composites demonstrated persistent flaming with material drips, earning them a classification of no rating, and they also experienced a decline in tensile strength.

The efficiency and use of these fillers depend on factors like cost, toxicity, color, and the effect they have on the polymers' physical properties [126]. Fire-retardant

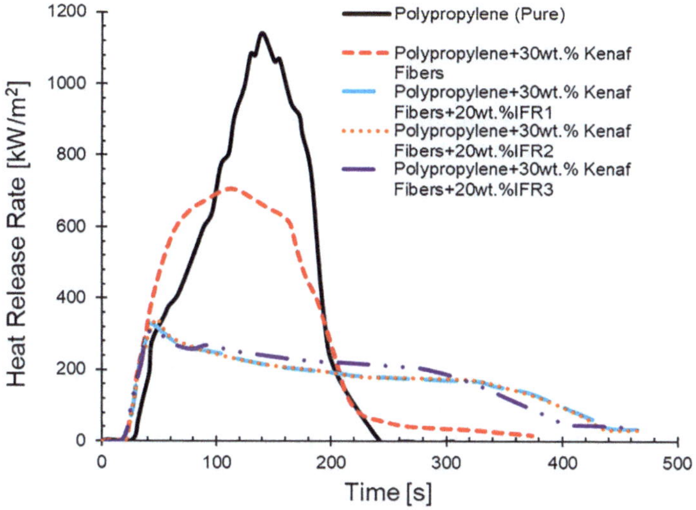

Fig. 10 Heat release rate curves for pure PP, kenaf-PP, and kenaf-PP-IFR composites [123]. Reproduced with permission from Swagata Dutta's Ph.D. thesis referenced in [123]

fillers offer several advantages when incorporated into hybrid natural fiber-reinforced composites [107]. They can significantly enhance the thermal stability, flame resistance, and overall safety profile of the composites. By effectively retarding the ignition and propagation of flames, these fillers contribute to creating a self-extinguishing and fire-resistant material, meeting stringent safety standards and regulatory requirements. Various types of fire-retardant fillers can be used in hybrid natural fiber-reinforced composites to enhance their fire retardancy. Some common examples of fire-retardant fillers include aluminum hydroxide, magnesium hydroxide, antimony trioxide, and organophosphorus compounds [127, 128]. These fillers operate through various mechanisms, including releasing water vapor, creating a protective char layer, or hindering the combustion process. Aluminum hydroxide and magnesium hydroxide are commonly used as fire-retardant fillers due to their ability to release water vapor when exposed to high temperatures. This released water vapor cools the composite material, delaying ignition and slowing fire spread. Furthermore, these hydroxides aid in forming a protective char layer during combustion, enhancing the composite's fire resistance. Antimony trioxide is another effective flame retardant that promotes char layer formation and acts synergistically with halogen-based flame retardants. It serves as a flame suppressant and can significantly lower the composite's flammability.

Organophosphorus compounds are known for their ability to prevent combustion by creating a protective and intumescent char layer when subjected to heat or flames. These compounds are essential for decreasing both the rate of heat release and smoke production during combustion, thereby enhancing the fire safety characteristics of the composite. Incorporating these fire-retardant fillers into hybrid natural fiber-reinforced composites offers a range of benefits, including improved fire resistance, reduced flammability, and compliance with industry-specific safety standards. The selection of the most suitable fire-retardant filler depends on the specific requirements of the application and the desired fire performance of the composite material. Moreover, fire-retardant fillers can significantly influence the combustion behavior of polymers by altering the polymer's thermal degradation and fire cycle in various ways. These fillers may interrupt the self-perpetuating combustion process by reducing the availability of fuel and reducing the rate at which oxygen and flammable pyrolysis products diffuse. They can also modify the melt flow properties of the polymer, influencing its tendency to drip. Fillers such as metal hydroxides, hydrates, and carbonates undergo endothermic breakdown, which cools the solid phase while releasing gases that reduce and cool flammable vapors in the vapor phase. Moreover, the inorganic residue left after the filler breaks down acts as a thermal insulator, protecting the underlying polymer from external heat and contributing to smoke reduction. Therefore, their presence in a polymer matrix is crucial for enhancing fire retardancy and reducing the overall risk of combustion. In the context of enhancing the fire-retardant properties of polymers, synergism refers to the combined effect of different additives leading to improved flame retardancy. Specifically, the article discusses how organic compounds, when used in conjunction with metal oxides as fire-retardant fillers, can significantly reduce the heat release rate and promote char formation during combustion. The synergy between these substances results in a composite material with superior fire resistance compared to what could be achieved by using either type of additive alone. For example, graphene oxide combined with titanium dioxide has been shown to not only enhance the fire resistance of a polymer but also improve its mechanical and thermal behavior when exposed to fire. This synergistic approach is a sophisticated method that allows the modification of existing polymer structures to achieve an exemplary balance between fire retardancy and mechanical performance [129].

Suriani et al. [130] investigated the potential of sugar palm fiber-reinforced with polyester yarn in an epoxy composite, enhanced with magnesium hydroxide to improve flame retardancy. It was found that the optimum content of SPF was 35%, which provided the lowest burning rate, thereby enhancing fire resistance, as well as the highest tensile strength, indicating improved mechanical stability. Conversely, a 50% SPF content resulted in reduced tensile strength due to weaker fiber-matrix bonding. These results are promising for the application of such biocomposites in industries that require materials with good mechanical properties and increased fire safety, such as automotive, construction, and aerospace. The use of SPF, an agricultural waste, aligns with the current trend toward sustainable and economical materials, presenting an eco-friendly alternative to synthetic, non-biodegradable fibers. In another study, Khalili et al. [131] found that the synergistic effect of combining fire retardants, specifically ammonium polyphosphate and alumina trihydrate, in natural fiber-reinforced epoxy composites, amplifies the material's resistance to fire. APP acts as an intumescent flame retardant, which upon heating, releases ammonia and water and forms a protective char barrier. This barrier helps in reducing combustion by limiting the transport of heat and fuel. On the other hand, ATH undergoes endothermic decomposition when exposed to fire, releasing water vapor which dilutes combustible gases and forms a thermally stable ceramic alumina layer that provides insulation. The combination of APP and ATH not only helps in reducing the flame time and preventing dripping but also enhances the char residue and self-extinguishing properties of the composite. This dual fire-retardant strategy leads to a fire-resistant material with improved thermal stability and a reduced spread of flames.

Data such as the duration of flame dripping, overall combustion time, and the percentage of weight loss were documented. The addition of alumina trihydrate to the epoxy composite reinforced with empty fruit bunch fiber did not enhance the composite's resistance to flammability compared to the control sample; this was evident in both the dripping and total flame durations. In fact, composites with higher ATH content showed pronounced burning with significant dripping. Nevertheless, higher amounts of ATH correlated with shorter overall flame durations but exhibited increased drip times. On the other hand, composites with increased levels of ammonium polyphosphate exhibited a slight reduction in total flame duration. Impressively, composites containing APP showed no dripping during burning, a stark contrast to other formulations. This improvement is attributed to how APP degrades when heated, forming a robust char layer that obstructs the transfer of heat and combusted material. This process is coupled with the emission of ammonia and steam, which serve to dilute flammable gases. Kim et al. [132] examined the impact of incorporating talc along with APP-based intumescent flame retardants in short flax fiber-reinforced polypropylene composites on their combustion and mechanical characteristics. Their results indicated that using 15 wt.% IFR on its own resulted in a 42% reduction in the peak heat release rate and an 8% decrease in the total heat release at an external heat flux of 50 kW/m^2. Adding 7 wt.% talc slightly improved the material's fire resistance, but the composite still failed the UL-94 vertical burning tests. Furthermore, although including IFR reduced the tensile and flexural strength, adding talc improved the mechanical properties compared to natural fiber composites without any flame-retardant treatment.

Hornsby [126] reported that certain fire-retardant fillers also have smoke suppression capabilities, which is an important aspect of safety during a fire. Smoke suppression is crucial because smoke often poses a greater risk to individuals during a fire than the flames themselves, due to its ability to obstruct visibility and its toxic nature. Fire-retardant fillers such as metal hydroxides can decompose endothermically, producing water vapor, and other gases that dilute flammable degradation products responsible for smoke. Additionally, the inorganic residue left after the decomposition of these fillers acts as a barrier, which not only insulates the underlying polymer from further heat exposure but also helps in reducing the volume and toxicity of the smoke generated. The choice of appropriate fillers contributes to minimizing smoke production, improving the chances of survivability and lessening property damage during a fire event. Numerous studies as reported by Hornsby [126] have shown that hydrated fillers effectively suppress smoke production in a range of polymers, such as ethylene-propylene-diene elastomers, polypropylene, polystyrene, modified polyphenylene oxide, polybutylene terephthalate, and ABS. These hydrated fillers not only decrease the total amount of smoke emitted but also can postpone the initial generation of smoke. This could provide critical additional moments for individuals to evacuate an area affected by fire, as illustrated in Fig. 11.

In addition, nanofillers in polymer composites are tiny particles that, when incorporated added to the polymer matrix, can substantially alter the characteristics of the base material. The addition of these nanofillers can lead to enhancements in mechanical strength, thermal stability, and fire resistance, among other traits. In the context of fire performance, specific types of nanofillers, such as graphitic carbon and inorganic nanoparticles, have been shown to enhance the polymer's resistance to ignition and combustion. When these nanofillers are incorporated into polymers, they can help in forming char layers that act as barriers to heat and mass transfer in a fire, thereby slowing down the spread of flames and decreasing the volume of smoke generated. Some examples mentioned in the paper are single-walled and multi-walled nanotubes, which can form compact network-like layers protecting the material by blocking heat and radiation from the flames [133–135]. The interaction between nanoparticles and polymers increases the viscosity of the polymer melt and can improve its fire performance if appropriately dispersed. Graphene, a two-dimensional nanofiller, is particularly noted for its high thermal stability and its ability to create a barrier effect in polymer matrices. Its inclusion can decrease the rate of heat release and delay ignition times due to the formation of a dense, continuous char layer [136–138]. The dispersion quality of these nanofillers within the polymer matrix is a critical factor that determines their effectiveness in enhancing fire performance. Ideally, they should be evenly distributed without agglomerating to achieve the best possible improvement in properties.

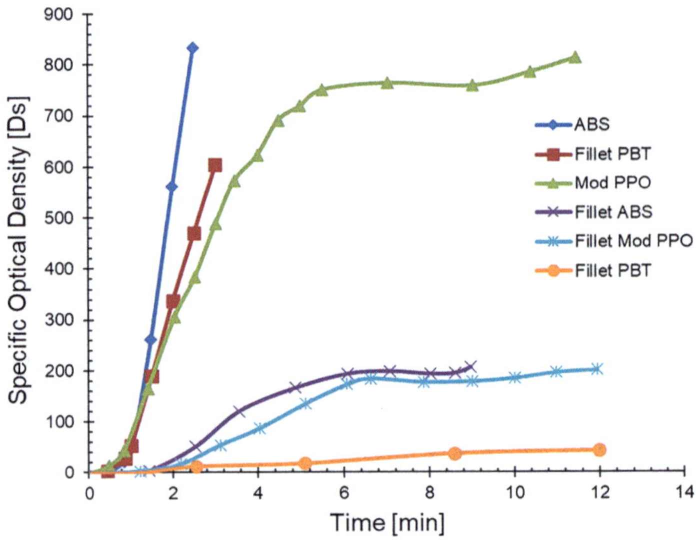

Fig. 11 ASTM E662 tests for smoke emission from selected polymers (with and without 40 wt. % MH filler) Reproduced with permission from P.R. Hornsby taken from his paper referenced in [126]

In a study by Kim et al. [139], the authors focused on how nanofillers can improve the fire safety of polymer composites. It outlines that appropriate nanofillers can significantly enhance the fire performance of polymers by elevating smoke suppression effects and contributing to better flame retardancy. Specifically, the paper discusses the role of one-dimensional nanofillers like single-walled nanotubes and multi-walled nanotubes, which form protective network-like layers that shield the polymer matrix from external heat and radiation during combustion. The effectiveness of nanofillers is largely dependent on their proper dispersion within the polymer matrix because this determines the nanocomposite's flammability characteristics. The paper also mentions two-dimensional graphene, which encourages the development of a thick char layer during decomposition, serving as a barrier to heat transfer and postponing the onset of ignition. Furthermore, the study includes the creation of novel flame-retardant systems by combining multiple types of flame retardants, which can produce synergistic effects and thus potentially offer improved fire safety solutions for polymers. Understanding these processes and researching new methodologies for flame retardants are vital for reducing the risk and damage associated with fires.

The synergistic effects of combining multiple flame-retardant fillers to improve the fire resistance of natural fiber-reinforced composites have garnered significant recent interest. Often, studies have focused on the use of ammonium polyphosphate alongside other flame-retardant agents to capitalize on this synergy. The literature outlines several reasons for employing this strategy. One main consideration is that single flame-retardant fillers typically need to be added in large quantities to be effective, but doing so can compromise the composite's structural integrity by reducing its mechanical properties. Additionally, there is the challenge of achieving good compatibility between the flame-retardant agents and the fiber-polymer matrix. Inadequate bonding between the fibers, matrix, and flame-retardant agents is known to negatively impact the mechanical strength of the composite. In certain instances, flame-retardant fillers are modified with reactive agents such as maleic anhydride, which interacts with the interface between the fiber and the polymer matrix. This modification process ensures that the integration of flame-retardant fillers does not detrimentally affect the composite's mechanical attributes. Alternatively, the inclusion of a different flame retardant with substantial mechanical, thermal, and chemical resilience can be considered for enhancing natural fiber hybrid composites.

It is important to recognize that the use of multiple flame-retardant fillers in a natural fiber and polymer matrix usually involves lesser quantities, preserving the composite's mechanical integrity. The intent here is to bolster the material's fire resistance without undermining its structural performance. Notably, the combination of flame retardants in natural fiber-reinforced composites is seen to reduce flammability more effectively than using a solitary flame retardant. Specifically, the simultaneous use of synthetic zeolites and ammonium polyphosphate (APP) has been studied for their impact on the flammability of wood flour/polypropylene composites [140]. APP is acknowledged for improving flame resistance in such composites, but at the cost of reduced mechanical properties due to its suboptimal compatibility with the composite matrix. To create a synergistic effect, zeolites, specifically types 4A and

13X, were utilized at varying concentrations (2, 4, 6, and 8 wt.%) in conjunction with APP. The reason for selecting zeolites is their high mechanical strength and their stability against heat and chemicals. Experiments, including limiting oxygen index and cone calorimetry tests, were conducted to assess the composites' fire performance. Results from these tests illustrated that the addition of zeolites in combination with APP provided a synergistic enhancement to the flame retardancy of the wood-plastic composites. The findings revealed that when zeolites were included with the ammonium polyphosphate in wood-plastic composites, there was an enhancement in the bending properties and resistance to impact, although tensile strength was found to diminish. The combination of zeolites (either type 4A or 13X) with APP in WPCs resulted in the formation of a dense and continuous char layer. In contrast, the presence of APP alone in WPCs produced a char layer that was less cohesive and fragmented. Consequently, the inclusion of zeolites with APP contributed to the creation of a superior, more protective char layer, shielding the material beneath from further thermal degradation. Such a robust and stable char acts as an efficient barrier, hindering heat penetration and the evolution of combustible volatiles, and thus achieves enhanced flame retardancy.

Additionally, the aspect of toxic gas emissions during combustion is a critical factor in assessing flame-retardant materials. In particular, carbon monoxide, a prevalent fire-emitted toxic gas, is considered a major contributor to fire-related fatalities. The application of APP in WPC was found to generate a higher yield of carbon monoxide, whereas integrating zeolites with the WPC/APP blend resulted in a lower CO yield, indicating that zeolites can indeed suppress the release of toxic gases during the combustion of WPCs. In a comparative analysis [141], the impacts of two disparate nanoparticles, namely halloysite nanotubes and montmorillonite nanoclay, on the flammability characteristics of a polypropylene/kenaf system using an intumescent APP formula were explored. Polypropylene on its own, without any inclusion of kenaf fiber and with the addition of ammonium polyphosphate as an intumescent flame retardant was found to combust swiftly, resulting in a substantial peak heat release rate (pk-HRR), as depicted in Fig. 9. The pk-HRR was somewhat diminished when kenaf was incorporated into the polymer matrix; however, this does not imply that kenaf serves as a flame retardant. Despite the reduction attributed to the presence of kenaf, the HRR of the PP/kenaf composite remained considerably high, with values ranging between 600 to 800 kW/m^2, underscoring the flammability of both PP and kenaf. The combination of kenaf, PP, and APP (termed KeC-IFR by the authors) led to a significant drop in pk-HRR by 67% compared to the unmodified PP matrix. This substantial reduction in pk-HRR is a result of both the transient nature of the lignocellulosic ash and the formation of more robust intumescent char. Additionally, a characteristic double peak was noted in the HRR curve (see Fig. 9) for the KeC-IFR system with APP. This second peak is often explained by the progressive breakdown of the protective char as the material continues to be subject to heat, followed by the establishment of a new protective char layer.

The incorporation of halloysite nanotubes and montmorillonite nanoclay contributed to improved resistance to flammability in the composite system, surpassing that of the APP-enhanced system (labeled as KeC-IFR), by diminishing the heat release rate as well as decreasing both the rate of smoke emission and the production of carbon monoxide, as illustrated in Fig. 9. The researchers emphasized that the efficient distribution of these particles was crucial for the development of a dense and resilient char layer, with the dispersion being more pronounced for MMT than for HNT. The enhancement in particle dispersion correlates with a more effective combative action during combustion, indicated by the lowered peak in HRR as seen in Fig. 12.

In hybrid nanocomposites, nanofillers are integrated into the polymer matrix, enhancing their mechanical properties. Introducing nanoscale additives such as oil palm nanofiller, montmorillonite, and organically modified montmorillonite at a 3% weight fraction into kenaf fiber-reinforced composites resulted in enhanced tensile strength, modulus, and impact strength compared to composites reinforced solely with kenaf fibers. This enhancement is due to the high surface area of the nanomaterials, which promotes a stronger bond between the kenaf fibers and the epoxy resin. However, it was noted that composites containing oil palm nanofiller exhibited lower tensile properties compared to those with montmorillonite and organically modified montmorillonite. In particular, composites enhanced with OMMT displayed notable tensile and impact strength due to the effective reinforcement from the organic modification. Additionally, findings from Arrakhiz et al. [142] indicated that combining pinecone fibers with clay increased material stiffness, and a higher clay content, up to 30%, enhanced stiffness without affecting tensile strength. Conversely, research by Piekarska et al. [143] and Essabir et al. [144] suggested that hybridization with fibers and nanomaterials could lead to a reduction in tensile properties. Moreover, Battegazzore et al. [145] reported that adding modified clay, like Cloisite 30B, to

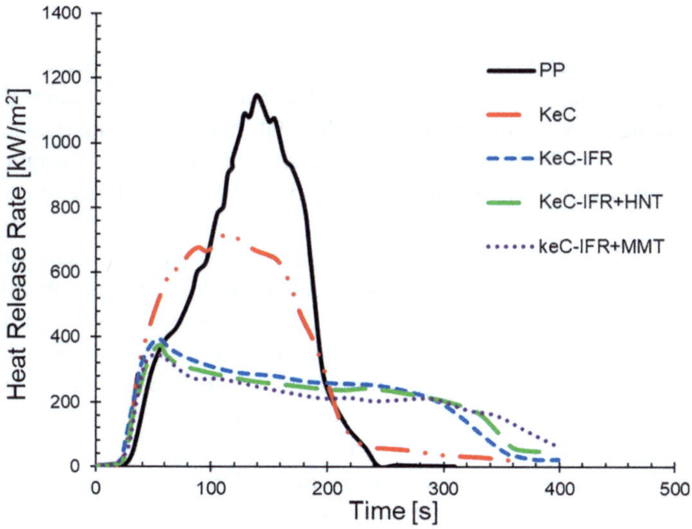

Fig. 12 Cone calorimeter test (HRR) of different materials [141]. Reproduced with permission from Aruna Dilshantha Lelwela Subasinghe taken from his paper referenced in [141]

rice husk ash in different polymers, increased their modulus but reduced elongation, noting a 94%, 85%, and 53% increase in modulus for polyamide and polylactic acid-based composites, respectively, with a 20% rice husk ash content. A general decrease in tensile strength was also seen, and upon applying Pukanszky's model [146, 147], there appeared to be a distinct interaction between the fillers and the matrix, with deviations in predicted behavior between 10 and 20 wt.% filler loading suggesting a potential compromise in matrix-filler bonding at higher concentrations of filler.

4.3 Testing Methods and Standards

To evaluate the flammability performance of hybrid natural fiber composites, various testing methods and standards are employed. These methods provide valuable insights into the ignition behavior, flame spread, smoke emission, and toxicity of the materials. The evaluation of the flammability of composite materials involves testing across various scales to ensure their safety and compliance with regulations. These tests are divided into scales, beginning from the material's flammability at a micro-scale, progressing to bench-scale testing, and then to large-scale testing. Each scale of testing provides valuable insights into the ignition behavior, flame spread, smoke emission, and toxicity of the materials. By systematically analyzing the behavior of the hybrid composites at different scales, researchers and industry professionals can gain a comprehensive understanding of their fire performance and make informed decisions regarding their application in critical sectors. This multistage approach is essential for tailoring the fire performance of natural fiber composites to meet specific safety standards and regulatory requirements. Also, the techniques of thermal degradation pathways can be classified into scientific techniques as well as commercial techniques for flammability evaluation. Micro-scale tests (scientific techniques) are crucial for analyzing the flammability and thermal behavior of materials. Scientific methods are employed to evaluate the thermal robustness of materials, assisting in the identification of substances released as degradation occurs. Insights gleaned from such analyses can provide an understanding of the underlying reasons for the degradation.

Thermogravimetric analysis (TGA) is another micro-scale test that evaluates the weight loss of a material as a function of temperature, offering valuable data on its decomposition and thermal stability [ASTM E1131]. As materials decompose, there is a measurable alteration in their mass. Thermogravimetric analysis can precisely quantify this mass change over temperature or time, as depicted in Fig. 13. Several key factors impact the reliability of the mass loss data collected during TGA, including the purge gas used, the starting mass of the sample, and the specific environment within the testing chamber. Gases that are inert—like helium, nitrogen, and argon—are typically used when assessing materials under conditions of thermal decomposition, known as pyrolysis. Conversely, oxidizing gases such as oxygen or air are utilized to gauge the thermo-oxidative decomposition of a material. The specimen's mass as it heats or cools is identified using a precision balance that holds the sample pan, coupled with a compensating signal. The choice of purge gas, which can be inert such as argon or nitrogen, or reactive like oxygen, helps maintain desired conditions within the sample space. Various elements play critical roles in securing the precision of both the instrument's function and the test results, including the specimen's preparation, its initial weight, the heating rate, the selected purge gas, as well as the pan and the thermal sensor involved. By subjecting the hybrid composites to a controlled temperature program while measuring their weight change, TGA provides valuable data on their thermal degradation kinetics, including onset temperature, degradation temperature range, and decomposition residues. This enables a comprehensive understanding of how the different constituents of hybrid composites contribute to their thermal stability and behavior under elevated temperatures, which is inherently linked to their flammability and fire resistance.

Differential Scanning Calorimetry (DSC) is commonly used to measure the heat flow in response to changes in the sample's temperature, providing insights into the material's thermal properties. Moreover, DSC and TGA are two fundamental techniques employed to quantitatively assess the thermal properties and behavior of hybrid natural fiber composites. DSC is used to measure the heat flow associated with thermal transitions in materials, offering insights into phenomena such as melting, crystallization, and glass transitions. In the context of hybrid composites, DSC can identify the specific temperature ranges and associated enthalpy changes related to the thermal degradation of natural and synthetic fibers, as well as the matrix material. This information is crucial for understanding the material's thermal stability and degradation mechanisms, which directly impact its fire performance. The combined use of DSC and TGA allows for a comprehensive assessment of the thermal properties and behavior of hybrid natural fiber composites, providing critical information for optimizing their fire performance through tailored material design, surface treatments, and flame-retardant strategies.

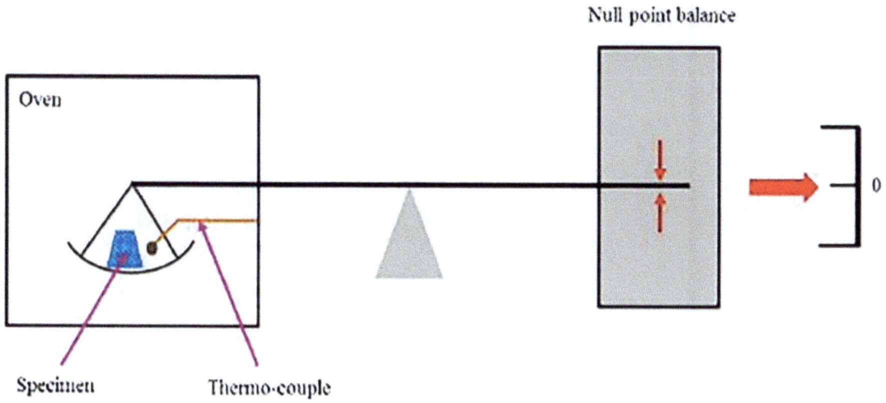

Fig. 13 Schematic diagram of a thermogravimetric analyzer [123]. Reproduced with permission from Swagata Dutta taken from his Ph.D thesis referenced in [123]

The UL-94 flammability test (see Fig. 14), which is widely used for evaluating the flammability of plastic materials, provides a classification based on the material's burning behavior. The UL-94 standard outlines a testing methodology to evaluate flammability, and it is organized into distinct procedures based on the orientation of the sample: either horizontal or vertical. Among these, the 20 mm vertical burn test, which includes classifications V-0, V-1, or V-2, is the most frequently applied. This test serves as a benchmark to ascertain whether plastic materials are suitable for incorporation into appliances or various types of equipment based on their burning characteristics under regulated lab conditions. Flammability classifications are derived from experimental procedures detailed in a specific set of experiments, as seen in Table 4. To achieve a rating, an average is taken from five samples that have been conditioned and tested as per the guidelines of the ASTM D 3801 standard. These ratings are designed to provide an initial evaluation of the materials' flammability and suitability for particular uses.

Hybrid natural fiber composites are also tested as per UL-94 testing specifications to evaluate their flammability and classification based on their burning behavior. The UL-94 flammability test provides valuable insights into the ignition behavior, flame spread, smoke emission, and toxicity of the materials. In the context of evaluating the flammability of hybrid natural fiber composites, the UL-94 test plays a crucial role in determining their fire performance and ensuring compliance with safety standards and regulatory requirements [148]. This test will provide essential data to assess

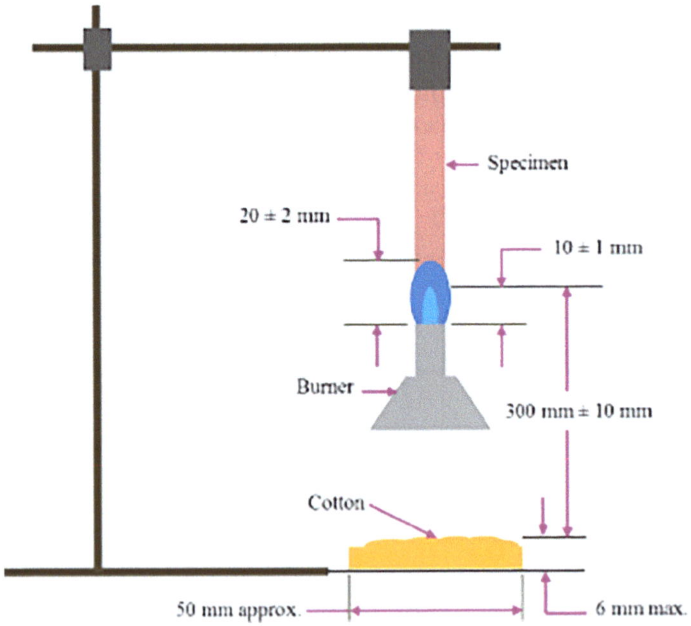

Fig. 14 Schematic diagram of a vertical burning setup—UL-94 test [123]. Reproduced with permission from Swagata Dutta taken from his Ph.D thesis referenced in [123]

Table 4 UL-94 test ratings [123]

Rating	Description
UL-94 V-0	The flame must extinguish within 10 s or less, with no glowing beyond 30 s, and no material should fall while burning
UL-94 V-1	The flame must extinguish within 30 s or less, with no glowing beyond 60 s, and no burning material should fall
UL-94 V-2	The flame must extinguish within 30 s or less, with no glowing beyond 60 s, but burning material is allowed to fall

the ignition characteristics and flame retardancy of the composites under specific fire conditions, contributing to the development of safer and more reliable materials for fire-sensitive applications in critical sectors. Furthermore, the results obtained from the UL-94 test will be instrumental in understanding the effectiveness of flame-retardant additives, surface treatments, and other modifications aimed at enhancing the fire resistance and thermal stability of hybrid natural fiber composites.

In addition, the CC test, described in the ASTM E1354 standard, is commonly used to evaluate the HRR, THR, SPR, production of toxic gases and other important fire behavior parameters of materials under controlled conditions [11]. This testing technique involves exposing the material sample to a controlled heat flux, typically in the form of a radiant heat source, and measuring various fire reaction properties. Cone calorimetry represents an advanced technique for fire testing, offering a scientific approach for plastic manufacturers to evaluate their materials and determine their appropriate uses. The test involves subjecting a 100 × 100 mm sample to a consistent heat flux while it is in a combustion-enhanced environment. If the sample's thickness is under 50 mm, extra insulating material is added to ensure the sample fits securely in its holder, facilitating uniform exposure of the specimen's surface to a controlled heat irradiance, with the potential to reach up to 100 kW/m^2 emitted by a conical heater. Initially, the surface temperature of the specimen climbs quickly and then stabilizes. The heat transferred from the specimen to its environment balances the heat applied to its surface. The test is expedient, cost-effective, and generates consistently reliable data. Nonetheless, its shortcomings include a disregard for the dynamic nature of fire growth. Additionally, its methodology often results in the complete consumption of the test sample, an outcome that does not always align with the variability of oxygen availability in actual fire circumstances, such as those in enclosed, unventilated spaces.

Another complication arises from the requirement for the samples to be flat sheets. While this is suitable for materials like cladding or panels, it does not apply to the wide array of composite products, which include variously shaped items, such as moldings, extruded profiles, and piping. The heat release data obtained from this test are crucial for analyzing the contributions of material combustibility to flame spread and heat release in a fire scenario. This test provides valuable data for understanding the fire performance of composite materials and assessing their suitability for specific applications where fire safety is a concern. It is essential for researchers, industry professionals, and regulatory bodies to adhere to these testing methods and standards to ensure that composite materials meet the necessary safety requirements. Additionally, the smoke production rate measured during cone calorimetry testing offers crucial information about the potential smoke hazards associated with the material's combustion, thus contributing to fire safety assessment [123]. Moreover, the cone calorimeter test can also reveal the effectiveness of flame-retardant additives or surface treatments in improving the fire performance of composite materials. By comparing the fire-related parameters of treated and untreated samples, researchers can evaluate the efficacy of various flame retardancy strategies and make informed decisions about the most suitable approaches for enhancing the material's resistance to ignition and fire spread. In the context of hybrid natural fiber composites, conducting cone calorimetry testing is essential for gaining a comprehensive understanding of the ignition behavior and overall flammability performance of these materials. It allows for the systematic evaluation of the interplay between different fiber types, resin matrices, and flame-retardant systems, aiding in the development of tailored fire safety solutions for specific applications in industries such as automotive, construction, and aerospace.

Moreover, micro-scale combustion calorimetry (MCC) is a sensitive technique used to measure the heat release and flammability properties of small material samples. This method is particularly valuable for assessing the fire performance of composite materials at a micro-scale level, providing detailed insights into their thermal behavior. Micro-scale combustion calorimetry [149–152] is a highly valuable technique for determining the effective heat of combustion of composite materials. By subjecting small material samples to controlled conditions and measuring their heat release characteristics, MCC provides detailed quantitative data on the flammability properties of the materials. This method allows for the precise evaluation of the heat release rate, peak heat release, and total heat release during combustion, enabling researchers and industry professionals to assess the materials' fire performance with high precision. Furthermore, MCC facilitates the assessment of the effectiveness of flame-retardant additives, surface treatments, and other modifications aimed at enhancing the fire resistance and thermal stability of composite materials. Through MCC analysis, the impact of these interventions on the heat release behavior of the materials can be quantitatively evaluated, providing essential information for optimizing fire safety strategies. Additionally, micro-combustion calorimetry is employed to study the flammability of small samples under controlled conditions, allowing researchers to assess their ignitability and burning behavior.

In addition to the aforementioned testing methods and standards, the limiting oxygen index (LOI) test [153, 154] is a crucial technique for assessing the flammability of composite materials. The LOI test measures the minimum concentration of oxygen required to sustain the combustion of a material, providing valuable insights into its fire resistance. By subjecting the sample to varying oxygen concentrations and observing its ignition and burning behavior, researchers can determine the material's susceptibility to ignition and flame spread under different oxygen conditions. For hybrid natural fiber composites, conducting LOI testing offers a comprehensive evaluation of their intrinsic flammability characteristics. The results obtained from the LOI test can aid in identifying the oxygen concentration at which the material becomes flammable, thus guiding the development of effective flame-retardant strategies to enhance its fire resistance. It is essential for researchers and industry professionals to include LOI testing in their comprehensive assessment of composite materials' fire performance. The insights gained from LOI testing can inform the design and formulation of hybrid composites with improved resistance to ignition and flame propagation, contributing to enhanced fire safety across various industrial applications.

Intermediate scale tests, such as the Single Burning Item test and the room corner test, are essential for simulating real-life fire scenarios and assessing the fire behavior of materials in larger settings. These tests offer insights into factors such as flame spread, heat release, and smoke production, contributing to a more thorough understanding of the materials' fire performance. The Euroclass system provides a framework for categorizing building materials based on their fire reaction behavior. Utilizing the Single Burning Item test, outlined in the EN 13823:2002 standard, the test mimics the scenario where a building product encounters thermal exposure from a solitary ignited item situated in a room's corner. The SBI procedure, born out of the European Construction Product Directive, is a modern methodology for assessing fire response. It is utilized for a vast range of construction materials, including composites, and facilitates the measurement of various characteristics in a single go: ignition timing, heat emission, smoke output, flame extension, and overall fire progress in true-to-life fire scenarios. The test calculates indices like FIGRA, which quantifies how rapidly a building product amplifies the heat of a fire, and SMOGRA, which pertains to the pace at which smoke is generated. Moreover, the process assesses incendiary particles and drips resulting from thermal degradation [155]. The SBI procedure is celebrated for being one of the most comprehensive fire test methods ever created.

Finally, large-scale tests, including the ISO 9705 Room Test and the NFPA 286 Standard Methods of Fire Tests for Evaluating Contribution of Wall and Ceiling Interior Finish to Room Fire Growth, are conducted to evaluate the fire behavior of materials in full-scale applications [156]. Large-scale fire experiments commonly employ the use of room corner tests globally as a primary method. There exists an array of established protocols for these room corner tests, which are incorporated into various standards and legal requirements to certify interior finishing materials. These tests simulate fire scenarios in actual environments, providing critical data on factors such as fire growth, smoke production, and toxicity. The spread of fire within a room can be influenced by numerous factors such as the contents of the room, the materials used in walls, ceilings, and floors, the layout and dimensions of the space, how well-ventilated the room is, whether there are automatic fire-extinguishing systems in place and the nature of potential ignition sources. Typically, the flammability of items within a room is not heavily restricted, with building regulations primarily concentrating on the structural design and the construction materials of the building itself. In the room corner test, which is quite sizable with dimensions of 2.4 m × 3.6 m × 2.4 m, the walls and ceiling are lined with the construction product under scrutiny, installed as it would be in a real setting, as shown in specific figure references. The material is ignited by placing a gas fuel source in a corner of the room, and the smoke and gases emitted are captured and analyzed to determine heat release rates among other metrics. The heat source intensity for the test starts at 100 kW for the initial ten minutes and is stepped up to 300 kW if no ignition has been observed. Based on the length of time to reach flashover as per the ISO 9705 standard, materials are categorized into five divisions ranging from Class A–E, with various peak heat release rate (pk-HRR) thresholds and timing criteria. Those materials that lead to flashover in under two minutes are typically not classified. There are also specific limits on average heat release and smoke production rates. Modern strides in flammability evaluation seek to bridge the gap between small-scale tests or material property assessments and full-scale fire behavior through tactics like using oxygen-enriched environments to better mimic large fire conditions or through the introduction of micro-calorimeters that assess heat release rates.

The flame spread test standard, often referred to as ASTM E84 or the Steiner tunnel test [157, 158], serves as an essential assessment for evaluating the surface burning properties of construction materials. This review of the ASTM Test for Surface Burning Characteristics of Building Materials (E 84) aims to serve as a foundational document regarding the 25-foot tunnel testing method and to facilitate its examination by ASTM Subcommittee E05.32 on Research. It outlines the historical context of ASTM E 84, detailing its evolution and recognition by regulatory bodies. The document describes the current testing methodology along with the operational and calculational procedures in use today. It also highlights the activities of the tunnel operator's task group, which is dedicated to the ongoing assessment of the method, conducting relevant research, and suggesting updates for potential adoption by ASTM Committee E-5 on Fire Standards. The review addresses how factors such as air velocity, burner gas flow, and other variables influence test outcomes and the consistency of results across different laboratories. Additionally, it presents the relationship between tunnel test results and those from room and corner tests. The current analytical status of the tunnel is discussed, along with efforts to quantify the total smoke produced during tests by enhancing the photometer system, mitigating stratification, and modifying calculation methods. The approach of using the oxygen consumption method to evaluate total heat contribution is explained, along with the significance of reporting both the rate of flame spread and the flame spread index (FSI). The advantages and disadvantages of these measures are examined. Finally, the document provides a list of facilities capable of conducting ASTM E 84 tests, along with contact information for relevant personnel, and includes a comprehensive bibliography related to the tunnel testing methodology. The ASTM Test for Surface Burning Characteristics of Building Materials (E 84) [2] is a crucial measure for assessing the surface burning characteristics of building materials. This test provides a flame spread index and a smoke-developed index, which are essential for understanding the fire performance of materials used in construction and building applications. By subjecting the material to controlled flame exposure and measuring its flame spread and smoke density, the flame spread test helps in evaluating the surface burning behavior and fire safety characteristics of composite materials. Adhering to the ASTM E84 standard is vital for ensuring the compliance of composite materials with regulatory requirements and industry safety standards. When assessing the fire performance of hybrid natural fiber composites, conducting the flame spread test according to ASTM E84 provides valuable data on the material's surface burning characteristics and its suitability for use in various applications. This test aids in the development of tailored fire safety solutions for industries such as automotive, construction, and aerospace, where surface burning behavior is a critical factor for ensuring fire safety and regulatory compliance. Additionally, the results of the flame spread test contribute to understanding the effectiveness of flame-retardant additives and surface treatments in improving the fire resistance of hybrid natural fiber composites, further guiding the development of enhanced fire safety strategies for these materials are designed to evaluate the surface burning characteristics of materials and assess their propensity to promote or resist flame spread. By subjecting the material to controlled ignition, this test helps in understanding its flame propagation behavior, which is critical for ensuring safety in various applications.

The radiant panel flame spread test, also known as the ASTM E162 test [159, 160], is a fundamental assessment for determining the surface burning characteristics of materials. This test method measures the flame spread index and smoke-developed index of the material, providing crucial data for evaluating its fire performance. By subjecting the material to controlled radiant heat exposure and analyzing its flame spread and smoke development, the radiant panel flame spread test aids in understanding the surface burning behavior and fire safety characteristics of composite materials. Conducting the radiant panel flame spread [161] test according to ASTM E162 is essential for ensuring the compliance of composite materials with regulatory requirements and industry safety standards. In the context of hybrid natural fiber composites, performing this test provides valuable insights into the material's propensity to promote or resist flame spread, enabling the development of tailored fire safety solutions for industries such as automotive, construction, and aerospace. Furthermore, the results of the radiant panel flame spread test contribute to evaluating the effectiveness of flame-retardant additives and surface treatments in enhancing the fire resistance of hybrid natural fiber composites. This data guides the development of enhanced fire safety strategies for these materials, ensuring their suitability for specific applications where flame spread behavior is a critical factor for regulatory compliance and ensuring fire safety.

Furthermore, assessing the smoke emission characteristics of composite materials is essential for understanding their potential impact on visibility and overall fire hazard. The smoke density test measures the darkness and density of smoke generated during material combustion, providing insights into its smoke production behavior. Toxicity assessment tests are vital for evaluating the potential release of toxic gases and compounds during the combustion of composite materials. These tests help in understanding the health hazards associated with the materials' decomposition products, guiding the development of safer and lower-toxicity materials.

Dreyer and Motoc [105] used testing methods focused on the fire behavior of CE/epoxy blend-based CF/FF and BF/FF hybrid composites materials were conducted using cone calorimetry, as outlined by the ISO 5660-1:2015 standard. This process involved exposing hybrid composite samples, composed of different blends of natural and synthetic fibers infused with a cyanate ester/epoxy polymer mixture, to an incident heat flux of 50 kW/m^2 to replicate a fully developed fire scenario. Critical fire risk factors, including heat release rate, total heat output, and ignition time, were evaluated alongside fire hazards such as fire load, flame spread, smoke, and CO production. The calorimeter also provided data on the peak of heat release rate and fire growth rate. These experiments were done in replicates for statistical reliability and to obtain a comprehensive understanding of the fire performance of these advanced material systems. The char formations, as well as the protective performance of the synthetic outermost layers of carbon or basalt fibers, were also analyzed during the exposure to the radiant heat.

5 Fire Resistance Properties of Hybrid NFRCs

Fire resistance is a critical attribute of materials and structures, determining their capability to inhibit fire spread while maintaining their structural integrity. This characteristic is gauged by two key factors: the prevention of heat transmission and resistance to burn-through. Another significant factor is the preservation of mechanical properties when the material is subject to fire. To evaluate the fire resistance of composite materials, several experimental methodologies are employed. The furnace test is a commonly used procedure to determine resistance to heat transmission. During this test, a composite panel is subjected to the intensity of a furnace fire over a prolonged duration. The progression of heat through the panel is tracked with thermocouples placed within the material. Composites that slow down the heat increase are deemed to have better fire resistance qualities. The furnace test is typically conducted under cellulosic fire conditions, with a peak temperature of approximately 850 °C for a standard room fire or hydrocarbon fire, and about 1100 °C for a more intense fuel fire. Research has extensively documented the fire resistance of laminate and sandwich composites under these conditions. An illustration provided in Fig. 15 displays the temperature progression on the rear side of both glass/polyester and glass/phenolic laminate samples as they face prolonged exposure to a cellulosic fire environment, showing a gradual temperature rise on the unexposed side over time.

Figure 15 indicates that initially, the phenolic laminate's back-face temperature escalates less sharply than that of the polyester composite, courtesy of the phenolic matrix's enhanced resistance to burning. Nonetheless, beyond the first forty minutes, the phenolic laminate's temperature spikes dramatically, a change attributed to the violent delamination damage resulting from the internal pressures formed by the water vapor encapsulated within the phenolic matrix transforming to steam. Conventionally, the interval needed for the reverse side temperature of a flammable material to hit 160 °C serves as an index of its fire resistance, marking the potential of fire spread to adjacent areas. In particular, the fire resistance of various laminates under hydrocarbon fire stress is shown in Fig. 16. The correlation between increased laminate thickness and the delayed attainment of 160 °C reflects the material's ability to withstand extensive temperatures for a substantial time. This durability is largely due

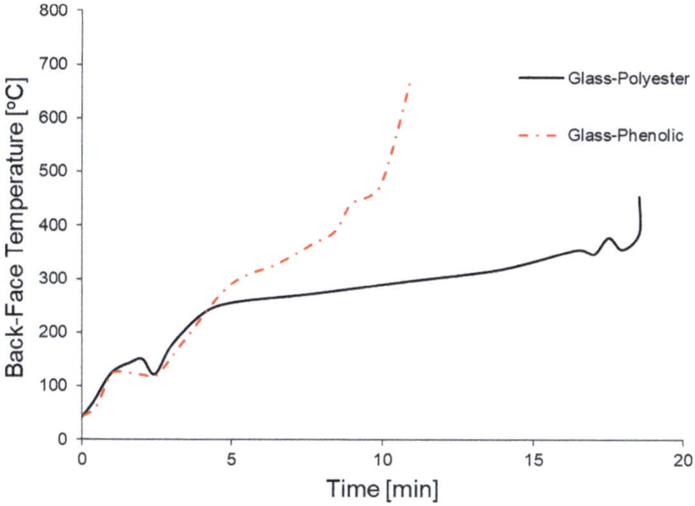

Fig. 15 Back-face temperature profile of different composites [123]. Reproduced with permission taken from Swagata Dutta taken from his Ph.D thesis referenced in [123]

to the material's low thermal conductance combined with the endothermic character of the resin's breakdown reaction, leading to the deduced inference that phenolic laminates exhibit superior fire resistance traits [22]. The inherent property of generating a robust char layer is a critical aspect of this material's performance, serving as a barrier to swift heat passage. In addition to this, another essential attribute of fire resistance is the material's ability to withstand penetration by fire or directly applied flame. This characteristic, known as burn-through resistance, is determined by the duration a composite withstands a jet flame before being breached. While the test employing a flame jet is one method, the furnace test is also commonly used to assess the resilience of composite materials against burn-through.

Fire resistance properties are crucial for assessing the safety and performance of hybrid natural fiber composites in various applications. To further explore this topic and gain deeper insights, it is essential to discover new references that offer comprehensive discussions on the fire resistance properties of these materials. This will provide an opportunity to broaden the understanding of the latest advancements, testing methodologies, and flame-retardant strategies aimed at enhancing the fire performance of hybrid natural fiber composites. When evaluating the fire performance of hybrid natural fiber composites, it is important to distinguish between their fire reaction and fire resistance properties. As mentioned before, fire reaction properties pertain to the material's behavior when exposed to fire, including its ignition, flame spread, and smoke emission characteristics. On the other hand, fire resistance properties focus on the material's ability to withstand fire and prevent its propagation, often influenced by flame-retardant additives and surface treatments.

Understanding the distinction between fire reaction and fire resistance properties is essential for developing comprehensive fire safety strategies for hybrid natural

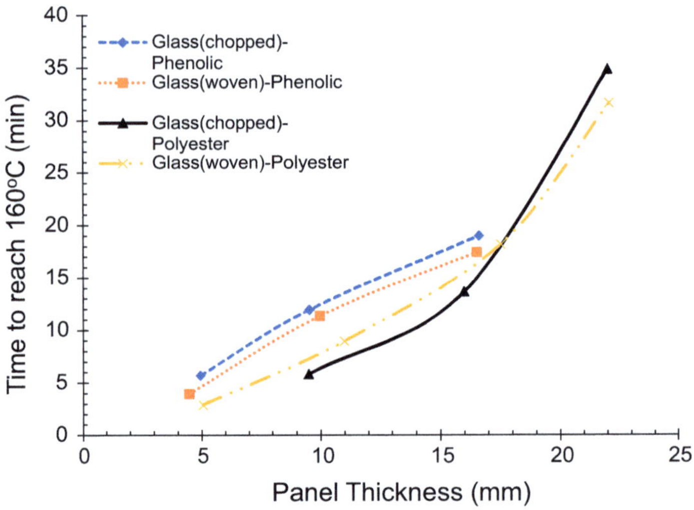

Fig. 16 Temperature profile of different composites [123]. Reproduced with permission taken from Swagata Dutta taken from his Ph.D referenced in [123]

fiber composites. While fire reaction properties provide insights into the material's initial response to fire, fire resistance properties enable the assessment of its ability to contain or withstand flames, contributing to overall fire safety in diverse industrial applications. By considering both aspects, researchers and industry professionals can aim to optimize the fire performance of hybrid natural fiber composites through tailored flame-retardant strategies and material formulations. The large-scale test, also known as ASTM E119, is a critical measure for evaluating the fire resistance of structural elements or assemblies, such as walls, floors, and roofs. This test is designed to simulate real-scale fire conditions and assess the ability of materials or constructions to withstand fire for a specified period. By subjecting the structural element to a standard fire exposure and observing its performance under these conditions, the large-scale test provides crucial data on the material's fire resistance and structural integrity during a fire event. Conducting the large-scale test according to ASTM E119 is essential for ensuring the compliance of structural elements with fire safety regulations and building codes. In the context of hybrid natural fiber composites used in construction and building applications, performing this test offers valuable insights into their fire resistance and structural performance under fire exposure. The results obtained from the large-scale test contribute to understanding how these materials behave in fire scenarios, guiding the development of fire safety strategies and the selection of appropriate materials for specific construction and building applications. Furthermore, the data from the large-scale test can also aid in evaluating the effectiveness of fire protection measures and structural design enhancements for hybrid natural fiber composites, ultimately contributing to improved fire safety and compliance with industry standards.

In a study by Mohammed et al. [162], the authors explored the thermal conductivity and fire resistance properties of fiber metal laminate composites, which are materials composed of a metal alloy and reinforced fibers within different types of polymer matrices. The study focuses on composites made from natural fibers like kenaf and flax as well as synthetic fibers like carbon, and it is particularly concerned with their application in areas where fire resistance is crucial, such as in the automotive and aerospace industries. The results indicated that natural fiber metal laminate composites showed a significant improvement in fire resistance properties, while pure synthetic fiber composites exhibited a slight decrease. Specifically, the flax composite demonstrated high thermal conductivity and was able to withstand flame temperature for 15 min according to the ISO 2685 standard, while the kenaf composite failed at 10 min and 30 s. The findings suggest that these fiber metal laminate composites can be suitably used as components in fire-designated zones of automotive, aerospace, and other machinery, due to their beneficial properties such as high strength, stiffness, resistance to corrosion, and reduced weight contributing to fuel efficiency—alongside their fire-resistant characteristics.

6 Applications and Future Directions

Interest in hybrid composite materials is growing as they show promise in surpassing traditional composites in performance. Notably, carbon fiber composites are commended for their high tensile strength and stiffness. However, the costliness and lower compressive-to-tensile strength ratio of carbon fibers limit their standalone use, prompting investigations into incorporating a secondary filler that excels in compression. Introducing glass fibers to carbon fiber composites has been researched to discover a more balanced performance in the hybrid material. Research by Irina et al. [163] explored substituting some carbon fibers with glass fibers, leading to lighter hybrid composites—beneficial for improving fuel efficiency and reducing environmental impacts in the automotive sector. The hybrid materials achieved promising mechanical properties, suggesting such mixtures could be employed for creating lighter, high-performing products. Adjusting the woven fabric stacking ratios of carbon and glass fibers could counteract the inherent challenges of using carbon fibers alone. Simultaneously, there is been a trend in exploring natural fibers like kenaf—whose mechanical properties are similar to glass fibers—for high-performance uses like automotive parts [164]. Utilizing nanoparticles such as $MgCO_3$ in these natural fiber composites could further enhance their mechanical attributes and fiber-matrix bonding. Studies have shown natural fiber composites outperform those with glass fibers, suggesting a potential shift in automotive materials. Nonetheless, hybridizing natural fibers with glass ones offers a compromise between strength and weight, providing a viable option for automotive applications. One such application, developed through the Analytical Hierarchy Process, found kenaf fibers as a top choice among several to be combined with glass fibers for designing automotive brake levers. In aerospace, the priority is on safety and minimizing weight.

Hybrid composites need to meet these criteria to be considered viable for aerospace applications. The quest to produce hybrid composites with high flame resistance is a dynamic field of research, stemming from the significant risk that fires pose in aircraft incidents. Traditional composites used in aircraft interiors, made from thermosets reinforced with carbon fibers, often exhibit high flammability. This has fueled in-depth studies on materials that can offer a balance between mechanical performance and flammability resistance. For instance, Alexander and Churchill's [165] research compared various hybrid composites, including combinations of sisal/glass and sisal/basalt, finding that sisal/basalt/epoxy composites.

Moreover, renewable energy sources, specifically wind and solar energies, have garnered significant attention as sustainable alternatives to fossil fuel-based energy. In solar energy applications, there has been a particular interest in utilizing hybrid composites as part of the structure for parabolic trough collectors, which are instrumental in gathering solar energy. Although the production of PTCs can be expensive due to the costs tied to their molds, the incorporation of natural fibers into these composites has come to light as a strategy to offset overall manufacturing expenses. Recent advancements have shown that the cost of trough collectors can be decreased by integrating natural fibers and synthetic fillers into the reinforcing phase of the collector material. In the realm of wind energy, the efficiency of power generation and cost is heavily influenced by key components, namely the wind turbine blades and rotors. These elements are crucial in determining the capacity of the wind turbines, especially under low wind conditions, typically ranging from 4 to 6 m per second. Commonly, carbon and glass fibers are employed as fillers in turbine blades due to their favorable properties. However, the drawbacks, notably the high cost and significant weight of these fibers, have spurred research into alternatives. Notably, a study suggested that replacing a proportion of carbon fibers with basalt fibers—around 20–30%—in the turbine blades does not substantially compromise the composite's properties. Moreover, an inventive composite incorporating woven silicon and mesoporous silica has displayed promising potential for wind turbine applications, achieving an excellent flexural modulus with just a 3% volume of mesoporous silica [166]. This efficiency gain is attributed to the robust interaction between the fillers and the matrix within the composite.

Furthermore, synthetic fibers like aramid, carbon, and glass are extensively utilized as reinforcement in resin-based systems for their potential use in armor. Armor systems typically comprise a dual-layer structure, with an outer metallic layer crafted from high-strength materials such as titanium or uranium, and an inner spall liner designed to shield against high-impact projectiles. Spall liners often harness polymers like ultra-high molecular weight polyethylene and aramid fibers, chosen for their high tensile strength, modulus, and elongation at break. Despite the efficacy of these materials, their high production costs have led researchers to seek more economical yet effective fibers. Exploratory studies have included glass, Kevlar, carbon fibers, and their hybrids as prospective alternatives. Of late, interest has surged in the use of natural fibers blended with synthetic fibers for ballistic use. Factors such as stacking sequence, thickness, fiber orientation, and areal density significantly influence ballistic resistance. Research by Yahaya et al. [167] thoroughly examined various sequences of Kevlar and nonwoven kenaf layers, including configurations with kenaf on the innermost, alternating, and outermost layers. They discovered that hybrids with kenaf positioned externally presented the highest V50—the velocity at which there is a 50% chance of penetration or prevention of projectile impact. However, it was noted that Kevlar-reinforced composites outperformed all other samples tested, demonstrating a negative hybrid effect. They also found that V50 values, which escalated alongside specimen thickness, attributed to the capacity for dissipating more surface energy due to the additional material the projectile must traverse. Further studies on woven kenaf-Kevlar hybrids showed improved performance over nonwoven kenaf configurations, with increases in both V50 and energy absorption correlating to thickness. The principal failure modes identified for these hybrid composites were shear plugging at the impact surface and fiber tension at the rear. It was deduced that thicker specimens necessitate higher velocities for penetration, given their augmented resistance, which in turn diminishes the projectile's residual velocity and heightens kinetic energy absorption. Additionally, increments in areal density were linked to greater energy absorption. Specifically, hybrids integrating two woven kenaf layers, situated third from the front and back surfaces, exhibited enhanced performance due to their increased thickness and areal density. These findings underscore the potential for integrating natural fibers with synthetic fibers in armor systems, offering a path toward cost-effective and efficient protection solutions.

The concept of hybridization has gained traction not merely for the enhanced performance of the final products but also for its ability to address the shortcomings that limit the use of natural fibers in structural applications. There is significant potential for a considerable amount of natural fibers to be integrated into traditional composites that are reinforced with synthetic materials, which is a progressive move both environmentally and economically. Nevertheless, the variable quality of natural fibers due to factors such as growth conditions and maturity level introduces variability in the composite properties. There is no question that further research into the modification and quality control of natural fibers could play a pivotal role in their commercial uptake. Additionally, the inclusion of a secondary filler, whether micro or nano-sized, could potentially mitigate some limitations associated with natural fiber-reinforced composites. Enhancing these materials with flame retardants could also boost thermal stability and reduce flammability, paving the way for the use of engineering polymers in high-end applications. It is important to note that natural fiber-based hybrid composites are positioning themselves to become a strong contender in various industrial markets.

Looking toward the future, the development of standardized methods for enhancing natural fiber quality will be critical to ensuring uniformity in composite properties. Advances in the interface engineering between natural fibers and polymers, optimization of fiber treatments, and exploration of new bio-based resins have the potential to substantially improve composite performance. The pursuit of greener manufacturing processes and the expansion of applications, especially in fields sensitive to environmental impact, holds the promise of eco-friendly advancements in composite technology. The ongoing research and development in this area are expected to yield materials that are not only cost-effective and high-performing but also aligned with sustainable practices and lower ecological footprints.

7 Summary

Natural fiber-reinforced composites (NFRCs) have traditionally been known to have issues with fiber-matrix adhesion, limited structural strength and stiffness, reduced performance and applications, moisture absorption and dimensional stability, durability and decay resistance, and flammability. Extensive studies have helped address some of these issues through chemical treatments of the fibers and systematic hybridization. Fiber modification techniques are known to help in improved interfacial adhesion due to increased fiber roughness, wettability, and even flammability. However, hybridization has recently emerged as a more potent and economical technique, allowing flexibility in fiber selection as per the matrix and the required mechanical and physical properties of the composite for the intended application. Thus, the possibility of applying and exploring new fibers with higher strengths and moduli and lower densities has the potential of both increasing the applicability of natural fiber composites and increasing the sustainability of the product by reducing the carbon footprint. General hybridizations are currently carried out among synthetic/mineral fibers and natural fibers, to make the composites lighter, more environmentally friendly, and sustainable, while ensuring the structural performances are not compromised. However, hybridization of two or more natural fibers in a single composite is also possible, which although might not be comparable in strengths to synthetic fibers can have great value in terms of sustainability. Although future studies are needed, hybridization has successfully helped in reducing many of the concerns in the NFRCs, where the dimensional stability, durability, decay resistance, flammability, and mechanical and structural integrity of the composites have been successfully improved by many folds. These have the potential to be used in applications where currently only synthetic fibers are being used, increasing the sustainability of those products. One such widely used example is the circuit board in various electronic applications, where glass fiber-reinforced epoxy composite is predominantly used in the current market.

References

1. Clyne TW, Hull D (2019) An introduction to composite materials. Cambridge University Press, Cambridge, New York
2. Chanda A, Bhattacharyya D (2018) Formability of wood veneers: a parametric approach for understanding some manufacturing issues. Holzforschung 72

3. Chanda A, Bhattacharyya D (2022) Introduction of the developable surface concept in fibrous composite materials. Compos Part Appl Sci Manuf 157
4. Currey J (2008) The structure and mechanical properties of bone. In: Bioceramics and their clinical applications. Elsevier, Netherlands, pp 3–27. https://doi.org/10.1533/9781845694227.1.3
5. Currey JD (2012) The structure and mechanics of bone. J Mater Sci 47:41–54
6. Begum K, Islam MA. Natural fiber as a substitute to synthetic fiber in polymer composites: a review. Res J Eng Sci 2:46–53
7. Daniel IM, Ishai O (2006) Engineering mechanics of composite materials. Oxford University Press, New York
8. Coutinho FMB, Costa THS (1999) Performance of polypropylene–wood fiber composites. Polym Test 18:581–587
9. Saheb DN, Jog JP (1999) Natural fiber polymer composites: a review. Adv Polym Technol 18:351–363
10. Bettini SHP, Uliana AT, Holzschuh D (2008) Effect of process parameters and composition on mechanical, thermal, and morphological properties of polypropylene/sawdust composites. J Appl Polym Sci 108:2233–2241
11. Chanda A (2021) Formability and flammability of thin walled veneer structures. University of Auckland, Auckland
12. Mohammadabadi M, Yadama V, Smith LV (2022) Wood-based prepreg for composite laminates: wood-based prepreg. Wood Fiber Sci 54
13. Mohammadabadi M, Yadama V (2020) Influence of a biaxially corrugated core geometry on flexural stiffness of wood-strand composite sandwich panels. Mater Today Commun 23
14. Biodegradable and biobased polymers for environmental and biomedical applications. Scrivener Publishing, Salem, Mass (2016)
15. Kim NK, Dutta S, Bhattacharyya D (2018) A review of flammability of natural fibre reinforced polymeric composites. Compos Sci Technol 162:64–78
16. Mohanty AK, Misra M, Hinrichsen G (2000) Biofibres, biodegradable polymers and biocomposites: an overview. Macromol Mater Eng 276–277:1–24
17. Kalia S, Kaith BS, Kaur I (2009) Pretreatments of natural fibers and their application as reinforcing material in polymer composites—a review. Polym Eng Sci 49:1253–1272
18. Keener TJ, Stuart RK, Brown TK (2004) Maleated coupling agents for natural fibre composites. Compos Part Appl Sci Manuf 35:357–362
19. Wongsriraksa P, Togashi K, Nakai A, Hamada H (2013) Continuous natural fiber reinforced thermoplastic composites by fiber surface modification. Adv Mech Eng 5:685104
20. Fire retardant materials. CRC Press [u.a.], Boca Raton, Fla. (2010)
21. Bhattacharyya D, Subasinghe A, Kim NK (2015) Natural fibers. In: Multifunctionality of polymer composites. Elsevier, Netherlands, pp 102–143. https://doi.org/10.1016/B978-0-323-26434-1.00004-0
22. Mouritz AP, Gibson AG (2010) Fire properties of polymer composite materials. Springer, Dordrecht
23. Sheng R (2019) Systems engineering fundamentals. In: Systems engineering for aerospace. Elsevier, Netherlands, pp 113–206. https://doi.org/10.1016/B978-0-12-816458-7.00007-7
24. Ching FDK, Winkel SR (2022) Building codes illustrated: a guide to understanding the 2021 international building code. Wiley, Hoboken
25. E05 Committee. Test methods for fire tests of building construction and materials. https://doi.org/10.1520/E0119-20
26. Arao Y et al (2014) Improvement on fire retardancy of wood flour/polypropylene composites using various fire retardants. Polym Degrad Stab 100:79–85
27. Azwa ZN, Yousif BF, Manalo AC, Karunasena W (2013) A review on the degradability of polymeric composites based on natural fibres. Mater Des 47:424–442
28. Saba N, Paridah MT, Abdan K, Ibrahim NA (2016) Dynamic mechanical properties of oil palm nano filler/kenaf/epoxy hybrid nanocomposites. Constr Build Mater 124:133–138

29. Mochane MJ et al (2019) Recent progress on natural fiber hybrid composites for advanced applications: a review. Express Polym Lett 13:159–198
30. Samanta S, Muralidhar M, Singh TJ, Sarkar S (2015) Characterization of mechanical properties of hybrid bamboo/GFRP and Jute/GFRP composites. Mater Today Proc 2:1398–1405
31. Mansor MR, Sapuan SM, Zainudin ES, Nuraini AA, Hambali A (2013) Hybrid natural and glass fibers reinforced polymer composites material selection using analytical hierarchy process for automotive brake lever design. Mater Des 51:484–492
32. Rodriguez-Castellanos W, Rodrigue D (2016) Production and characterization of hybrid polymer composites based on natural fibers. In: Poletto M (ed) Composites from renewable and sustainable materials. InTech. https://doi.org/10.5772/64995
33. The Paris Agreement on Climate Change (2017) Analysis and commentary. Oxford University Press, Oxford, United Kingdom
34. Pötschke P (2004) Dispersion of carbon nanotubes into thermoplastic polymers using melt mixing. In: AIP conference proceedings, vol 723. AIP, Kirchberg, Tirol (Austria), pp 478–484
35. Pötschke P et al (2005) Melt mixing as method to disperse carbon nanotubes into thermoplastic polymers. Fuller Nanotub Carbon Nanostruct 13:211–224
36. Luo J, Krause B, Pötschke P (2016) Melt-mixed thermoplastic composites containing carbon nanotubes for thermoelectric applications. AIMS Mater Sci 3:1107–1116
37. Sormunen P, Kärki T (2019) Compression molded thermoplastic composites entirely made of recycled materials. Sustainability 11:631
38. Léger É, Landry B, LaPlante G (2020) High flow compression molding for recycling discontinuous long fiber thermoplastic composites. J Compos Mater 54:3343–3350
39. Song Y et al (2018) CAE method for compression molding of carbon fiber-reinforced thermoplastic composite using bulk materials. Compos Part Appl Sci Manuf 114:388–397
40. Kim J-W, Lee D-G (2014) Study on the fiber orientation during compression molding of reinforced thermoplastic composites. Int J Precis Eng Manuf Green Technol 1:335–339
41. Edhirej A, Sapuan SM, Jawaid M, Zahari NI (2017) Cassava/sugar palm fiber reinforced cassava starch hybrid composites: Physical, thermal and structural properties. Int J Biol Macromol 101:75–83
42. Bengtsson M, Oksman K, Stark NM (2006) Profile extrusion and mechanical properties of crosslinked wood–thermoplastic composites. Polym Compos 27:184–194
43. Li H, Englund K (2017) Recycling of carbon fiber-reinforced thermoplastic composite wastes from the aerospace industry. J Compos Mater 51:1265–1273
44. Mamanpush SH, Li H, Englund K, Tavousi Tabatabaei A (2020) Extruded fiber-reinforced composites manufactured from recycled wind turbine blade material. Waste Biomass Valorization 11:3853–3862
45. Yao S-S, Jin F-L, Rhee KY, Hui D, Park S-J (2018) Recent advances in carbon-fiber-reinforced thermoplastic composites: a review. Compos Part B Eng 142:241–250
46. Zhang X et al (2022) Nano silicon carbide-treated wheat straw fiber reinforced high-density polyethylene composites. Ind Crops Prod 182:114834
47. Maldas D, Kokta BV, Daneault C (1989) Thermoplastic composties of polystyrene: effect of different wood species on mechanical properties. J Appl Polym Sci 38:413–439
48. Zhao H, Gao Z, Zhai D, Zhao G (2023) Enhanced mechanical property of continuous carbon fiber/polyamide thermoplastic composites by combinational treatments of carbon fiber fabric. Compos Commun 38:101508
49. Khalil HA et al (2013) Natural fiber reinforced poly(vinyl chloride) composites: a review. J Reinf Plast Compos 32:330–356
50. Hamerton I, Mooring L (2012) The use of thermosets in aerospace applications. In: Thermosets. Elsevier, Netherlands, pp 189–227. https://doi.org/10.1533/9780857097637.2.189
51. Rudd CD (2000) Composites for automotive applications. Rapra Technology Ltd, Shawbury, Shrewsbury
52. Brøndsted P, Lilholt H, Lystrup A (2005) Composite materials for wind power turbine blades. Annu Rev Mater Res 35:505–538

53. Miranda Campos B, Bourbigot S, Fontaine G, Bonnet F (2022) Thermoplastic matrix-based composites produced by resin transfer molding: a review. Polym Compos 43:2485–2506
54. Gartner B, Yadama V, Smith L (2022) Resin transfer molding of wood strand composite panels. Forests 13
55. Chaturvedi SK, Sun CT, Sierakowski RL (1983) Mechanical characterization of sheet molding compound composites. Polym Compos 4:167–171
56. Andresen FR (2001) Open molding: hand lay-up and spray-up. In: Miracle DB, Donaldson SL (eds) Composites. ASM International, pp 450–456. https://doi.org/10.31399/asm.hb.v21.a0003406
57. Minchenkov K et al (2023) Pultrusion of thermoplastic composites with mechanical properties comparable to industrial thermoset profiles. Compos Commun 44:101766
58. Dittenber DB, GangaRao HVS (2012) Critical review of recent publications on use of natural composites in infrastructure. Compos Part Appl Sci Manuf 43:1419–1429
59. Chanda A, Bhattacharyya D (2021) A parametric study to minimise spring-back while producing plywood channels. J Clean Prod 304
60. Das O, Sarmah AK (2015) Mechanism of waste biomass pyrolysis: effect of physical and chemical pre-treatments. Sci Total Environ 537:323–334
61. Horrocks AR (1983) An introduction to the burning behaviour of cellulosic fibres. J Soc Dye Colour 99:191–197
62. Ferdous D, Dalai AK, Bej SK, Thring RW (2002) Pyrolysis of lignins: experimental and Kinetics Studies. Energy Fuels 16:1405–1412
63. Kozłowski R, Władyka-Przybylak M (2008) Flammability and fire resistance of composites reinforced by natural fibers. Polym Adv Technol 19:446–453
64. Galaska ML, Horrocks AR, Morgan AB (2017) Flammability of natural plant and animal fibers: a heat release survey. Fire Mater 41:275–288
65. Bradbury AGW, Sakai Y, Shafizadeh F (1979) A kinetic model for pyrolysis of cellulose. J Appl Polym Sci 23:3271–3280
66. Wilkie CA, Morgan AB (2010) Fire retardancy of polymeric materials. CRC Press, Boca Raton
67. Benisek L (1975) Flame retardance of protein fibers. In: Lewin M, Atlas SM, Pearce EM (eds) Flame-retardant polymeric materials. Springer US, Boston, MA, pp 137–191. https://doi.org/10.1007/978-1-4684-2148-4_3
68. Kandola BK (2012) Flame retardant characteristics of natural fibre composites. RSC Publ, Cambridge
69. Guan J-P, Chen G-Q (2006) Flame retardancy finish with an organophosphorus retardant on silk fabrics. Fire Mater 30:415–424
70. Gualtieri AF, Tartaglia A (2000) Thermal decomposition of asbestos and recycling in traditional ceramics. J Eur Ceram Soc 20:1409–1418
71. Kusiorowski R, Zaremba T, Piotrowski J, Adamek J (2012) Thermal decomposition of different types of asbestos. J Therm Anal Calorim 109:693–704
72. Karvanis K, Rusnáková S, Krejčí O, Žaludek M (2020) Preparation, thermal analysis, and mechanical properties of basalt fiber/epoxy composites. Polymers 12:1785
73. Maxineasa SG, Taranu N (2018) Life cycle analysis of strengthening concrete beams with FRP. In: Eco-efficient repair and rehabilitation of concrete infrastructures. Elsevier, Netherlands, pp 673–721. https://doi.org/10.1016/B978-0-08-102181-1.00024-1
74. Fiore V, Di Bella G, Valenza A (2011) Glass–basalt/epoxy hybrid composites for marine applications. Mater Des 32:2091–2099
75. Zhang Y et al (2012) Mechanical and thermal properties of basalt fiber reinforced poly(butylene succinate) composites. Mater Chem Phys 133:845–849
76. Deák T, Czigány T (2009) Chemical composition and mechanical properties of basalt and glass fibers: a comparison. Text Res J 79:645–651
77. Derradji M et al (2019) Cost effective surface-modified basalt fibers-reinforced phthalonitrile composites with improved mechanical properties and advanced nuclear shielding efficiency. Polym Compos 40

78. Keszei S et al (2005) Progress in interface modifications: from compatibilization to adaptive and smart interphases. Eur Polym J 41:697–705
79. Jones FR, Huff NT (2018) The structure and properties of glass fibers. In: Handbook of properties of textile and technical fibres. Elsevier, Netherlands, pp 757–803. https://doi.org/10.1016/B978-0-08-101272-7.00019-5
80. Ren B, Li Z, Zhang Y, Feng S, Wang L (2023) New interpretation of glass formation in isomeric substances: shifting from melting-point to melting-entropy. Adv Sci 10:2206389
81. Kirk-Othmer encyclopedia of chemical technology. Wiley, USA. https://doi.org/10.1002/0471238961
82. Sun K (1947) Fundamental condition of glass formation. J Am Ceram Soc 30:277–281
83. Thomason JL, Nagel U, Yang L, Bryce D (2019) A study of the thermal degradation of glass fibre sizings at composite processing temperatures. Compos Part Appl Sci Manuf 121:56–63
84. Price D, Anthony G, Carty P (2001) Introduction: polymer combustion, condensed phase pyrolysis and smoke formation. In: Fire Retardant Mater. Elsevier, Netherlands, pp 1–30. https://doi.org/10.1533/9781855737464.1
85. Horrocks AR, Kandola BK (2005) Flammability and fire resistance of composites. In: Design and manufacture of textile composites. Elsevier, Netherlands, pp 330–363. https://doi.org/10.1533/9781845690823.330
86. Parodi E, Govaert LE, Peters GWM (2017) Glass transition temperature versus structure of polyamide 6: a flash-DSC study. Thermochim Acta 657:110–122
87. Sun L, Yang J-T, Lin G-Y, Zhong M-Q (2007) Crystallization and thermal properties of polyamide 6 composites filled with different nanofillers. Mater Lett 61:3963–3966
88. Grebowicz J, Lau S-F, Wunderlich B (1984) The thermal properties of polypropylene. J Polym Sci Polym Symp 71:19–37
89. Godovsky YK (2012) Thermophysical properties of polymers. Springer Science & Business Media
90. Seleem HEH (2006) The effect of inorganic fillers on the mechanical and thermal properties of polyester. Polym-Plast Technol Eng 45:585–590
91. Bakar M, Duk R, Przybyłek M, Kostrzewa M (2009) Mechanical and thermal properties of epoxy resin modified with polyurethane. J Reinf Plast Compos 28:2107–2118
92. Wu J et al (2022) Effect of the addition of thermoplastic resin and composite on mechanical and thermal properties of epoxy resin. Polymers 14:1087
93. Monk JD et al (2015) Computational and experimental study of phenolic resins: thermal-mechanical properties and the role of hydrogen bonding. Macromolecules 48:7670–7680
94. Sweileh BA, Al-Hiari YM, Kailani MH, Mohammad HA (2010) Synthesis and characterization of polycarbonates by melt phase interchange reactions of alkylene and arylene diacetates with alkylene and arylene diphenyl dicarbonates. Molecules 15:3661–3682
95. Shi H et al (2020) Bismuth oxychloride nanosheets for improvement of flexible poly (vinyl chloride) flame retardancy. J Mater Sci 55:631–643
96. Xiao S, Akinyi C, Longun J, Iroh JO (2022) Polyimide copolymers and nanocomposites: a review of the synergistic effects of the constituents on the fire-retardancy behavior. Energies 15:4014
97. Hachani SE et al (2018) Effects of magnesia incorporation on properties of polystyrene/magnesia composites. Acta Chim Slov 65:646–651
98. Martins JN, Klohn TG, Bianchi O, Fiorio R, Freire E (2010) Dynamic mechanical, thermal, and morphological study of ABS/textile fiber composites. Polym Bull 64:497–510
99. Ao X, Vázquez-López A, Mocerino D, González C, Wang D-Y (2024) Flame retardancy and fire mechanical properties for natural fiber/polymer composite: a review. Compos Part B Eng 268:111069
100. Prabhakar MN, Naga Kumar C, Dong Woo L, Jung-IL S (2022) Hybrid approach to improve the flame-retardant and thermal properties of sustainable biocomposites used in outdoor engineering applications. Compos Part Appl Sci Manuf 152:106674
101. Neto J et al (2022) A review of recent advances in hybrid natural fiber reinforced polymer composites. J Renew Mater 10:561–589

102. Suriani MJ et al (2021) Flammability, tensile, and morphological properties of oil palm empty fruit bunches fiber/pet yarn-reinforced epoxy fire retardant hybrid polymer composites. Polymers 13:1282
103. Suriani MJ et al (2021) Kenaf fiber/pet yarn reinforced epoxy hybrid polymer composites: morphological, tensile, and flammability properties. Polymers 13:1532
104. Neto JSS, De Queiroz HFM, Aguiar RAA, Banea MD (2021) A review on the thermal characterisation of natural and hybrid fiber composites. Polymers 13:4425
105. Dreyer C, Motoc DL (2020) Reaction-to-fire parameters of CE/epoxy blend-based CF/FF and BF/FF hybrid composites. J Therm Anal Calorim 142:661–669
106. Mohammed L, Ansari MNM, Pua G, Jawaid M, Islam MS (2015) A review on natural fiber reinforced polymer composite and its applications. Int J Polym Sci 2015:1–15
107. Chapple S, Anandjiwala R (2010) Flammability of natural fiber-reinforced composites and strategies for fire retardancy: a review. J Thermoplast Compos Mater 23:871–893
108. Affan Usmani M, Anas M (2018) Study of natural fibre reinforced composites. IOP Conf Ser Mater Sci Eng 404:012048
109. Shah AUR, Prabhakar MN, Song J-I (2017) Current advances in the fire retardancy of natural fiber and bio-based composites—a review. Int J Precis Eng Manuf Green Technol 4:247–262
110. Kureemun U et al (2018) Effects of hybridization and hybrid fibre dispersion on the mechanical properties of woven flax-carbon epoxy at low carbon fibre volume fractions. Compos Part B Eng 134:28–38
111. Mazur K, Kuciel S, Salasinska K (2019) Mechanical, fire, and smoke behaviour of hybrid composites based on polyamide 6 with basalt/carbon fibres. J Compos Mater 53:3979–3991
112. Dutta S, Kim NK, Das R, Bhattacharyya D (2021) Evaluating orientation effects on the fire reaction properties of flax-polypropylene composites. Polymers 13:2586
113. Dutta S, Kim NK, Das R, Bhattacharyya D (2019) Effects of sample orientation on the fire reaction properties of natural fibre composites. Compos Part B Eng 157:195–206
114. Kim NK, Dutta S, Bhattacharyya D (2019) Heat and smoke production of flax fibre reinforced composites under horizontal and vertical orientations. Compos Part B Eng 178:107467
115. Ghazzawi YM, Osorio AF, Heitzmann MT (2020) The effect of fibre length and fibre type on the fire performance of thermoplastic composites: The behaviour of polycarbonate as an example of a charring matrix. Constr Build Mater 234:117889
116. Mouritz AP, Mathys Z, Gibson AG (2006) Heat release of polymer composites in fire. Compos Part Appl Sci Manuf 37:1040–1054
117. Sałasińska K et al (2022) The Effect of Manufacture Process on Mechanical Properties and Burning Behavior of Epoxy-Based Hybrid Composites. Materials 15:301
118. Sorathia U, Rollhauser CM, Hughes WA (1992) Improved fire safety of composites for naval applications. Fire Mater 16:119–125
119. Sorathia U, Beck C, Dapp T (1993) Residual strength of composites during and after fire exposure. J Fire Sci 11:255–270
120. Sorathia U, Long G, Gracik T, Blum M, Ness J (2001) Screening tests for fire safety of composites for marine applications. Fire Mater 25:215–222
121. Mouritz AP, Mathys Z (1999) Post-fire mechanical properties of marine polymer composites. Compos Struct 47:643–653
122. Khuntia T, Biswas S (2022) An investigation on the flammability and dynamic mechanical behavior of coir fibers reinforced polymer composites. J Ind Text 51:1616–1640
123. Dutta S (2020) Understanding and modelling the thermo-mechanical response of FRP composites under fire. The University of Auckland, Auckland, NZ
124. Subasinghe A (2015) Characterisation of halogen free fire retardant polypropylene composites reinforced by short natural fibres. The University of Auckland
125. Subasinghe A, Bhattacharyya D (2014) Performance of different intumescent ammonium polyphosphate flame retardants in PP/kenaf fibre composites. Compos Part Appl Sci Manuf 65:91–99
126. Hornsby PR (2001) Fire retardant fillers for polymers. Int Mater Rev 46:199–210

127. Sain M, Park SH, Suhara F, Law S (2004) Flame retardant and mechanical properties of natural fibre–PP composites containing magnesium hydroxide. Polym Degrad Stab 83:363–367
128. Stark NM, White RH, Mueller SA, Osswald TA (2010) Evaluation of various fire retardants for use in wood flour–polyethylene composites. Polym Degrad Stab 95:1903–1910
129. Nithiyapathi C, Thirunavukkarasu K, Daniel Das A, Tamilvendan D (2021) Progression in fire retardant properties of polymer composites – a review. IOP Conf Ser Mater Sci Eng 1059:012058
130. Suriani M, Sapuan S, Ruzaidi C, Nair D, Ilyas R (2021) Flammability, morphological and mechanical properties of sugar palm fiber/polyester yarn-reinforced epoxy hybrid biocomposites with magnesium hydroxide flame retardant filler. Text Res J 91:2600–2611
131. Khalili P, Tshai KY, Hui D, Kong I (2017) Synergistic of ammonium polyphosphate and alumina trihydrate as fire retardants for natural fiber reinforced epoxy composite. Compos Part B Eng 114:101–110
132. Kim NK, Lin RJT, Bhattacharyya D (2017) Flammability and mechanical behaviour of polypropylene composites filled with cellulose and protein based fibres: a comparative study. Compos Part Appl Sci Manuf 100:215–226
133. Wu Q, Zhu W, Zhang C, Liang Z, Wang B (2010) Study of fire retardant behavior of carbon nanotube membranes and carbon nanofiber paper in carbon fiber reinforced epoxy composites. Carbon 48:1799–1806
134. Dittrich B, Wartig K-A, Hofmann D, Mülhaupt R, Schartel B (2013) Flame retardancy through carbon nanomaterials: carbon black, multiwall nanotubes, expanded graphite, multi-layer graphene and graphene in polypropylene. Polym Degrad Stab 98:1495–1505
135. Marney DCO et al (2008) The suitability of halloysite nanotubes as a fire retardant for nylon 6. Polym Degrad Stab 93:1971–1978
136. Dhand V, Rhee KY, Ju Kim H, Ho Jung D (2013) A comprehensive review of graphene nanocomposites: research status and trends. J Nanomater 2013:1–14
137. Vengatesan MR, Mittal V (2015) Polyolefin/graphene nanocomposite materials. In: Handbook of nanoceramic and nanocomposite coatings and materials. Elsevier, Netherlands, pp 129–154. https://doi.org/10.1016/B978-0-12-799947-0.00006-7
138. Sang B, Li Z, Li X, Yu L, Zhang Z (2016) Graphene-based flame retardants: a review. J Mater Sci 51:8271–8295
139. Kim Y, Lee S, Yoon H (2021) Fire-safe polymer composites: flame-retardant effect of nanofillers. Polymers 13:540
140. Wang W, Zhang W, Chen H, Zhang S, Li J (2015) Synergistic effect of synthetic zeolites on flame-retardant wood-flour/polypropylene composites. Constr Build Mater 79:337–344
141. Subasinghe A, Das R, Bhattacharyya D (2016) Study of thermal, flammability and mechanical properties of intumescent flame retardant PP/kenaf nanocomposites. Int J Smart Nano Mater 7:202–220
142. Arrakhiz FZ, Benmoussa K, Bouhfid R, Qaiss A (2013) Pine cone fiber/clay hybrid composite: mechanical and thermal properties. Mater Des 50:376–381
143. Piekarska K, Sowinski P, Piorkowska E, Haque MdM-Ul, Pracella M (2016) Structure and properties of hybrid PLA nanocomposites with inorganic nanofillers and cellulose fibers. Compos Part Appl Sci Manuf 82:34–41
144. Essabir H, Bensalah MO, Rodrigue D, Bouhfid R, Qaiss A (2016) Structural, mechanical and thermal properties of bio-based hybrid composites from waste coir residues: fibers and shell particles. Mech Mater 93:134–144
145. Battegazzore D et al (2016) Thermo-mechanical properties enhancement of bio-polyamides (PA10.10 and PA6.10) by using rice husk ash and nanoclay. Compos Part Appl Sci Manuf 81:193–201
146. Zare Y (2016) "a" interfacial parameter in Nicolais-Narkis model for yield strength of polymer particulate nanocomposites as a function of material and interphase properties. J Colloid Interface Sci 470:245–249
147. Pregi E, Kun D, Faludi G, Móczó J, Pukánszky B (2022) Modeling the mechanical properties of polypropylene/lignin/flax hybrid composites. Mater Des 220:110833

148. Izquierdo V (2018) Testing fabrics for flammability and fire safety. In: Advanced characterization and testing of textiles. Elsevier, Netherlands, 189–209. https://doi.org/10.1016/B978-0-08-100453-1.00009-X
149. Ceylan Ö, Alongi J, Van Landuyt L, Frache A, De Clerck K (2013) Combustion characteristics of cellulosic loose fibres. Fire Mater 37:482–490
150. Rojas-Aguilar A, Valdés-Ordoñez A (2004) Micro-combustion calorimetry employing a Calvet heat flux calorimeter. J Chem Thermodyn 36:619–626
151. Rabe S, Schartel B (2017) The rapid mass calorimeter: Understanding reduced-scale fire test results. Polym Test 57:165–174
152. Lyon RE, Walters RN (2004) Pyrolysis combustion flow calorimetry. J Anal Appl Pyrolysis 71:27–46
153. Anioł B, Jankowski D (2020) Flame retardants of polymeric materials—calculation of the oxygen index. Czas Tech 9
154. Abdul Rashid MK, Ramli Sulong NH, Alengaram UJ (2020) Fire resistance performance of composite coating with geopolymer-based bio-fillers for lightweight panel application. J Appl Polym Sci 137:49558
155. International handbook of structural fire engineering. Springer, Cham (2021). https://doi.org/10.1007/978-3-030-77123-2
156. Wade C (2001) Building fire safety and hazard assessment methods for combustible surface finishes
157. Abrams M et al (1985) Review of ASTM test for surface burning characteristics of building materials (E 84) (an ASTM white paper). J Test Eval 13:89–114
158. ASTM International (2023) ASTM E84-23: standard test method for surface burning characteristics of building materials. ASTM International. https://doi.org/10.1520/E0084-23
159. Wade C (2003) Future directions in fire testing of building products
160. Settle GE (1979) Assessment of draft ISO radiant panel spread-of-flame test. Fire Mater 3:171–181
161. Miriel J, Serres L, Trombe A (2002) Radiant ceiling panel heating–cooling systems: experimental and simulated study of the performances, thermal comfort and energy consumptions. Appl Therm Eng 22:1861–1873
162. Mohammed I (2022) Fire properties of hybrid composites. J Technol Innov Energy 1:4–8
163. Irina MMW, Azmi AI, Tan CL, Lee CC, Khalil ANM (2015) Evaluation of mechanical properties of hybrid fiber reinforced polymer composites and their architecture. Proc Manuf 2:236–240
164. Xia C, Shi SQ, Wu Y, Cai L (2016) High pressure-assisted magnesium carbonate impregnated natural fiber-reinforced composites. Ind Crops Prod 86:16–22
165. Alexander J, Elphej Churchill SJ (2017) Mechanical characterization of baslat based natural hybrid composites for aerospace applications. IOP Conf Ser Mater Sci Eng 197:012008
166. Reddy KS, Singla H (2017) Optimization of woven jute/glass fibre-reinforced polyester hybrid composite solar parabolic trough collector. IOP Conf Ser Mater Sci Eng 222:012016
167. Yahaya R, Sapuan SM, Jawaid M, Leman Z, Zainudin ES (2014) Quasi-static penetration and ballistic properties of kenaf–aramid hybrid composites. Mater Des 63:775–782

Post-Fire Response of the Composite Facesheet Used in the Main Rotor Blades: A Case Study

S. I. Dhaniya Lakshmi, G. Samiksha, G. Ramesh Babu, Muthukumar Chandrasekar, Naveen Jesuarockiam, and N. Karthikeyan

Abstract A case study on the post-fire performance of the facesheet extracted from the sandwich structure of a helicopter rotor blade is presented. The facesheet of the composite sandwich was exposed to heating through oxyacetylene flame only on its top face replicating a real-life fire scenario. Load bearing ability under tension and bending were affected such that the post-fire specimens possessed inferior residual tensile strength. The flame exposure created rougher surface morphology and increased oxygen elemental concentration from the EDX spectra. As the flame exposure temperature was increased, a significant increase in glass transition temperature and shifting of loss modulus as well as tan delta peaks were observed as per the dynamic mechanical analysis (DMA). The flame exposed specimens were also marked by the major decline storage modulus indicating the severity of fire exposure.

Keywords Composite · Simulated fire exposure · Post-fire testing · Mechanical properties · DMA

1 Introduction

The composite sandwich structure consisting of a core and facesheet has been prominently used in various structural applications, such as space, marine, construction. The facesheet effectively sustains in-plane loads, while the core sustains the shear load when a typical sandwich material is subjected to bending [15]. Both organic and non-organic materials have been utilized as core depending on the application needs.

S. I. Dhaniya Lakshmi · G. Samiksha · M. Chandrasekar (✉)
SIMCRASH Centre, Department of Aerospace Engineering, Hindustan Institute of Technology and Science, Padur, Chennai, Tamil Nadu, India
e-mail: chandrasekar.25j@gmail.com

G. Ramesh Babu
Advanced Composite Division, Hindustan Aeronautics Limited (HAL), Bengaluru, India

N. Jesuarockiam · N. Karthikeyan
School of Mechanical Engineering, Vellore Institute of Technology, Vellore, India

C. Muthukumar et al. (eds.), *Flammability Performance of Biocomposites and Bionanocomposites*, Composites Science and Technology,
https://doi.org/10.1007/978-981-95-2389-4_14

Organic materials such as biopolymer foams, syntactic foam, Nomex, and balsa wood while synthetic polymer foam and metallic core come under the category of non-organic material [6, 9].

The main rotor blade used in the helicopter is a typical composite sandwich material constructed using carbon fibre, glass fibre, and copper wire mesh and rohacell foam or honeycomb structure as the core. The characteristics of foam core such as acoustic insulation, low thermal conductivity, excellent ductility, and customizable density have made them an attractive choice for their utilization in the sandwich structure [8]. The FRP facesheets provide a notable advantage due to their exceptional strength-to-weight ratio. Furthermore, the primary rotor blade consists of one or few layers of the copper mesh layer inserted between the composite fabrics to enhance the structural rigidity [7].

Fire poses a substantial threat to the structural integrity and durability of the sandwich material during its service life. The fire testing of composites on a large scale structure is time consuming and expensive. Thus, researchers have utilized small-scale burners and flame torch to induce fire on the materials at a range of heat fluxes and temperatures [13].

Post-fire response of the sandwich explored by some researchers revealed that fire exposure or exposure to heat flux causes reduction in the mechanical properties. In a study, sandwich consisting fibreglass/vinylester facesheet with PVC foam core and phenolic foam core was subjected to heat flux from 10,100 to 100 kW/m^2 with a heating time between 200 and 1800 s. The surface blackening occurred in both the sandwich structures due to thermal degradation followed by char formation. The extent of char formation on the surface as well as in the transverse direction along its depth was dependent on duration of the fire exposure [10]. Recently, E-glass/vinylester/balsa core sandwich to constant thermal flux of 35 kW/m^2 and exposure time varied between 60 and 1200 s. The tensile strength and stiffness of the post-fire exposed specimens declined with the increase in exposure time. Compared to tensile properties, stiffness and strength deterioration under compression was more prominent. The char layer formed on the front skin due to fire exposure caused softening and weakening of the interfacial bonding such that the facesheet failed due to buckling at the exposure time as low as 60 s. Another interesting observation was that as the heat flux and exposure time was increased, sandwich failed by buckling with facesheet delaminating from the core as opposed to the core shear failure in the unexposed specimens [2] exposed. In another work, the post-fire E-glass/vinylester/balsa core subjected to compression failed by buckling in which facesheet exposed to fire delaminated from the core while the core and undamaged facesheet (the surface which was not exposed to fire) remained intact. They also indicated that one of their specimens displayed delamination between facesheet and core during the heating [3]. These studies indicate the detrimental effects of the fire on failure behaviour of the sandwich and particularly highlight the vulnerability of the facesheet as it is being exposed and damaged under fire as compared to the core which undergoes least damage or almost no damage. In a study, the tensile properties, compression properties, and inter-laminar shear strength (ILSS) of the post-fire fibreglass/polyester specimens dropped significantly with the increase in fire exposure time. They also indicated

Table 1 Layup sequence of the composite facesheet employed in this study

Fibreglass 0/90 fabric
Fibreglass 0 unidirectional tape
Fibreglass 0 unidirectional tape
Carbon 0/90 fabric
Carbon ± 45 fabric
Carbon ± 45 fabric
Fibreglass 0/90 fabric
Carbon ± 45 fabric
Carbon ± 45 fabric
Carbon ± 45 fabric
Carbon ± 45 fabric
Carbon ± 45 fabric
Copper mesh layer
Copper mesh layer

that surface of the fire exposed specimens examined under the scanning electron microscope (SEM) revealed the presence of char layer and fibre damage. Furthermore, the extent of damage propagation and extensive delamination was observed in the unburnt section of the composite from the transverse section [11]. The investigations carried out in the literature were primarily on the coupon level testing while the testing on the real-time structure was scarcely reported. Thus, this work focuses on determining the mechanical properties and morphological characteristics of the FRP facesheet from the sandwich structure removed from the propeller blade before and after exposure to fire. The microstructure and damage to the facesheet was examined using SEM and energy dispersive x-ray spectroscopy (EDX).

2 Materials and Methods

2.1 Materials

The composite facesheet used in this research work was extracted from the main rotor blade with layup sequence as shown in Table 1.

2.2 Fire Exposure

The composite facesheet along with its core is exposed to oxyacetylene flame in two cycles. Initially, the sample was exposed to a temperature of 800 °C for 20 min and

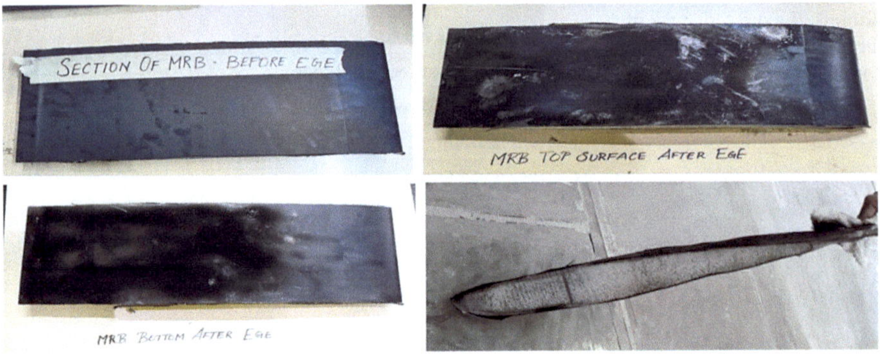

Fig. 1 Main rotor blade before fire and post-fire

cooled to room temperature. Subsequently, after 2 h, the sample was re-heated to 800 °C for 20 min. Figure 1 depicts the main rotor blade before fire and after fire. The specimen was exposed to fire only on the outer surface.

2.3 *Characterization*

The surface exposed to fire was placed under the loading nose while the bottom surface without fire exposure was placed under support during the flexural test, short beam test, and DMA.

2.3.1 DMA

DMA was performed on the specimens after fire exposure as per ASTM D4065. The specimens were heated between 30 °C and 200 °C in the nitrogen atmosphere using NETZSCH machine at Indian Institute of Technology, Madras.

2.3.2 Mechanical Tests

Tensile test was performed as per ASTM D3039 on rectangular specimens at a crosshead speed of 2 mm/min. Flexural test was performed in 3-point bending mode as per the ASTM D790 with a crosshead displacement of 1 mm/min. Short beam test was performed as per ASTM D2344 where peak load from the load–displacement curve was used to determine Interlaminar Shear Strength (ILSS) as shown in the Eq. (1).

$$\text{ILSS(MPa)} = \frac{3F}{4bt} \tag{1}$$

where F is the peak load (N), b is the breadth (mm), and t is the thickness (mm).

2.3.3 Morphological Analysis

Surface morphology and microstructure of the tensile fracture specimens were analysed through the Scanning electron microscope equipped with EDX for composition.

3 Results and Discussion

3.1 Surface Morphology

The surface morphology of the composite before and after fire along with EDX analysis is shown in Fig. 2a–d. It is evident from Fig. 2b that exposing the composite to fire has led to scorching of the surface as observed by the rougher texture over the smooth texture before the exposure (Fig. 2a). EDX analysis (Fig. 2c, d) further revealed the changes in composition as represented by the increase in oxygen element and nitrogen element. According to Vieira et al.[14], the increase in oxygen concentration in the burnt area represents the presence of char layer.

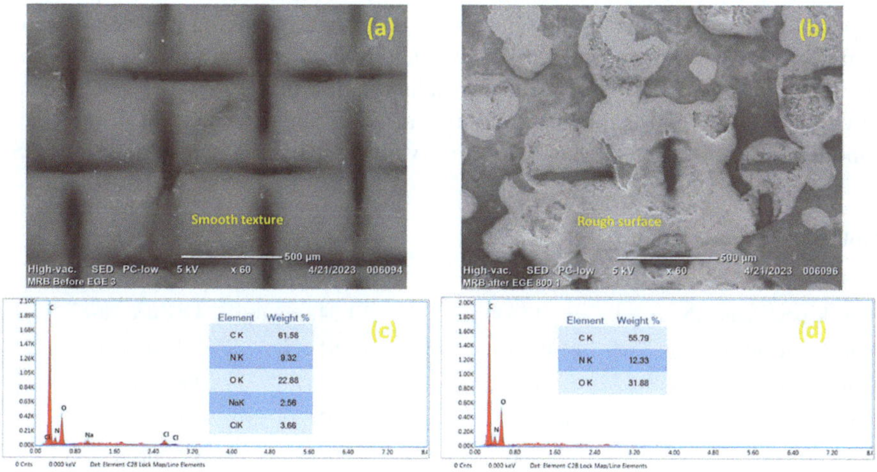

Fig. 2 Morphological and composition changes. **a** Smooth texture before fire, **b** rough surface post-fire, **c** EDX analysis before fire, and **d** EDX analysis post-fire

3.2 Mechanical Properties

The results from various mechanical tests have been highlighted in Table 2. The detrimental effects of fire are clearly visible from the substantial drop in the load bearing ability under tension and bending. It is interesting to note that tensile strength of the post-fire specimens dropped by 47% as compared to infinitesimal change in the flexural strength of the post-fire specimens. Moreover, the interfacial bonding characteristics were found to have affected as highlighted by the 35% decline in ILSS for the post-fire specimens. As the composite is exposed to a flame temperature of 800 °C, it not only affects the top layers exposed to fire, the heat generated due to flame exposure transfers to the adjacent layers. Once the flame temperature exceeds the glass transition temperature of the matrix, it causes matrix degradation such that the interfacial bonding is affected. The decreasing ILSS with the increase in exposure temperature was reported on fibreglass/vinylester and fibreglass/phenolic composites [14]. However, the magnitude of drop in ILSS was not significant as the specimens were only exposed to elevated temperature unlike fire in the present study. Thus, fire exposure is more detrimental to the material which is further evident from the inferior load bearing ability and lower ILSS for the post-fire specimens. In another work, a major decrease in flexural properties, tensile properties, and ILSS were reported for the glass/polyester exposed to fire. The reason for inferior mechanical properties of the fire exposed specimen was due to the delamination and micro-cracking in the char layer leading to premature failure at lower load than the unexposed specimen [11].

Figure 3a, b presents the microstructural images of the tensile fractured specimens. Typical failure pattern such as the fibre breakage, matrix cracking, and delamination could be observed. However, the extent of delamination was prominent in the post-fire specimens (Fig. 3b). The extensive delamination in the post-fire specimens is an indication of the structural weakening caused by the deterioration of interfacial bonding between various layers. This observation substantiates with drop in ILSS.

Table 2 Mechanical properties before fire and post-fire

Property	Before fire	Post-fire	Inference
Tensile peak load (kN)	2.554	1.342	47% decrease
Tensile strength (MPa)	146.04	103.51	29% decrease
Bending peak load (kN)	0.0572	0.0457	20% decrease
Flexural strength (MPa)	474.53	471.92	0.5% decrease
ILSS (MPa)	24.06	15.66	35% decrease

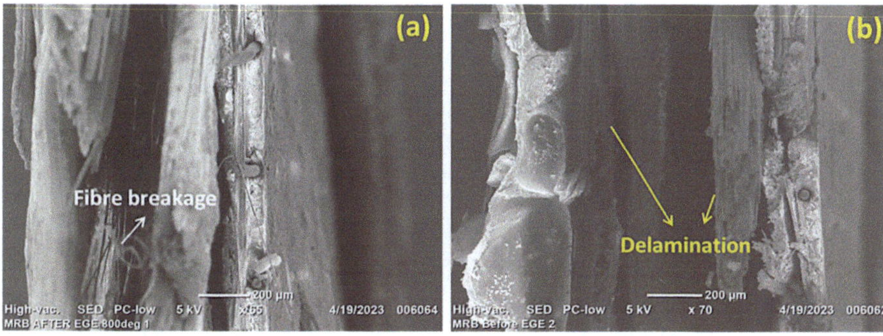

Fig. 3 Microstructural images of the tensile fractured specimens **a** before fire and **b** post-fire

3.3 DMA

The previous section highlighted the effects of flame exposure on the load bearing ability and interfacial bonding characteristics of the material. This section sheds light on the viscoelastic properties of the material investigated through DMA. The specimens exposed to 800 and 1200 °C were examined and the plots from DMA are illustrated in Fig. 4a–c.

E' plot consists of three distinctive zones as represented in (Fig. 4a). Initially at room temperature, E' is higher, and as the chamber temperature was increased, E' tends to drop during which the material undergoes transition from the solid state to rubbery state. The slope of the E' curve in the transition zone represents T_g [1]. It can be noticed that the E' curve flattens for the flame temperature of 1200 °C indicating major changes in T_g over its counterpart. At temperature above 150 °C, on surpassing the transition zone, the E' plot tends to coalesce in the rubbery region. The rapid change in E' with the temperature in the transition zone and rubbery region is due to the increase in mobility of the polymeric chains within the matrix [4]. Apart from the changes in T_g, detrimental effects of flame exposure can also be noticed from the Peak E' tabulated in Table 3 where the E' for the 800 °C flame exposed specimen was 13,483.74 MPa while a lower value of 5484.66 MPa was observed for the 1200 °C flame exposed specimens. The decrease in E' value is a clear indication of loss in stiffness.

A decreasing trend was also observed for E'' (Fig. 4b) and tan delta (Fig. 4c) as the flame temperature was increased from 800 to 1200 °C. Another interesting observation was the shift in T_g corresponding to the peak values of E'' and tan delta evident from Fig. 4b, c. Specimen exposed to 1200 °C displayed higher values of T_g over the specimen exposed to 800 °C (Table 3). This phenomenon of curve shifting in the loss modulus and tan delta curve as well as increase in T_g for the post-fire specimens was identical to the trend reported in literature [5]. Higher T_g for the fire exposed specimen could have been caused by residual curing of the matrix [12].

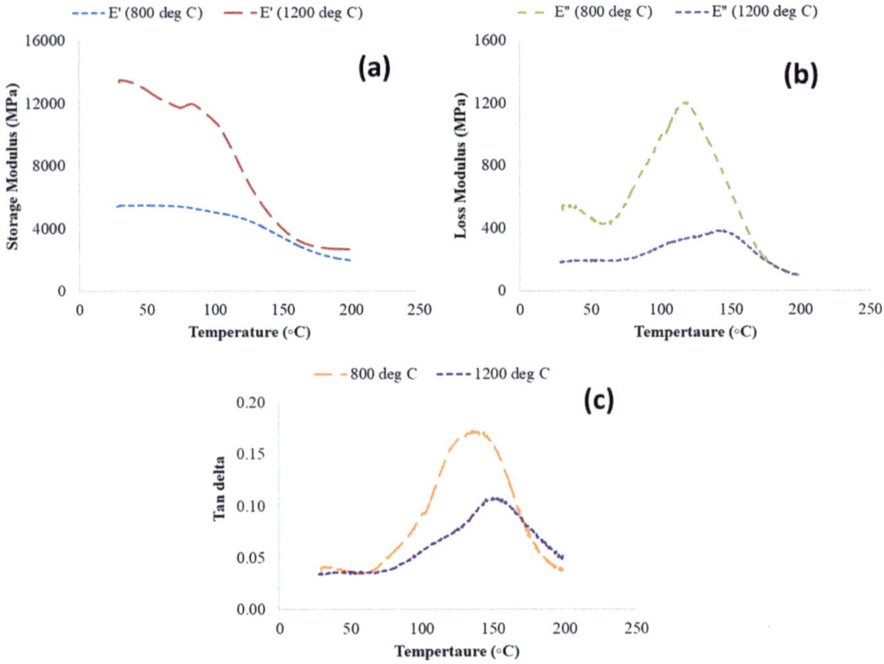

Fig. 4 DMA plots. **a** Storage modulus-temperature plot, **b** loss modulus-temperature, and **c** tan delta-temperature plot

Table 3 Inferences from DMA

Parameter	800 °C	1200 °C
Peak E' (MPa)	13,483.74	5484.66
Peak E'' (MPa)	1204.37	386.047
T_g corresponding to peak E''	118.94	145.07
Peak Tan delta	0.172	0.108
T_g corresponding to peak Tan delta	135.96	151.12

4 Conclusion

In this work, a main rotor blade comprising of rohacell foam and composite facesheet was subjected to simulated fire exposure and the post-fire response was investigated. Following were the conclusions:

- The flame exposure caused changes in the surface morphology and composition as reflected by the rough surface texture and elemental variation through EDX analysis.

- The load bearing ability of the composite post-fire under bending and tension was significantly affected. Tensile strength declined by 30% while infinitesimal changes were observed for the flexural strength. The fractured specimens exposed to fire displayed higher degree of delamination indicating structural weakening. This observation corroborates with the decrease in ILSS and the degradation of interfacial bonding leading to inferior strength for the post-fire specimens.
- DMA results revealed drastic drop in storage modulus, loss modulus, and tan delta on increasing the flame exposure from 800 to 1200 °C. Furthermore, a shift in T_g represented by the change in slope of the storage modulus plot and peak values of loss modulus and tan delta were observed.

Acknowledgements The authors would like to thank the HAL, Composite Division for supporting this work by providing the materials and testing facility required for this research.

References

1. Alothman OY, Jawaid M, Senthilkumar K et al (2020) Thermal characterization of date palm/epoxy composites with fillers from different parts of the tree. J Mater Res Technol 9:15537–15546
2. Anjang A, Chevali VS, Lattimer BY et al (2015) Post-fire mechanical properties of sandwich composite structures. Compos Struct 132:1019–1028
3. Asaro RJ, Lattimer B, Ramroth W (2009) Structural response of FRP composites during fire. Compos Struct 87:382–393
4. Asim M, Paridah MT, Chandrasekar M et al (2020) Thermal stability of natural fibers and their polymer composites. Cellulose 174:175
5. Bai Y, Keller T (2007) Modeling of post-fire stiffness of E-glass fiber-reinforced polyester composites. Compos Part A Appl Sci Manuf 38:2142–2153
6. Galos J, Das R, Sutcliffe MP, Mouritz AP (2022) Review of balsa core sandwich composite structures. Mater Des 221:111013
7. Haentzsche E, Mueller R, Huebner M et al (2016) Manufacturing technology of integrated textile-based sensor networks for in situ monitoring applications of composite wind turbine blades. Smart Mater Struct 25:105012
8. Joseph C, Muthukumar C, Ng LF et al (2023) The effect of nanoclay on the performance of basalt-epoxy facesheet and foam core sandwich panels. J Sandw Struct Mater 25:730–746
9. Kausar A, Ahmad I, Rakha SA et al (2023) State-of-the-art of sandwich composite structures: manufacturing—to—high performance applications. J Compos Sci 7:102
10. Mouritz AP, Gardiner CP (2002) Compression properties of fire-damaged polymer sandwich composites. Compos Part A Appl Sci Manuf 33:609–620
11. Mouritz AP, Mathys Z (2001) Post-fire mechanical properties of glass-reinforced polyester composites. Compos Sci Technol 61:475–490
12. Toubia E, Morgan A, Elmushyakhi A (2017) Post-fire residual mechanical properties of a composite laminate—experimental study. FL, Dec 11–14
13. Ulven CA, Vaidya UK (2008) Impact response of fire damaged polymer-based composite materials. Compos Part B Eng 39:92–107
14. Vieira PSC, de Souza FS, Cardoso DCT et al (2020) Influence of moderate/high temperatures on the residual flexural behavior of pultruded GFRP. Compos Part B Eng 200:108335
15. Yang B, Wang Z, Zhou L et al (2015) Study on the low-velocity impact response and CAI behavior of foam-filled sandwich panels with hybrid facesheet. Compos Struct 132:1129–1140

MIX
Papier aus verantwortungsvollen Quellen
Paper from responsible sources
FSC® C105338

If you have any concerns about our products,
you can contact us on
ProductSafety@springernature.com

In case Publisher is established outside the EU,
the EU authorized representative is:
Springer Nature Customer Service Center GmbH
Europaplatz 3, 69115 Heidelberg, Germany

Printed by Libri Plureos GmbH
in Hamburg, Germany